A COMPUTATIONAL INTRODUCTION
to DIGITAL IMAGE PROCESSING

SECOND EDITION

Alasdair McAndrew

Victoria University
Melbourne, Australia

CRC Press
Taylor & Francis Group
Boca Raton London New York

CRC Press is an imprint of the
Taylor & Francis Group, an **informa** business

A CHAPMAN & HALL BOOK

MATLAB® is a trademark of The MathWorks, Inc. and is used with permission. The MathWorks does not warrant the accuracy of the text or exercises in this book. This book's use or discussion of MATLAB® software or related products does not constitute endorsement or sponsorship by The MathWorks of a particular pedagogical approach or particular use of the MATLAB® software.

CRC Press
Taylor & Francis Group
6000 Broken Sound Parkway NW, Suite 300
Boca Raton, FL 33487-2742

First issued in paperback 2021

© 2016 by Taylor & Francis Group, LLC
CRC Press is an imprint of Taylor & Francis Group, an Informa business

No claim to original U.S. Government works

Version Date: 20150824

ISBN 13: 978-0-367-78333-4 (pbk)
ISBN 13: 978-1-4822-4732-9 (hbk)

Library of Congress Cataloging-in-Publication Data

McAndrew, Alasdair.
  A computational introduction to digital image processing / Alasdair McAndrew.
    pages cm
  ISBN 978-1-4822-4732-9 (alk. paper)
  1. Image processing--Digital techniques--Mathematics. 2. MATLAB. I. Title.

TA1632.M3638 2016
006.6--dc23   2015018371

**Visit the Taylor & Francis Web site at
http://www.taylorandfrancis.com**

**and the CRC Press Web site at
http://www.crcpress.com**

To my dear wife, Felicity, for all her love, support and understanding during my writing of this book.

Contents

Preface

Human beings are predominantly visual creatures, and our computing environments reflect this. We have the World Wide Web, filled with images of every possible type and provenance; and our own computers are crammed with images from the operating system, downloaded from elsewhere, or taken with our digital cameras. Then there are the vast applications which use digital images: remote sensing and satellite imaging; astronomy; medical imaging; microscopy; industrial inspection and so on.

This book is an introduction to *digital image processing*, from a strictly elementary perspective. We have selected only those topics that can be introduced with simple mathematics; however, these topics provide a very broad introduction to the discipline. The initial text was published in 2004; the current text is a rewritten version, with many changes.

This book is based on some very successful image processing subjects that have been taught at Victoria University in Melbourne, Australia, as well as in Singapore and Kuala Lumpur, Malaysia. The topics chosen, and their method of presentation, are the result of many hours talking to, and learning from, the hundreds of students who have taken these subjects.

There are a great many books on the subject of image processing, and this book differs from all of them in several important respects:

Not too much mathematics. Some mathematics is necessary for the explanation and discussion of image processing algorithms. But this book attempts to keep the mathematics at a level commensurate with elementary undergraduate computer science. The level of mathematics required is about one year's study at a tertiary level, including calculus and linear algebra.

A discrete approach. Since digital images are discrete entities, we have adopted an approach using mainly discrete mathematics. Although calculus is required for the development of some image processing topics, we attempt to connect the calculus-based (continuous) theory with its discrete implementation.

A strong connection between theory and practice. Since we have used a mainly discrete approach to explain the material, it becomes much easier to extend the theory into practice. This becomes particularly significant when we develop our own functions for implementing specific image processing algorithms.

Software based. There are image processing books that are based around programming languages: generally C or Java. The problem is that to use such software, a specialized image processing library must be used, and at least for C, there is no standard for this. And a problem with the Java image processing libraries is they are not really suitable for beginning students.

This book is based entirely around three different systems: MATLAB® and its Image Processing Toolbox, GNU Octave and its Image Processing Toolbox, and Python with various scientific and imaging libraries. Each system provides a complete environment for image processing that is easy to use, easy to explain, and easy to extend.

Plenty of examples. *All* the example images in this text are accompanied by commands in each of the three systems. Thus, if you work carefully through the book, you can create the images as given in the text.

Exercises. Most chapters finish off with a selection of exercises enabling the student to consolidate and extend the material. Some of the exercises are "pencil-and-paper"; designed for better understanding of the material; others are computer based to explore the algorithms and methods of that chapter.

What Is in the Book

The first three chapters set the scene for much of the rest of the book: exploring the nature and use of digital images, and how they can be obtained, stored, and displayed. Chapter 1 provides a brief introduction to the field of image processing, and attempts to give some idea as to its scope and areas of practice. We also define some common terms. Chapter 2 shows how images can be handled as matrices, and how the manipulation of these matrices forms the background of all subsequent work. Chapter 3 investigates aspects of image display, and looks at resolution and quantization and how these affect the appearance of the image. Appendix A provides a background in the use of MATLAB and Octave, and gives a brief introduction to programming in them.

Chapter 4 looks at some of the simplest, yet most powerful and widely used, of all image processing algorithms. These are the *point operations*, where the value of a pixel (a single "dot" in a digital image) is changed according to a single function of its value.

Chapter 5 introduces spatial filtering. Spatial filtering can be used for a vast range of image processing operations: from removing unnecessary detail, to sharpening edges and removing noise.

Chapter 6 looks at the geometry of an image: its size and orientation. Resizing an image may be necessary for inclusion in a web page or printed text; we may need to reduce it to fit or enlarge it.

Chapter 7 introduces the Fourier transform. This is possibly the single most important transform for image processing. To get a feeling for how the Fourier transform "works" and what information it provides, we need to spend some time exploring its mathematical foundations. This is the heaviest mathematics in this book, and requires some knowledge of complex numbers. In keeping with our philosophy, we use discrete mathematics only. We then show how images can be processed with great efficiency using the Fourier transform, and how various operations can be performed only using the Fourier transform.

Chapter 8 discusses the restoration of an image from different forms of degradation. Among these is the problem of *noise*, or "errors" in an image. Such errors are a natural consequence of electronic transmission of image signals, and although error correction of the signal can go a long way to ensuring that the image arrives "clean," we may still receive images with noise. We also look at the removal of blur.

Chapter 9 addresses the problems of thresholding and of finding edges in an image. Edges are a vital aspect of object recognition: we can classify the size, shape, and type of object by an analysis of its edges. As well, edges form a vital aspect of human visual interpretation, and so the sharpening of edges is often an important part of image enhancement.

Chapter 10 introduces *morphology* or *mathematical morphology*, which is an area of image processing very much wedded to set theory. Historically, morphology developed from the need for granulometry, or the measurement of grains in ore samples. It is now a very powerful method for investigating shapes and sizes of objects. Morphology is generally defined in terms of binary images (which is what we do here) and then can be extended to grayscale images. With the latter, we can also perform edge detection and some noise reduction.

Chapter 11 investigates the *topology* of digital images. This is concerned with the neighborhoods of pixels, and how the exploration of different neighborhoods leads to an understanding of the structure of image objects.

We continue the investigation of shapes in Chapter 12, but from a more spatial viewpoint; we look at traversing the edges of an object, and how the traversal can be turned into descriptors of the size and shape of the object.

Chapter 13 looks at color. Color is one of most important aspects of human interpretation. We look at the definition of color, from physical and digital perspectives, and how a color image can be processed using the techniques we have developed so far.

Chapter 14 discusses some basic aspects of image compression. Image files tend to be large, and their compression can be a matter of some concern, especially if there are many of them. We distinguish two types of compression: *lossless*, where there is no loss of information, and *lossy*, where higher compression rates can be obtained at the cost of losing some information.

Chapter 15 introduces wavelets. These have become a very hot topic in image processing; in some places they are replacing the use of the Fourier transform. Our treatment is introductory only; we show how wavelets and waves differ; how wavelets can be defined; how they can be applied to images; and the affects that can be obtained. In particular, we look at image compression, and show how wavelets can be used to obtain very high rates of lossy compression, with no apparent loss of image quality.

The final chapter is intended to be a bit more light-hearted than the others: here we look at some "special effects" on images. These are often provided with image editing programs—if you have a digital camera, chances are the accompanying software will allow this—our treatment attempts to provide an understanding into the nature of these algorithms.

What This Book Is Not

This book is *not* an introduction to MATLAB, Octave, Python, or their image processing tools. We have used only a small fraction of the many commands and functions available in each system—only those useful for an elementary text such as this. There is an enormous number of books on MATLAB available; fine general introductions are provided by the texts by Attaway [2] and Palm [20]. Python is equally well represented with texts; the texts by Langtangen [28], McKinney [30], and Howse [19] are all excellent basic references, even if Howse uses a different imaging library to the one in this text. Octave is less well represented by printed material, but the text by Quarteroni et al. [36] is very good. To really get to grips with any of these systems or their image processing tools and libraries, you can browse the excellent manuals online or through the system's own help interface.

How to Use This Book

This book can be used for two separate streams of image processing; one very elementary, another a little more advanced. A first course consists of the following:

- Chapter 1

- Chapter 2 except for Section 2.5

- Chapter 3 except for Section 3.5

- Chapter 4

- Chapter 5

- Chapter 7

- Chapter 8

- Chapter 9 except for Sections 9.4, 9.5 and 9.9

- Chapter 10 except for Sections 10.8 and 10.9

- Chapter 13

- Chapter 14, Sections 14.2 and 14.3 only

A second course fills in the gaps:

- Section 2.5

- Section 3.5

- Chapter 6

- Sections 9.4, 9.5 and 9.9

- Sections 10.8 and 10.9

- Chapter 11

- Chapter 12

- Section 14.5

- Chapter 15

As taught at Victoria University, for the first course we concentrated on introducing students to the principles and practices of image processing, and did not worry about the implementation of the algorithms. In the second course, we spent much time discussing programming issues, and encouraged the students to write their own programs, and edit programs already written. Both courses have been very popular since their inception.

Changes Since the First Edition

The major change is that MATLAB, the sole computing environment in the first edition, has been joined by two other systems: GNU Octave [10], which in many ways is an open-source version of MATLAB, and Python [12], which is a deservedly popular general programming language. Python has various libraries for the handling of matrices and images, and is fast becoming recognized as a viable alternative to systems such as MATLAB.

At the time of writing the first edition, MATLAB was the the software of choice: its power, ease of use, and its functionality gave it few, if any, competitors. But the world of software has expanded greatly since then, and the beginning user is now spoiled for choice. In particular, many open-source systems (such as GNU Octave) have reached a maturity that makes them genuine alternatives to MATLAB for scientific programming. For users who prefer the choice that open-source provides, this has been a welcome development. MATLAB still has an edge in its enormous breadth: its image processing toolbox has over 300 commands and functions—far more than any competitor—as well as a very mature graphics interface library. However, for basic and fundamental image processing—such as discussed in this text—MATLAB is equalled by both Octave and Python, and in some respects even surpassed by them.

Octave has been designed to be a "drop-in" replacement for MATLAB: standard MATLAB functions and scripts should run unchanged in Octave. This philosophy has been extended to the Octave Image Processing Toolbox, which although less comprehensive than that of MATLAB, is equal in depth. Thus, all through this text, except in a very few places, MATLAB and Octave are treated as though they are one and the same.

Python has several imaging libraries; I have chosen to use scikit-image [53]. This is one of several toolkits that are designed to extend and deepen the functionality in the standard Python scientific library SciPy ("Scikit" is a shortened version of "SciPy Toolkit"). It is designed to be easy to use: an image is just an array, instead of a specialized data-type unique to the toolkit. It is also implemented in Python, with the static library Cython used when fast performance is necessary. Other imaging libraries, such as Mahotas [9] and OpenCV [4, 19], while excellent, are implemented in C++. This makes it hard for the user (without a good knowledge of C++) to understand or extend the code. The combination of scikit-image with the functionality already available in SciPy provides very full-featured and robust image processing.

Many teachers, students, and practitioners are now attempting to find the best software for their needs. One of the aims of this text is to provide several different options, or alternatively enable users to migrate from one system to another.

There are also many changes throughout the text—some minor and typographical, some major. All the diagrams have been redrawn using the amazing TiKZ package [31]; some have been merely rewritten using TiKZ; others have been redrawn for greater symmetry, clarity, and synthesis of their graphics elements. Programs (in all three systems) have been written to be as modular as possible: in the first edition they tended to be monolithic and large. Modularity means greater flexibility, code reuse, and often conciseness—one example in the first edition which took most of a page, and 48 lines of small type, has been reduced to 9 lines, and includes greater functionality! Such large programs as are left have been relegated to the ends of the chapters.

All chapters have been written to be "system neutral": as far as possible, imaging techniques are implemented in MATLAB and Octave, and again in Python. In Chapter 2, the section on PNG images has been expanded to provide more information about this format. Chapter 5 has been extended with some material about the Kuwahara and bilateral filters—both methods for blurring an image while maintaining edge sharpness—as well as a rewritten section on region of interest processing. In Chapter 6 an original section on anamorphosis has been removed, on account of difficulties obtaining a license to use a particular image; this section has been replaced with a new section containing an example of image warping. For readers who may pine for anamorphosis, the excellent recent text by Steven Tanimoto [50] contains much wonderful material on anamorphosis, as well as being written in a beautifully approachable style. Chapter 7 has had a few new diagrams inserted, hopefully to clarify some of the material. Chapter 9 includes an extended discussion on Otsu's method of thresholding, as well as new material about the ISODATA method. This chapter also contains a new section about corner detection, with discussions of both the Moravec and Harris-Stephens operators. The section on the Hough transform has been completely rewritten to include the Radon transform. In Chapter 11, the algorithms for the Zhang-Suen and Guo-Hall skeletonization methods have been rewritten to be simpler. Chapter 12 contains a new discussion on Fourier shape descriptors, with different examples.

Chapter 13 contains a new section at the end introducing one version of the retinex algorithm. Chapter 14 implements a simpler and faster method for run-length encoding, and includes a new section on LZW compression. Chapter 15 has been simplified, and more connections made between the discrete wavelet transform and digital filtering. Also in this chapter a new wavelet toolbox is used, one that amazingly has been written to be used in each of MATLAB, Octave, and Python. In Chapter 16 the greater modularity of the programs has allowed a greater conciseness, with (I hope) increased clarity.

Finally, new images have been used to ameliorate licensing or copyright difficulties. All images in this book are either used with permission of their owners, or are the author's and may be used freely and without permission. They are listed at the end.

Acknowledgments

The first edition of this book began as set of notes written to accompany an undergraduate introductory course to image processing; these notes were then extended for a further course. Thanks must go to the many students who have taken these courses, and who have discussed the subject material and its presentation with me.

I would like to thank my colleagues at Victoria University for many stimulating conversations about teaching and learning. I would also like to thank Associate Professor Robyn Pierce at the University of Melbourne, for providing me with a very pleasant environment for six months, during which much of the initial programming and drafting of this new edition was done.

Sunil Nair and Sarfraz Khan, of Taylor and Francis Publishing Company, have provided invaluable help and expert advice, as well as answering all my emails with admirable promptness. Their helping hands when needed have made this redrafting and rewriting a very pleasant and enjoyable task.

I am indebted to the constant hard work of David Miguel Susano Pinto (maintainer of the Octave Forge image package under the alias of Carnë Draug), and his predecessor Søren Hauberg, the original author of the image package; also Stefan van der Walt, lead developer of the Python scikit-image package. They have made lasting and valuable contributions to open-source imaging software, and also have answered many of my questions.

I would also like to thank the reviewers for their close and careful reading of the initial proposal, and for their thoughtful, detailed, and constructive comments.

Finally, heartfelt thanks to my long suffering wife Felicity, and my children Angus, Edward, Fenella, William, and Finlay, who have put up, for far too long, with absent-mindedness, tables and benches covered with scraps of papers, laptops, and books, and a husband and father who was more concentrated on his writing than their needs.

A Note on the Images

Some of the images used in this book are the author's and may be used freely and without restriction. They are:

```
arch.png
backyard.png
blocks.png
bot_gardens.png
buffalo.png
car.png
cat.png
chickens.png
circles.png
circles2.png
color_sunset.png
emu.png
engineer.png
handmade.png
iris.png
koala.png
meet_text.png
monkey.png
morph_text.png
newborn.png
nicework.png
paperclips.png
pelicans.png
pinenuts.png
rings.png
seagull.png
stairs.png
street.png
sunset.png
twins.png
venice.png
wombats.png
```

MIT Press has kindly allowed the use of the standard test image

```
cameraman.png
```

The following images have been provided courtesy of `shutterstock.com`:

```
blood.png
paramecium.png
```

The image

```
caribou.tif
```

has been cropped from the NOAA image `anim0614.jpg`, a photograph of a caribou (*Rangifer tarandus*) by Capt. Budd Christman of the NOAA Corps.

The image

`monastery.png`

has been taken from `http://www.public-domain-image.com`; more particularly from `http://bit.ly/18V2Z57`: the photographer is Andrew McMillan.

The image

`daisies.png`

has been taken from `www.pixabay.com` as an image in the public domain; the photographer is Marianne Langenbach.

The iconic image

`thylacine.png`

showing the last known living thylacine, or Tasmanian tiger, at the Hobart zoo in 1933, is now in the public domain.

The x-ray image

`xray.png`

has been taken from Wikimedia Commons, as an image in the public domain. It can be found at `http://bit.ly/1Az0Tzk`; the original image has been provided by Diego Grez from Santa Cruz, Chile.

Chapter 1

Introduction

1.1 Images and Pictures

As we mentioned in the Preface, human beings are predominantly visual creatures: we rely heavily on our vision to make sense of the world around us. We not only look at things to identify and classify them, but we can scan for differences, and obtain an overall rough "feeling" for a scene with a quick glance.

Humans have evolved very precise visual skills: we can identify a face in an instant; we can differentiate colors; we can process a large amount of visual information very quickly.

However, the world is in constant motion: stare at something for long enough and it will change in some way. Even a large solid structure, like a building or a mountain, will change its appearance depending on the time of day (day or night); amount of sunlight (clear or cloudy), or various shadows falling upon it.

We are concerned with single images: snapshots, if you like, of a visual scene. Although image processing can deal with changing scenes, we shall not discuss it in any detail in this text.

For our purposes, an *image* is a single picture that represents something. It may be a picture of a person, people, or animals, an outdoor scene, a microphotograph of an electronic component, or the result of medical imaging. Even if the picture is not immediately recognizable, it will not be just a random blur.

1.2 What Is Image Processing?

Image processing involves changing the nature of an image in order to either

1. Improve its pictorial information for human interpretation

2. Render it more suitable for autonomous machine perception

We shall be concerned with *digital image processing*, which involves using a computer to change the nature of a *digital image* (see Section 1.4). It is necessary to realize that these two aspects represent two separate but equally important aspects of image processing. A procedure that satisfies condition (1)—a procedure that makes an image "look better"—may be the very worst procedure for satisfying condition (2). Humans like their images to be sharp, clear, and detailed; machines prefer their images to be simple and uncluttered.

Examples of (1) may include:

- Enhancing the edges of an image to make it appear sharper; an example is shown in Figure 1.1. Note how the second image appears "cleaner"; it is a more pleasant image. Sharpening edges is a vital component of printing: in order for an image to appear "at its best" on the printed page; some sharpening is usually performed.

(a) The original image   (b) Result after "sharpening"

FIGURE 1.1: Image sharpening

- Removing "noise" from an image; noise being random errors in the image. An example is given in Figure 1.2. Noise is a very common problem in data transmission: all sorts of electronic components may affect data passing through them, and the results may be undesirable. As we shall see in Chapter 8, noise may take many different forms, and each type of noise requires a different method of removal.

(a) The original image   (b) After removing noise

FIGURE 1.2: Removing noise from an image

- Removing motion blur from an image. An example is given in Figure 1.3. Note that in the deblurred image (b) it is easier to read the numberplate, and to see the spikes on the fence behind the car, as well as other details not at all clear in the original image (a). Motion blur may occur when the shutter speed of the camera is too long for the speed of the object. In photographs of fast moving objects—athletes, vehicles for example—the problem of blur may be considerable.

(a) The original image   (b) After removing the blur

FIGURE 1.3: Image deblurring

Examples of (2) may include:

- Obtaining the edges of an image. This may be necessary for the measurement of objects in an image; an example is shown in Figures 1.4. Once we have the edges we can measure their spread, and the area contained within them. We can also use edge detection algorithms as a first step in edge enhancement, as we saw previously.

From the edge result, we see that it may be necessary to enhance the original image slightly, to make the edges clearer.

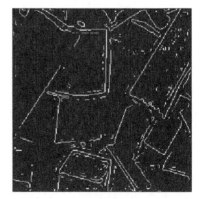

(a) The original image   (b) Its edge image

FIGURE 1.4: Finding edges in an image

- Removing detail from an image. For measurement or counting purposes, we may not be interested in all the detail in an image. For example, a machine inspected items on an assembly line, the only matters of interest may be shape, size, or color. For such cases, we might want to simplify the image. Figure 1.5 shows an example: in image (a) is a picture of an African buffalo, and image (b) shows a blurred version in which extraneous detail (like the logs of wood in the background) have been removed. Notice that in image (b) all the fine detail is gone; what remains is the coarse structure of the image. We could for example, measure the size and shape of the animal without being "distracted" by unnecessary detail.

(a) The original image   (b) Blurring to remove detail

FIGURE 1.5: Blurring an image

1.3   Image Acquisition and Sampling

Sampling refers to the process of digitizing a continuous function. For example, suppose we take the function

$$y = \sin(x) + \frac{1}{3}\sin(3x)$$

and sample it at ten evenly spaced values of x only. The resulting sample points are shown in Figure 1.6. This shows an example of *undersampling*, where the number of points is not sufficient to reconstruct the function. Suppose we sample the function at 100 points, as shown in Figure 1.7. We can clearly now reconstruct the function; all its properties can be determined from this sampling. In order to ensure that we have enough sample points, we require that the sampling period is not greater than one-half the finest detail in our function. This is known as the *Nyquist criterion*, and can be formulated more precisely in terms of "frequencies," which are discussed in Chapter 7. The Nyquist criterion can be stated as the *sampling theorem*, which says, in effect, that a continuous function can be reconstructed from its samples provided that the sampling frequency is at least twice the maximum frequency in the function. A formal account of this theorem is provided by Castleman [7].

Sampling an image again requires that we consider the Nyquist criterion, when we consider an image as a continuous function of two variables, and we wish to sample it to produce a digital image.

An example is shown in Figure 1.8 where an image is shown, and then with an under-sampled version. The jagged edges in the undersampled image are examples of *aliasing*.

FIGURE 1.6: Sampling a function—undersampling

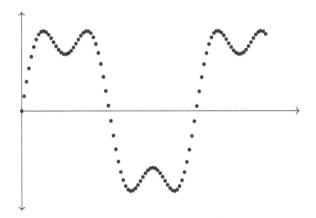

FIGURE 1.7: Sampling a function with more points

Correct sampling; no aliasing   An undersampled version with aliasing

FIGURE 1.8: Effects of sampling

The sampling rate will of course affect the final resolution of the image; we discuss this in Chapter 3. In order to obtain a sampled (digital) image, we may start with a continuous representation of a scene. To view the scene, we record the energy reflected from it; we may use visible light, or some other energy source.

Using Light

Light is the predominant energy source for images; simply because it is the energy source that human beings can observe directly. We are all familiar with photographs, which are a pictorial record of a visual scene.

Many digital images are captured using visible light as the energy source; this has the advantage of being safe, cheap, easily detected, and readily processed with suitable hardware. Two very popular methods of producing a digital image are with a digital camera or a flat-bed scanner.

CCD camera. Such a camera has, in place of the usual film, an array of *photosites*; these are silicon electronic devices whose voltage output is proportional to the intensity of light falling on them.

For a camera attached to a computer, information from the photosites is then output to a suitable storage medium. Generally this is done on hardware, as being much faster and more efficient than software, using a *frame-grabbing card*. This allows a large number of images to be captured in a very short time—in the order of one ten-thousandth of a second each. The images can then be copied onto a permanent storage device at some later time.

This is shown schematically in Figure 1.9.

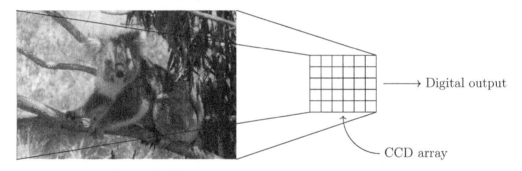

FIGURE 1.9: Capturing an image with a CCD array

The output will be an array of values; each representing a sampled point from the original scene. The elements of this array are called *picture elements*, or more simply *pixels*.

Digital still cameras use a range of devices, from floppy discs and CDs, to various specialized cards and "memory sticks." The information can then be downloaded from these devices to a computer hard disk.

Flat bed scanner. This works on a principle similar to the CCD camera. Instead of the entire image being captured at once on a large array, a single row of photosites is moved across the image, capturing it row-by-row as it moves. This is shown schematically in Figure 1.10.

Since this is a much slower process than taking a picture with a camera, it is quite reasonable to allow all capture and storage to be processed by suitable software.

FIGURE 1.10: Capturing an image with a CCD scanner

Other Energy Sources

Although light is popular and easy to use, other energy sources may be used to create a digital image. Visible light is part of the *electromagnetic spectrum*: radiation in which the energy takes the form of waves of varying wavelength. These range from cosmic rays of very short wavelength, to electric power, which has very long wavelength. Figure 1.11 illustrates this. For microscopy, we may use x-rays or electron beams. As we can see from Figure 1.11,

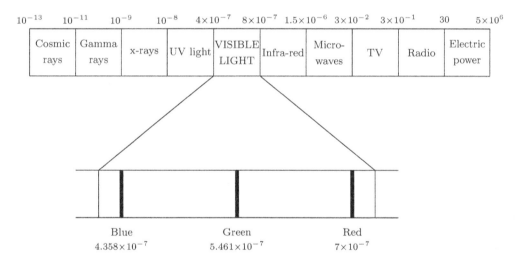

FIGURE 1.11: The electromagnetic spectrum

x-rays have a shorter wavelength than visible light, and so can be used to resolve smaller objects than are possible with visible light. See Clark [8] for a good introduction to this. X-rays are of course also useful in determining the structure of objects usually hidden from view, such as bones.

A further method of obtaining images is by the use of *x-ray tomography*, where an object is encircled by an x-ray beam. As the beam is fired through the object, it is detected on the other side of the object, as shown in Figure 1.12. As the beam moves around the object, an image of the object can be constructed; such an image is called a *tomogram*. In a CAT (Computed Axial Tomography) scan, the patient lies within a tube around which x-ray beams are fired. This enables a large number of tomographic "slices" to be formed, which can then be joined to produce a three-dimensional image. A good account of such systems (and others) is given by Siedband [48].

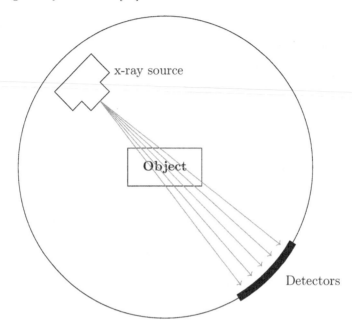

FIGURE 1.12: X-ray tomography

1.4   Images and Digital Images

Suppose we take an image, a photo, say. For the moment, let's make things easy and suppose the photo is monochromatic (that is, shades of gray only), so no color. We may consider this image as being a two-dimensional function, where the function values give the brightness of the image at any given point, as shown in Figure 1.13. We may assume that in such an image brightness values can be any real numbers in the range 0.0 (black) to 1.0 (white). The ranges of x and y will clearly depend on the image, but they can take all real values between their minima and maxima.

Such a function can of course be plotted, as shown in Figure 1.14. However, such a plot is of limited use to us in terms of image analysis. The concept of an image as a function, however, will be vital for the development and implementation of image processing techniques.

A *digital image* differs from a photo in that the x, y, and $f(x, y)$ values are all *discrete*. Usually they take on only integer values, so the image shown in Figure 1.13 will have x and y ranging from 1 to 256 each, and the brightness values also ranging from 0 (black) to 255 (white). A digital image, as we have seen above, can be considered a large array of sampled points from the continuous image, each of which has a particular quantized brightness; these points are the *pixels*, which constitute the digital image. The pixels surrounding a given pixel constitute its *neighborhood*. A neighborhood can be characterized by its shape in the same way as a matrix: we can speak, for example, of a 3×3 neighborhood or of a 5×7

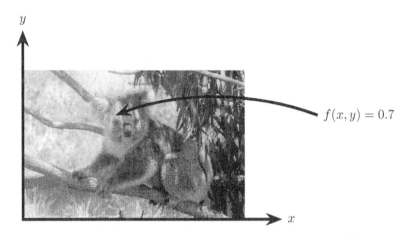

FIGURE 1.13: An image as a function

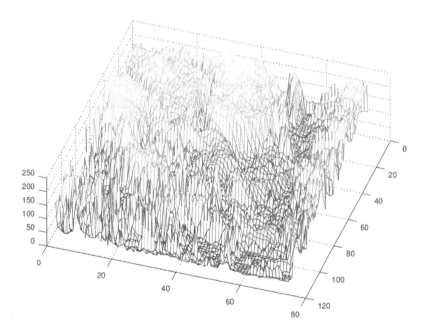

FIGURE 1.14: The image of Figure 1.13 plotted as a function of two variables

neighborhood. Except in very special circumstances, neighborhoods have odd numbers of rows and columns; this ensures that the current pixel is in the center of the neighborhood. An example of a neighborhood is given in Figure 1.15. If a neighborhood has an even number of rows or columns (or both), it may be necessary to specify which pixel in the neighborhood is the "current pixel."

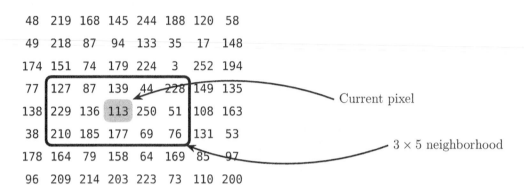

FIGURE 1.15: Pixels, with a neighborhood

1.5   Some Applications

Image processing has an enormous range of applications; almost every area of science and technology can make use of image processing methods. Here is a short list just to give some indication of the range of image processing applications.

1. Medicine

    - Inspection and interpretation of images obtained from x-rays, MRI, or CAT scans
    - Analysis of cell images, of chromosome karyotypes

2. Agriculture

    - Satellite/aerial views of land, for example to determine how much land is being used for different purposes, or to investigate the suitability of different regions for different crops
    - Inspection of fruit and vegetables—distinguishing good and fresh produce from old

3. Industry

    - Automatic inspection of items on a production line
    - Inspection of paper samples

4. Law enforcement

    - Fingerprint analysis
    - Sharpening or deblurring of speed-camera images

1.6   Image Processing Operations

It is convenient to subdivide different image processing algorithms into broad subclasses. There are different algorithms for different tasks and problems, and often we would like to distinguish the nature of the task at hand.

Image enhancement.   This refers to processing an image so that the result is more suitable for a particular application. Examples include:

- Sharpening or deblurring an out of focus image
- Highlighting edges
- Improving image contrast, or brightening an image
- Removing noise

Image restoration.   This may be considered as reversing the damage done to an image by a known cause, for example:

- Removing of blur caused by linear motion
- Removal of optical distortions
- Removing periodic interference

Image segmentation.   This involves subdividing an image into constituent parts, or isolating certain aspects of an image:

- Finding lines, circles, or particular shapes in an image
- In an aerial photograph, identifying cars, trees, buildings, or roads

Image registration.   This involves "matching" distinct images so that they can be compared, or processed together. The initial images must all be joined to share the same coordinate system. In this text, registation as such is not covered, but some tasks that are vital to registration are discussed, for example corner detection.

These classes are not disjoint; a given algorithm may be used for both image enhancement or for image restoration. However, we should be able to decide what it is that we are trying to do with our image: simply make it look better (enhancement) or removing damage (restoration).

1.7   An Image Processing Task

We will look in some detail at a particular real-world task, and see how the above classes may be used to describe the various stages in performing this task. The job is to obtain, by an automatic process, the postcodes from envelopes. Here is how this may be accomplished:

Acquiring the image. First we need to produce a digital image from a paper envelope. This can be done using either a CCD camera or a scanner.

Preprocessing. This is the step taken before the "major" image processing task. The problem here is to perform some basic tasks in order to render the resulting image more suitable for the job to follow. In this case, it may involve enhancing the contrast, removing noise, or identifying regions likely to contain the postcode.

Segmentation. Here is where we actually "get" the postcode; in other words, we extract from the image that part of it that contains just the postcode.

Representation and description. These terms refer to extracting the particular features which allow us to differentiate between objects. Here we will be looking for curves, holes, and corners, which allow us to distinguish the different digits that constitute a postcode.

Recognition and interpretation. This means assigning labels to objects based on their descriptors (from the previous step), and assigning meanings to those labels. So we identify particular digits, and we interpret a string of four digits at the end of the address as the postcode.

1.8 Types of Digital Images

We shall consider four basic types of images:

Binary. Each pixel is just black or white. Since there are only two possible values for each pixel, we only need one bit per pixel. Such images can therefore be very efficient in terms of storage. Images for which a binary representation may be suitable include text (printed or handwriting), fingerprints, or architectural plans.

An example was the image shown in Figure 1.4(b). In this image, we have only the two colors: white for the edges, and black for the background. See Figure 1.16.

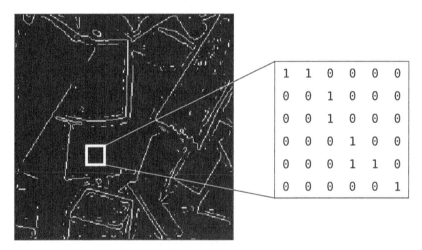

FIGURE 1.16: A binary image

Grayscale. Each pixel is a shade of gray, normally from 0 (black) to 255 (white). This range means that each pixel can be represented by eight bits, or exactly one byte. This is a very natural range for image file handling. Other grayscale ranges are used, but generally they are a power of 2. Such images arise in medicine (X-rays), images of printed works, and indeed 256 different gray levels is sufficient for the recognition of most natural objects.

An example is the street scene shown in Figure 1.1, and in Figure 1.17.

230	229	232	234	235	232	148
237	236	236	234	233	234	152
255	255	255	251	230	236	161
99	90	67	37	94	247	130
222	152	255	129	129	246	132
154	199	255	150	189	241	147
216	132	162	163	170	239	122

FIGURE 1.17: A grayscale image

True color, or RGB. Here each pixel has a particular color; that color being described by the amount of red, green, and blue in it. If each of these components has a range 0–255, this gives a total of $255^3 = 16,777,216$ different possible colors in the image. This is enough colors for any image. Since the total number of bits required for each pixel is 24, such images are also called 24-*bit color images*.

Such an image may be considered as consisting of a "stack" of three matrices; representing the red, green, and blue values for each pixel. This means that for every pixel there are three corresponding values.

An example is shown in Figure 1.18.

Indexed. Most color images only have a small subset of the more than sixteen million possible colors. For convenience of storage and file handling, the image has an associated *color map*, or *color palette*, which is simply a list of all the colors used in that image. Each pixel has a value which does not give its color (as for an RGB image), but an *index* to the color in the map.

It is convenient if an image has 256 colors or less, for then the index values will only require one byte each to store. Some image file formats (for example, Compuserve GIF) allow only 256 colors or fewer in each image, for precisely this reason.

49	55	56	57	52	53
58	60	60	58	55	57
58	58	54	53	55	56
83	78	72	69	68	69
88	91	91	84	83	82
69	76	83	78	76	75
61	69	73	78	76	76

Red

64	76	82	79	78	78
93	93	91	91	86	86
88	82	88	90	88	89
125	119	113	108	111	110
137	136	132	128	126	120
105	108	114	114	118	113
96	103	112	108	111	107

Green

66	80	77	80	87	77
81	93	96	99	86	85
83	83	91	94	92	88
135	128	126	112	107	106
141	129	129	117	115	101
95	99	109	108	112	109
84	93	107	101	105	102

Blue

FIGURE 1.18: **SEE COLOR INSERT** A true color image

Figure 1.19 shows an example. In this image the indices, rather then being the gray values of the pixels, are simply indices into the color map. Without the color map, the image would be very dark and colorless. In the figure, for example, pixels labelled 5 correspond to `0.2627 0.2588 0.2549`, which is a dark grayish color.

Indices

Color map

FIGURE 1.19: **SEE COLOR INSERT** An indexed color image

1.9   Image File Sizes

Image files tend to be large. We shall investigate the amount of information used in different image types of varying sizes. For example, suppose we consider a 512×512 binary image. The number of bits used in this image (assuming no compression, and neglecting, for the sake of discussion, any header information) is

$$512 \times 512 \times 1 = 262{,}144$$
$$= 32{,}768 \text{ bytes}$$
$$= 32.768 \text{ Kb}$$
$$\approx 0.033 \text{ Mb}.$$

(Here we use the convention that a kilobyte is 1000 bytes, and a megabyte is one million bytes.)

A grayscale image of the same size requires:

$$512 \times 512 \times 1 = 262{,}144 \text{ bytes}$$
$$= 262.14 \text{ Kb}$$
$$\approx 0.262 \text{ Mb}.$$

If we now turn our attention to color images, each pixel is associated with 3 bytes of color information. A 512×512 image thus requires

$$521 \times 512 \times 3 = 786{,}432 \text{ bytes}$$
$$= 786.43 \text{ Kb}$$
$$\approx 0.786 \text{ Mb}.$$

Many images are of course larger than this; satellite images may be of the order of several thousand pixels in each direction.

1.10   Image Perception

Much of image processing is concerned with making an image appear "better" to human beings. We should therefore be aware of the limitations of the human visual system. Image perception consists of two basic steps:

1. Capturing the image with the eye

2. Recognizing and interpreting the image with the *visual cortex* in the brain

The combination and immense variability of these steps influences the ways in which we perceive the world around us.

There are a number of things to bear in mind:

1. *Observed intensities* vary as to the background. A single block of gray will appear darker if placed on a white background than if it were placed on a black background. That is, we don't perceive gray scales "as they are," but rather as they differ from their surroundings. In Figure 1.20, a gray square is shown on two different backgrounds. Notice how much darker the square appears when it is surrounded by a light gray. However, the two central squares have exactly the same intensity.

2. We may observe non-existent intensities as bars in continuously varying gray levels. See, for example, Figure 1.21. This image varies continuously from light to dark as we travel from left to right. However, it is impossible for our eyes not to see a few horizontal edges in this image.

3. Our visual system tends to undershoot or overshoot around the boundary of regions of different intensities. For example, suppose we had a light gray blob on a dark gray background. As our eye travels from the dark background to the light region, the boundary of the region appears lighter than the rest of it. Conversely, going in the other direction, the boundary of the background appears *darker* than the rest of it.

FIGURE 1.20: A gray square on different backgrounds

FIGURE 1.21: Continuously varying intensities

Exercises

1. Watch the TV news, and see if you can observe any examples of image processing.

2. If your TV set allows it, turn down the color as far as you can to produce a monochromatic display. How does this affect your viewing? Is there anything that is hard to recognize without color?

3. Look through a collection of old photographs. How can they be enhanced or restored?

4. For each of the following, list five ways in which image processing could be used:

   - Medicine
   - Astronomy
   - Sport
   - Music
   - Agriculture
   - Travel

5. Image processing techniques have become a vital part of the modern movie production process. Next time you watch a film, take note of all the image processing involved.

6. If you have access to a scanner, scan in a photograph, and experiment with all the possible scanner settings.

   (a) What is the smallest sized file you can create which shows all the detail of your photograph?

   (b) What is the smallest sized file you can create in which the major parts of your image are still recognizable?

   (c) How do the color settings affect the output?

7. If you have access to a digital camera, again photograph a fixed scene, using all possible camera settings.

   (a) What is the smallest file you can create?

   (b) How do the light settings affect the output?

8. Suppose you were to scan in a monochromatic photograph, and then print out the result. Then suppose you scanned in the printout, and printed out the result of that, and repeated this a few times. Would you expect any degradation of the image during this process? What aspects of the scanner and printer would minimize degradation?

9. Look up *ultrasonography*. How does it differ from the image acquisition methods discussed in this chapter? What is it used for? If you can, compare an ultrasound image with an x-ray image. How to they differ? In what ways are they similar?

10. If you have access to an image viewing program (other than MATLAB, Octave or Python) on your computer, make a list of the image processing capabilities it offers. Can you find imaging tasks it is unable to do?

Chapter 2

Images Files and File Types

We shall see that matrices can be handled very efficiently in MATLAB, Octave and Python. Images may be considered as matrices whose elements are the pixel values of the image. In this chapter we shall investigate how the matrix capabilities of each system allow us to investigate images and their properties.

2.1 Opening and Viewing Grayscale Images

Suppose you are sitting at your computer and have started your system. You will have a prompt of some sort, and in it you can type:

```
>> w = imread('wombats.png');
```

MATLAB/Octave

or from the io module of skimage

```
In :  import skimage.io as io
In :  w = io.imread('wombats.png')
```

Python

This takes the gray values of all the pixels in the grayscale image wombats.png and puts them all into a matrix w. This matrix w is now a system variable, and we can perform various matrix operations on it. In general, the imread function reads the pixel values from an image file, and returns a matrix of all the pixel values.

Two things to note about this command:

1. If you are using MATLAB or Octave, end with a *semicolon*; this has the effect of not displaying the results of the command to the screen. As the result of this particular command is a matrix of size 256×256, or with 65,536 elements, we do not really want all its values displayed. Python, however, does not automatically display the results of a computation.

2. The name wombats.png is given in quotation marks. Without them, the system would assume that wombats.png was the name of a variable, rather than the name of a file.

Now we can display this matrix as a grayscale image. In MATLAB:

```
>> figure,imshow(w),impixelinfo
```

MATLAB

This is really three commands on one line. MATLAB allows many commands to be entered on the same line, using commas to separate the different commands. The three commands we are using here are:

figure, which creates a *figure* on the screen. A figure is a window in which a graphics object can be placed. Objects may include images or various types of graphs.

imshow(g), which displays the matrix **g** as an image.

impixelinfo, which turns on the pixel values in our figure. This is a display of the gray values of the pixels in the image. They appear at the bottom of the figure in the form

        Pixel info: $(c, r)\ p$

where c is the column value of the given pixel; r is its row value, and p is its gray value. Since **wombats.png** is an 8-bit grayscale image, the pixel values appear as integers in the range 0–255.

This is shown in Figure 2.1.

FIGURE 2.1: The wombats image with **impixelinfo**

In Octave, the command is:

```
>> figure,imshow(w)
```
Octave

and in Python one possible command is:

```
In: io.imshow(w)
```
Python

However, an interactive display is possible using the **ImageViewer** method from the module with the same name:

```
In :   from skimage.viewer import ImageViewer as IV
In :   viewer = IV(w)
In :   viewer.show()
```

                                                                              Python

This is shown in Figure 2.2.

FIGURE 2.2: The wombats image with **ImageViewer**

At present, Octave does not have a built-in method for interactively displaying the pixel indices and value comparable to MATLAB's **impixelinfo** or Python's **ImageViewer**.

If there are no figures open, then in Octave or MATLAB an **imshow** command, or any other command that generates a graphics object, will open a new figure for displaying the object. However, it is good practice to use the **figure** command whenever you wish to create a new figure.

We could display this image directly, without saving its gray values to a matrix, with the command

```
>> imshow('wombats.png')
```

                                                                         MATLAB/Octave

or

```
In: io.imshow('wombats.png')
```

                                                                              Python

However, it is better to use a matrix, seeing as these are handled very efficiently in each system.

2.2   RGB Images

As we shall discuss in Chapter 13, we need to define colors in some standard way, usually as a subset of a three-dimensional coordinate system; such a subset is called a *color model*. There are, in fact, a number of different methods for describing color, but for image display and storage, a standard model is RGB, for which we may imagine all the colors sitting inside a "color cube" of side 1 as shown in Figure 2.3. The colors along the black-white diagonal,

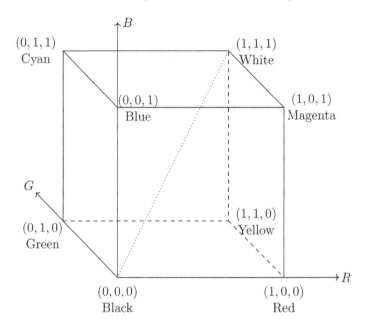

FIGURE 2.3: The color cube for the RGB color model

shown in the diagram as a dotted line, are the points of the space where all the R, G, B values are equal. They are the different intensities of gray. We may also think of the axes of the color cube as being discretized to integers in the range 0–255.

RGB is the standard for the *display* of colors on computer monitors and TV sets. But it is not a very good way of *describing* colors. How, for example, would you define light-brown using RGB? As we shall see also in Chapter 13, some colors are not realizable with the RGB model in that they would require negative values of one or two of the RGB components. In general, 24-bit RGB images are managed in much the same way as grayscale. We can save the color values to a matrix and view the result:

```
>> b = imread('backyard.png');
>> figure,imshow(b),impixelinfo
```

MATLAB

Note now that the pixel values consist of a list of three values, giving the red, green, and blue components of the color of the given pixel.

An important difference between this type of image and a grayscale image can be seen by the command

```
>> size(b)
```

MATLAB/Octave

which returns *three* values: the number of rows, columns, and "pages" of b, which is a three-dimensional matrix, also called a *multidimensional array*. Each system can handle arrays of any dimension, and b is an example. We can think of b as being a stack of three matrices, each of the same size.

In Python:

```
In: b.shape
```

Python

again returns a triplet of values.

To obtain any of the RGB values at a given location, we use indexing methods similar to above. For example,

```
>> b(100,200,2)
ans =

   126
```

MATLAB/Octave

returns the second color value (green) at the pixel in row 100 and column 200. If we want all the color values at that point, we can use

```
>> b(100,200,1:3)
ans(:,:,1) =

   114

ans(:,:,2) =

   126

ans(:,:,3) =

   58
```

MATLAB/Octave

However, MATLAB and Octave allow a convenient shortcut for listing all values along a particular dimension just using a colon on its own:

```
>> b(100,200,:)
```

MATLAB/Octave

This can be done similarly in Python:

```
In: print b[99,199,:]
[114 126   58]
```

Python

(Recall that indexing in Python starts at zero, so to obtain the same results as with MAT-LAB and Octave, the indices need to be one less.)

A useful function for obtaining RGB values is `impixel`; the command

```
>> impixel(b,200,100)
ans =

   114   126    58
```

MATLAB/Octave

returns the red, green, and blue values of the pixel at column 200, row 100. Notice that the order of indexing is the same as that which is provided by the `impixelinfo` command. This is opposite to the row, column order for matrix indexing. This command also applies to grayscale images:

```
>> impixel(w,100,200)
ans =

   189   189   189
```

MATLAB/Octave

Again, three values are returned, but since w is a single two-dimensional matrix, all three values will be the same.

Note that in Octave, the command `impixel` will in fact produce a 3×3 matrix of values; the three columns however will be the same. If we just want the values once, then:

```
impixel(b,200,100)(:,1)'
ans =

   114   126    58
```

Octave

2.3 Indexed Color Images

The command

```
>> figure,imshow('emu.png'),impixelinfo
```

MATLAB/Octave

produces a nice color image of an emu. However, the pixel values, rather than being three integers as they were for the RGB image above, are three fractions between 0 and 1, for example:

Pixel info: $(61, 109)$ $\langle 37 \rangle$ $[0.58\ 0.54\ 0.51]$

What is going on here?

If we try saving to a matrix first and then displaying the result:

```
>> em = imread('emu.png');
>> figure,imshow(em),impixelinfo
```

we obtain a dark, barely distinguishable image, with single integer gray values, indicating that em is being interpreted as a single grayscale image.

In fact the image emu.png is an example of an *indexed image*, consisting of two matrices: a *color map*, and an *index* to the color map. Assigning the image to a single matrix picks up only the index; we need to obtain the color map as well:

```
>> [em,emap] = imread('emu.png');
>> figure,imshow(em,emap),impixelinfo
```

MATLAB and Octave store the RGB values of an indexed image as values of type double, with values between 0 and 1. To obtain RGB values at any value:

```
>> v = em(109,61)

v =

    37
```

To find the color values at this point, note that the color indexing is done with eight bits, hence the first index value is zero. Thus, index value 37 is in fact the 38th row of the color map:

```
>> >> emap(38,:)

ans =

    0.5801    0.5410    0.5098
```

Python automatically converts an indexed image to a true-color image as the image file is read:

```
In:  em = io.imread('emu.png')
In:  em.shape
Out: (384, 331, 3)
```

To obtain values at a pixel : :

```
In :  print em[108,60,:]
[148, 138, 130]
In :  np.set_printoptions(precision = 4)
In :  print em[108,60,:].astype(float)/255.0
[ 0.5804 0.5412  0.5098]
```

This last shows how to obtain information comparable to that obtained with MATLAB and Octave.

Information About Your Image

Using MATLAB or Octave, a great deal of information can be obtained with the `imfinfo` function. For example, suppose we take our indexed image `emu.png` from above. The MATLAB and Octave outputs are in fact very slightly different; here is the Octave output:

```
>> imfinfo('emu.png')

ans =

  scalar structure containing the fields:

    Filename = /home/amca/Images/emu.png
    FileModDate =  2-Jan-2015 15:12:52
    FileSize =  71116
    Format = PNG
    FormatVersion =
    Width =  331
    Height =  384
    BitDepth =  8
    ColorType = indexed
    DelayTime = 0
    DisposalMethod =
    LoopCount = 0
    ByteOrder = undefined
    Gamma = 0
    Chromaticities = [](1x0)
    Comment =
    Quality =  75
    Compression = undefined
    Colormap =
```
Octave

(For saving of space, the colormap, which is given as part of the output, is not listed here.)

Much of this information is not useful to us; but we can see the size of the image in pixels, the size of the file (in bytes), the number of bits per pixel (this is given by `BitDepth`), and the color type (in this case "indexed").

For comparison, let's look at the output of a true color file (showing only the first few lines of the output), this time using MATLAB:

```
>> imfinfo('backyard.png')

ans =

              Filename: 'backyard.png'
           FileModDate: '02-Jan-2015_05:45:17'
              FileSize: 264103
                Format: 'png'
         FormatVersion: []
                 Width: 482
                Height: 643
              BitDepth: 24
             ColorType: 'truecolor'
       FormatSignature: [73 73 42 0]
             ByteOrder: 'little-endian'
        NewSubFileType: 0
         BitsPerSample: [8 8 8]
           Compression: 'JPEG'
```

MATLAB

Now we shall test this function on a binary image, in Octave:

```
>> imfinfo('circles.png')

ans =
  scalar structure containing the fields:

    Filename = /home/amca/Images/circles.png
    FileModDate =  2-Jan-2015 21:37:54
    FileSize =  1268
    Format = PNG
    FormatVersion =
    Width =  256
    Height =  256
    BitDepth =  1
    ColorType = grayscale
```

Octave

In MATLAB and Octave, the value of `ColorType` would be `GrayScale`.

What is going on here? We have a binary image, and yet the color type is given as either "indexed" or "grayscale." The fact is that the `imfinfo` command in both MATLAB nor Octave has no value such as "Binary", "Logical" or "Boolean" for `ColorType`. A binary image is just considered to be a special case of a grayscale image which has only two intensities. However, we can see that `circles.png` is a binary image since the number of bits per pixel is only one.

In Chapter 3 we shall see that a binary image *matrix*, as distinct from a binary image *file*, can have the numerical data type `Logical`.

To obtain image information in Python, we need to invoke one of the libraries that supports the metadata from an image. For TIFF and JPEG images, such data is known as EXIF data, and the Python `exifread` library may be used:

```
In :  import exifread
In :  f = open('backyard.tif')
In :  tags = exifread.process_file(f,details=False)
In :  for x in tags:
...:      print x.ljust(32),":",tags[x]

Image Orientation                : Horizontal (normal)
Image Compression                : JPEG
Image FillOrder                  : 1
Image ImageLength                : 643
Image PhotometricInterpretation  : 2
Image ImageWidth                 : 482
Image DocumentName               : /home/amca/Images/backyard.tif
Image BitsPerSample              : [8, 8, 8]
Image XResolution                : 72
Image YResolution                : 72
Image SamplesPerPixel            : 3
```

Python

Some of the output is not shown here. For images of other types, it is easiest to invoke a system call to something like "ExifTool"[1]:

```
In :  ! exiftool cassowary.png
ExifTool Version Number   : 9.46
File Name                 : cassowary.png
File Size                 : 21 MB
MIME Type                 : image/png
Image Width               : 4000
Image Height              : 2248
Bit Depth                 : 8
Color Type                : RGB
Compression               : Deflate/Inflate
Background Color          : 255 255 255
Pixels Per Unit X         : 7086
Pixels Per Unit Y         : 7086
Pixel Units               : Meters
Exif Byte Order           : Little-endian (Intel, II)
Orientation               : Horizontal (normal)
X Resolution              : 180
Y Resolution              : 180
Resolution Unit           : inches
```

Python

Again most of the information is not shown.

2.4 Numeric Types and Conversions

Elements in an array representing an image may have a number of different numeric data types; the most common are listed in Table 2.1. MATLAB and Octave use "double"

---

[1]Available from http://www.sno.phy.queensu.ca/~phil/exiftool/

Data type	Description	Range
logical	Boolean	0 or 1
int8	8-bit integer	$-128 - 127$
uint8	8-bit unsigned integer	$0 - 255$
int16	16-bit integer	$-32768 - 32767$
uint16	16-bit unsigned integer	$0 - 65535$
double or float	Double precision real number	Machine specific

TABLE 2.1: Some numeric data types

and Python uses "float." There are others, but those listed will be sufficient for all our work with images. Notice that, strictly speaking, logical is not a numeric type. However, we shall see that some image use this particular type. These data types are also functions, and we can convert from one type to another. For example:

```
>> a = 23;
>> b = uint8(a);
>> b

b =

    23

>> whos a b
  Name      Size            Bytes  Class

    a       1x1                 8  double
    b       1x1                 1  uint8
```
MATLAB/Octave

Similarly in Python:

```
In :  a = 23
In :  b = uint8(a)
In :  %whos int uint8
Variable   Type    Data/Info
--------------------------------
a          int        23
b          uint8      23
```
Python

Note here that %whos is in fact a magic function in the IPython shell. If you are using another interface, this function will not be available.

Even though the variables a and b have the same numeric value, they are of different data types.

A grayscale image may consist of pixels whose values are of data type uint8. These images are thus reasonably efficient in terms of storage space, because each pixel requires only one byte.

We can convert images from one image type to another. Table 2.2 lists all of MATLAB's functions for converting between different image types. Note that there is no gray2rgb

Function	Use	Format
ind2gray	Indexed to grayscale	y=ind2gray(x,map);
gray2ind	Grayscale to indexed	[y,map]=gray2ind(x);
rgb2gray	RGB to grayscale	y=rgb2gray(x);
rgb2ind	RGB to indexed	[y,map]=rgb2ind;
ind2rgb	Indexed to RGB	y=ind2rgb(x,map);

TABLE 2.2: Converting images in MATLAB and Octave

function. But a gray image can be turned into an RGB image where all the R, G, and B matrices are equal, simply by stacking the grayscale array three deep. This can done using several different methods, and will be investigated in Chapter 13.

In Python, there are the `float` and `uint8` functions, but in the skimage library there are methods from the `util` module; these are listed in Table 2.3.

Function	Use	Format
img_as_bool	Convert to boolean	y=sk.util.img_as_bool(x)
img_as_float	Convert to 64-bit floating point	y=sk.util.img_as_bool(x)
img_as_bool	Convert to 16-bit integer	y=sk.util.img_as_int(x)
img_as_bool	Convert to 8-bit unsigned integer	y=sk.util.img_as_ubyte(x)
img_as_bool	Convert to 16-bit unsigned integer	y=sk.util.img_as_uint(x)}

TABLE 2.3: Converting images in Python

2.5 Image Files and Formats

We have seen in Section 1.8 that images may be classified into four distinct types: binary, grayscale, colored, and indexed. In this section, we consider some of the different image file formats, their advantages and disadvantages. You can use MATLAB, Octave or Python for image processing very happily without ever really knowing the difference between GIF, TIFF, PNG, and all the other formats. However, some knowledge of the different graphics formats can be extremely useful to be able to make a reasoned decision as to which file type to use and when.

There are a great many different formats for storing image data. Some have been designed to fulfill a particular need (for example, to transmit image data over a network); others have been designed around a particular operations system or environment.

As well as the gray values or color values of the pixels, an image file will contain some *header information*. This will, at the very least, include the size of the image in pixels (height and width); it may also include the color map, compression used, and a description of the image. Each system recognizes many standard formats, and can read image data from them and write image data to them. The examples above have mostly been images with extension `.png`, indicating that they are PNG (Portable Network Graphics) images. This is a particularly general format, as it allows for binary, grayscale, RGB, and indexed color images, as well as allowing different amounts of transparency. PNG is thus a good format for transmitting images between different operating systems and environments. The

PNG standard only allows one image per file. Other formats, for example TIFF, allow for more than one image per file; a particular image from such a multi-image file can be read into MATLAB by using an optional numeric argument to the `imread` function.

Each system can read images from a large variety of file types, and save arrays to images of those types. General image file types, all of which are handled by each system, include:

JPEG	Images created using the Joint Photographics Experts Group compression method. We will discuss this more in Chapter 14.
TIFF	Tagged Image File Format: a very general format which supports different compression methods, multiple images per file, and binary, grayscale, truecolor and indexed images.
GIF	Graphics Interchange Format: This venerable format was designed for data transfer. It is still popular and well supported, but is somewhat restricted in the image types it can handle.
BMP	Microsoft Bitmap: This format has become very popular, with its use by Microsoft operating systems.
PNG	Portable Network Graphics: This is designed to overcome some of the disadvantages of GIF, and to become a replacement for GIF.

We will discuss some of these briefly below.

A Hexadecimal Dump Function

To explore binary files, we need a simple function that will enable us to list the contents of the file as hexadecimal values. If we try to list the contents of a binary file directly to the screen, we will see masses of garbage. The trouble is that any file printing method will interpret the file's contents as ASCII characters, and this means that most of these will either be unprintable or nonsensical as values.

A simple hexadecimal dump function, called "hexdump," is given at the end of the chapter. It is called with the name of the file, and the number of bytes to be listed:

```
>> hexdump('backyard.tif',64)
000010   4949 2a00 4c01 0400 ffd8 ffdb 0043 0008   II*.L........C..
000020   0606 0706 0508 0707 0709 0908 0a0c 140d   ...............
000030   0c0b 0b0c 1912 130f 141d 1a1f 1e1d 1a1c   ...............
000040   1c20 242e 2720 222c 231c 1c28 3729 2c30   . $.' ",#..(7),0
```
MATLAB/Octave

and in Python:

```
In :  hexdump('backyard.tif',64)
000000:   4949 2a00 4c01 0400 ffd8 ffdb 0043 0008   |II*.L........C..|
000010:   0606 0706 0508 0707 0709 0908 0a0c 140d   |...............|
000020:   0c0b 0b0c 1912 130f 141d 1a1f 1e1d 1a1c   |...............|
000030:   1c20 242e 2720 222c 231c 1c28 3729 2c30   |. $.'␣",#..(7),0|
```
Python

The result is a listing of the bytes as hexadecimal characters in three columns: the first gives the hexadecimal value of the index of the first byte in that row, the second column consists of the bytes themselves, and the third column lists those that are representable as ASCII text.

Vector versus Raster Images

We may store image information in two different ways: as a collection of lines or vectors, or as a collection of dots. We refer to the former as *vector* images; the latter as *raster* images. The great advantage of vector images is that they can be magnified to any desired size without losing any sharpness. The disadvantage is that are not very good for the representation of natural scenes, in which lines may be scarce. The standard vector format is Adobe PostScript; this is an international standard for page layout. PostScript is the format of choice for images consisting mostly of lines and mathematically described curves: architectural and industrial plans, font information, and mathematical figures. The reference manual [21] provides all necessary information about PostScript.

The great bulk of image file formats store images as raster information; that is, as a list of the gray or color intensities of each pixel. Images captured by digital means–digital cameras or scanners–will be stored in raster format.

A Simple Raster Format

As well as containing all pixel information, an image file must contain some *header information*; this must include the size of the image, but may also include some documentation, a color map, and compression used. To show the workings of a raster image file, we shall briefly describe the ASCII PGM format. This was designed to be a generic format used for conversion between other formats. Thus, to create conversion routines between, say, 40 different formats, rather than have $40 \times 39 = 1560$ different conversion routines, all we need is the $40 \times 2 = 80$ conversion routines between the formats and PGM.

```
P2
# CREATOR: The GIMP's PNM Filter Version 1.0
256 256
255
 41  53  53  53  53  49  49  53  53  56  56  49  41  46  53  53  53
 53  41  46  56  56  56  53  53  46  53  41  41  53  56  49  39  46
```

FIGURE 2.4: The start of a PGM file

Figure 2.4 shows the beginning of a PGM file. The file begins with **P2**; this indicates that the file is an ASCII PGM file. The next line gives some information about the file: any line beginning with a hash symbol is treated as a comment line. The next line gives the number of columns and rows, and the following line gives the number of grayscales. Finally we have all the pixel information, starting at the top left of the image, and working across and down. Spaces and carriage returns are delimiters, so the pixel information could be written in one very long line or one very long column.

Note that this format has the advantage of being very easy to write to and to read from; it has the disadvantage of producing very large files. Some space can be saved by using "raw" PGM; the only difference is that the header number is **P5**, and the pixel values are stored one per byte. There are corresponding formats for binary and colored images (PBM and PPM, respectively); colored images are stored as three matrices; one for each of red, green, and blue; either as ASCII or raw. The format does not support color maps.

Binary, grayscale, or color images using these formats are collectively called PNM images. MATLAB and Octave can read PNM images natively; Python can't, but it is not hard to write a file to read a PNM image to an ndarray.

Microsoft BMP

The Microsoft Windows BMP image format is a fairly simple example of a binary image format, noting that here binary means non-ascii, as opposed to Boolean. Like the PGM format above, it consists of a header followed by the image information. The header is divided into two parts: the first 14 bytes (bytes number 0 to 13), is the "File Header", and the following 40 bytes (bytes 14 to 53), is the "Information Header". The header is arranged as follows:

Bytes	Information	Description
0–1	Signature	"BM" in ASCII = 42 4D in hexadecimal.
2–5	FileSize	The size of the file in bytes.
6–9	Reserved	All zeros.
10–13	DataOffset	File offset to the raster data.
14–17	Size	Size of the information header = 40 bytes.
18–21	Width	Width of the image in pixels.
22–25	Height	Height of the image in pixels.
26–27	Planes	Number of image planes ($= 1$).
28–29	BitCount	Number of bits per pixel: 1: Binary images; two colors 4: $2^4 = 16$ colors (indexed) 8: $2^8 = 256$ colors (indexed) 16: 16-bit RGB; $2^{16} = 65,536$ colors 24: 24-bit RGB; $2^{24} = 17,222,216$ colors
30–33	Compression	Type of compression used: 0: no compression (most common) 1: 8-bit RLE encoding (rarely used) 2: 4-bit RLE encoding (rarely used)
34–37	ImageSize	Size of the image. If compression is 0, then this value may be 0.
38–41	HorizontalRes	The horizontal resolution in pixels per meter.
42–45	VerticalRes	The vertical resolution in pixels per meter.
46–49	ColorsUsed	The number of colors used in the image. If this value is zero, then the number of colors is the maximum obtainable with the bits per pixel, that is 2^{BitCount}.
50–53	ImportantColors	The number of important colors in the image. If all the colors are important, then this value is set to zero.

After the header comes the Color Table, which is only used if BitCount is less than or equal to 8. The total number of bytes used here is 4 × ColorsUsed. This format uses the Intel "least endian" convention for bytes ordering, where in each word of four bytes the *least* valued byte comes *first*. To see an example of this, consider a simple example:

```
>> hexdump('backyard.bmp',64)
000000   424d 8235 0e00 0000 0000 8a00 0000 7c00   BM.5..........|.
000010   0000 e201 0000 8302 0000 0100 1800 0000   ...............
000020   0000 f834 0e00 120b 0000 120b 0000 0000   ...4...........
000030   0000 0000 0000 0000 ff00 00ff 0000 ff00   ...............
```

The image width is given by bytes 18–21; they are in the second row:

`e201 0000`

To find the actual width; we re-order these bytes back to front:

`0000 01e2`

Now we can convert to decimal:

$$(1 \times 16^2) + (14 \times 16^1) + (2 \times 16^0) = 482$$

which is the image width in pixels. We can do the same thing with the image height; bits 22–25:

`8302 0000`

Re-ordering and converting to hexadecimal:

$$(2 \times 16^2) + (8 \times 16^1) + (3 \times 16^0) = 16 + 15 = 643.$$

Recall that hexadecimal symbols a, b, c, d, e, f have decimal values 10 to 15, respectively.

GIF and PNG

Compuserve GIF (pronounced "jif") is a venerable image format that was first proposed in the late 1980s as a means for distributing images over networks. Like PGM, it is a raster format, but it has the following properties:

1. Colors are stored using a color map; the GIF specification allows a maximum of 256 colors per image.

2. GIF doesn't allow for binary or grayscale images; except as can be produced with red, green, and blue values.

3. The pixel data is compressed using LZW (Lempel-Ziv-Welch) compression. This works by constructing a "codebook" of the data: the first time a pattern is found, it is placed in the codebook; subsequent times the encoder will output the code for that pattern. LZW compression can be used on any data; until relatively recently, it was a patented algorithm, and legal use required a license from Unisys. LZW is described in Chapter 14.

4. The GIF format allows for multiple images per file; this aspect can be used to create "animated GIFs."

A GIF file will contain a header including the image size (in pixels), the color map, the color *resolution* (number of bits per pixel), a flag indicating whether the color map is ordered, and the color map size.

The GIF format is greatly used; it has become one of the standard formats supported by the World Wide Web, and by the Java programming language. Full descriptions of the GIF format can be found in [5] or [26].

The PNG (pronounced "ping") format has been more recently designed to replace GIF, and to overcome some of its disadvantages. Specifically, PNG was not to rely on any patented algorithms, and it was to support more image types than GIF. PNG supports grayscale, true-color, and indexed images. Moreover, its compression utility, zlib, *always* results in genuine compression. This is not the case with LZW compression; it can happen that the result of an LZW compression is larger than the original data. PNG also includes support for *alpha channels*, which are ways of associating variable transparencies with an image, and *gamma correction*, which associates different numbers with different computer display systems, to ensure that a given image will appear the same independently of the system.

PNG is described in detail by [38]. PNG is certainly to be preferred to GIF; it is now well supported by every system.

A PNG file consists of what are called "chunks"; each is referenced by a four-letter name. Four of the chunks must be present: IHDR, giving the image dimensions and bit depth, PLTE is the color palette, IDAT contains the image data, IEND marks the end. These four chunks are called "Critical Chunks." Other chunks, known as "Ancillary Chunks," may or may not be present; they include pHYS: the pixel size and aspect ratio; bKGD the background color; sRGB, indicating the use of the standard RGB color space; gAMA, specifying gamma; cHRM, which describes the primary chromaticites and the white point; as well as others.

For example, here are the hexadecimal dumps of the first few bytes of three PNG images, first `cameraman.png`:

```
000000:  8950 4e47 0d0a 1a0a 0000 000d 4948 4452  |.PNG........IHDR|
000010:  0000 0100 0000 0100 0800 0000 0079 19f7  |.............y..|
000020:  ba00 0000 0970 4859 7300 000b 1200 000b  |.....pHYs.......|
000030:  1201 d2dd 7efc 0000 8000 4944 4154 78da  |....~.....IDATx.|
```

Next `iguana.png`:

```
000000:  8950 4e47 0d0a 1a0a 0000 000d 4948 4452  |.PNG........IHDR|
000010:  0000 0267 0000 017d 0800 0000 0064 7c90  |...g...}.....d|.|
000020:  ad00 0000 0262 4b47 4400 ff87 8fcc bf00  |.....bKGD.......|
000030:  0000 0970 4859 7300 000b 1200 000b 1201  |...pHYs.........|
000040:  d2dd 7efc 0000 8000 4944 4154 78da 3cfd  |..~.....IDATx.<.|
```

Finally `backyard.png`:

```
000000:  8950 4e47 0d0a 1a0a 0000 000d 4948 4452  |.PNG........IHDR|
000010:  0000 01e2 0000 0283 0802 0000 0049 cb9e  |.............I..|
000020:  9600 0000 0467 414d 4100 00b1 8f0b fc61  |.....gAMA......a|
000030:  0500 0000 0173 5247 4200 aece 1ce9 0000  |.....sRGB.......|
000040:  0020 6348 524d 0000 7a26 0000 8084 0000  |. cHRM..z&......|
```

Each chunk consists of four bytes giving the length of the chunk's data, four bytes giving its type, then the data of the chunk, and finally four bytes of a checksum.

For example, consider the last file. The IHDR chunk has its initial four bytes of length `0000 000d`, which has decimal value 13, and the four bytes of its name **IHDR**. The 13 bytes of data start with four bytes each for width and height, followed by one byte each for bit depth, color type, compression method, filter method, and interlace method. So, for the example given:

Object	Bytes	Decimal value	Meaning
Width	`0000 01e2`	482	
Height	`0000 0283`	683	
Bit depth	`08`	8	Bits per sample or per palette index
Color type	`02`	2	RGB color space used
Compression	`00`	0	No compression
Interlacing	`00`	0	No interlacing

Note that the "bit depth" refers not to bit depth per pixel, but bit depth (in this case) for each color plane.

As an example of an ancillary chunk, consider pHYS, which is always nine bytes. From the cameraman image we can read:

Object	Bytes	Decimal value	Meaning
Pixels per unit, X axis	`00 000b 12`	2834	
Pixels per unit, Y axis	`00 000b 12`	2834	
Unit specifier	`01`	1	Unit is meter.

If the unit specifier was 0, then the unit is not specified. The cameraman image is thus expected to have 2834 pixels per meter in each direction, or 71.984 pixels per inch.

JPEG

The compression methods used by GIF and PNG are *lossless*: the original information can be recovered completely. The JPEG (Joint Photographics Experts Group) algorithm uses *lossy* compression, in which not all the original data can be recovered. Such methods result in much higher compression rates, and JPEG images are in general much smaller than GIF or PNG images. Compression of JPEG images works by breaking the image into 8×8 blocks, applying the discrete cosine transform (DCT) to each block, and removing small values. JPEG images are best used for the representation of natural scenes, in which case they are to be preferred.

For data with any legal significance, or scientific data, JPEG is less suitable, for the very reason that not all data is preserved. However, the mechanics of the JPEG transform ensures that a JPEG image, when restored from its compression routine, will *look* the same as the original image. The differences are in general too small to be visible to the human eye. JPEG images are thus excellent for display.

We shall investigate the JPEG algorithm in Section 14.5. More detailed accounts can be found in [13] and [37].

A JPEG image then contains the compression data with a small header providing the image size and file identification. We can see a header by using our **hexdump** function applied to the **greentreefrog.jpg** image:

```
000000:  ffd8 ffe0 0010 4a46 4946 0001 0101 00b4  |......JFIF......|
000010:  00b4 0000 ffe1 35fe 4578 6966 0000 4949  |......5.Exif..II|
000020:  2a00 0800 0000 0a00 0e01 0200 2000 0000  |*........... ...|
000030:  8600 0000 0f01 0200 0600 0000 a600 0000  |................|
```

An image file containing JPEG compressed data is usually just called a "JPEG image." But this is not quite correct; such an image should be called a "JFIF image" where JFIF stands for "JPEG File Interchange Format." The JFIF definition allows for the file to contain a *thumbnail* version of the image; this is reflected in the header information:

Bytes	Information	Description
0–1	Start of image marker	Always `ffd8`.
2–3	Application marker	Always `ffe0`.
4–5	Length of segment	
6–10	`JFIF\ 0`	ASCII "JFIF."
11–12	JFIF version	In our example above `01 01`, or version 1.1.
13	Units	Values are 0: Arbitrary units; 1: pixels/in; 2: pixels/cm.
14–15	Horizontal pixel density	
16–17	Vertical pixel density	
18	Thumbnail width	If this is 0, there is no thumbnail.
19	Thumbnail height	If this is 0, there is no thumbnail.

In this image, both horizontal and vertical pixel densities have hexadecimal value `b4`, or decimal value 180. After this would come the thumbnail information (stored as 24-bit RGB values), and further information required for decompressing the image data. See [5] or [26] for further information.

TIFF

The *Tagged Image File Format*, or TIFF, is one of the most comprehensive image formats. It can store multiple images per file. It allows different compression routines (none at all, LZW, JPEG, Huffman, RLE); different byte orderings (little-endian, as used in BMP, or big-endian, in which the bytes retain their order within words); it allows binary, grayscale, truecolor or indexed images; it allows for opacity or transparency.

For that reason, it requires skillful programming to write image reading software that will read all possible TIFF images. But TIFF is an excellent format for data exchange.

The TIFF header is in fact very simple; it consists of just eight bytes:

Bytes	Information	Description
0–1	Byte order	Either `4d4d`: ASCII "MM" for big-endian, or `49 49`: ASCII "II" for little endian.
2–3	TIFF version	Always `002a` or `2a00` (depending on the byte order) = 42.
4–8	Image offset	Pointer to the position in the file of the data for the first image.

We can see this with looking at the `newborn.tif` image:

```
000000:  4949 2a00 e001 0100 327c 5b2d 2319 0e15  |II*.....2|[-#...|
000010:  100e 0d0f 100f 0e10 1111 0e12 1312 1017  |................|
000020:  101d 708e 99a0 aeb5 bbba c2c6 c6cb d3d0  |..p.............|
000030:  d2d1 cadb dede e1e5 e6df e4e9 feeb 0bed  |................|
```

This particular image uses the little-endian byte ordering. The first image in this file (which is in fact the only image) begins at byte `e001 0100`.

Since this is a little-endian file, we reverse the order of the bytes: `00 01 01 e0`; this works out to 66016.

As well as the previously cited references, Baxes [3] provides a good introduction to TIFF, and the formal specification [1] is also available.

Writing Image Files

In Python, an image matrix may be written to an image file with the `imsave` function; its simplest usage is

```
In :  io.imsave(x,'filename.abc')
```
Python

where `abc` may be any of the image file types recognized by the system: `gif`, `jpg`, `tif`, `bmp`, for example.

MATLAB and Octave have an `imwrite` function:

```
>> imwrite(x,'filename.abc')
```
MATLAB/Octave

An indexed image can be written by including its colormap as well:

```
>> imwrite(x,map,'filename.abc')
```
MATLAB/Octave

If we wish to be doubly sure that the correct image file format is to be used, a string representing the format can be included; in MATLAB (but at the time of writing not in Octave), a list of supported formats can be obtained with the `imformats` function. Octave's image handling makes extensive use of the ImageMagick library;[2] these however support over 100 different formats, as long as your system is set up with the appropriate format handling libraries.

For example, given the matrix `c` representing the cameraman image, it can be written to a PNG image with

```
In :  io.imsave(c,'cameraman.png')
```
Python

or with

```
>> imwrite(c,'cameraman.png')
```
MATLAB/Octave

2.6 Programs

Here are the programs for the hexadecimal dump. First in MATLAB/Octave:

---

[2]`http://www.imagemagick.org/`

```matlab
function hexdump(filenm, n)
% hexdump(filenm, n)
% Print the first n bytes of a file in hex and ASCII.
fid = fopen(filenm, 'r');
if (fid<0) disp(['Error opening ',filenm]); return; end;
nread = 0;
while (nread < n)
    width = 16;
    [A,count] = fread(fid, width, 'uchar');
    nread = nread + count;
    if (nread>n) count = count - (nread-n); A = A(1:count); end;
    hexstring = repmat(' ',1,width*2);
    hexstring(1:2*count) = sprintf('%02x',A);
    hexdisp = repmat(' ',1,40);
    for i = 1:idivide(count,2)
        hexdisp(5*i-4:5*i-1) = hexstring(4*i-3:4*i);
    end
    ascstring = repmat('.',1, count);
    idx = find(double(A)>=32 & double(A)<=126);
    ascstring(idx) = char(A(idx));
    fprintf('%s: %s |%s|\n', num2str(dec2hex(nread-16,6)), hexdisp,
        ascstring);
end;
fclose(fid);
```

MATLAB/Octave

Now Python:

```python
def hexdump(file,num):
    from math import ceil
    f = open(file)
    h = f.read(num)
    hl = [hex(ord(c))[2:].zfill(2) for c in h] # all the bytes as 2
        character hex strings
    hl2 = hl +(-len(hl)%16)*['  ']  # fill up to a multiple of width 16
    asc = ['.']*num
    for i in range(num):                # creates ascii values (if printable)
        of bytes
        ii = int(hl[i],16)
        if ii>=32 and ii<=126:
            asc[i] = chr(ii)
    asc = asc + (-len(hl)%16)*[' '] # fill up to a multiple of width 16
    for n in range(int(ceil(num/16.0))):
        print hex(n*16)[2:].zfill(6)+': ',
        print ' '.join([hl2[16*n+2*i]+hl2[16*n+2*i+1] for i in range(8)]),
        print ' |'+''.join([asc[16*n+i] for i in range(16)])+'|'
```

Python

Exercises

1. Dig around in your system and see if you can find what image formats are supported.

2. If you are using MATLAB or Octave, read in an RGB image and save it as an indexed image.

3. If you are using Python, several images are distributed with the **skimage** library: enter **from skimage import data** at the prompt, and then open up some of the images. This can be done, for example, with

   ```
   In :  x = data.clock()
   In :  io.imshow(x)
   ```
   Python

   List the data images and their types (binary, grayscale, color).

4. If you are using MATLAB, there are a lot of sample images distributed with the Image Processing Toolbox, in the **imdata** subdirectory. Use your file browser to enter that directory and check out the images.

   For each of the images you choose:

   (a) Determine its type (binary, grayscale, true color or indexed color)

   (b) Determine its size (in pixels)

   (c) Give a brief description of the picture (what it looks like; what it seems to be a picture of)

5. Pick a grayscale image, say **cameraman.png** or **wombats.png**. Using the **imwrite** function, write it to files of type JPEG, PNG, and BMP.

   What are the sizes of those files?

6. Repeat the above question with

   (a) A binary image,

   (b) An indexed color image,

   (c) A true color image.

7. The following shows the hexadecimal dump of a PNG file:

   ```
   000000  8950 4e47 0d0a 1a0a 0000 000d 4948 4452   .PNG........IHDR
   000010  0000 012c 0000 00f6 0800 0000 0049 c4e5   ...,.........I..
   000020  5400 0000 0774 494d 4507 d209 1314 1f0c   T....tIME.......
   000030  035d c49d 0000 0027 7445 5874 436f 7079   .].....'tEXtCopy
   ```

   Determine the height and width of this image (in pixels), and whether it is a grayscale or color image.

8. Repeat the previous question with this BMP image:

```
000000  424d 3603 0000 0000 0000 3600 0000 2800   BM6.......6...(.
000010  0000 1000 0000 1000 0000 0100 1800 0000   ................
000020  0000 0003 0000 c40e 0000 c40e 0000 0000   ................
000030  0000 0000 0000 e1f5 ffe1 f5ff e1f5 ffe1   ................
```

Chapter 3

Image Display

3.1 Introduction

We have touched briefly in Chapter 2 on image display. In this chapter, we investigate this matter in more detail. We look more deeply at the use of the image display functions in our systems, and show how spatial resolution and quantization can affect the display and appearance of an image. In particular, we look at image quality, and how that may be affected by various image attributes. Quality is of course a highly subjective matter: no two people will agree precisely as to the quality of different images. However, for human vision in general, images are preferred to be sharp and detailed. This is a consequence of two properties of an image: its spatial resolution, and its quantization.

An image may be represented as a matrix of the gray values of its pixels. The problem here is to display that matrix on the computer screen. There are many factors that will effect the display; they include:

1. Ambient lighting

2. The monitor type and settings

3. The graphics card

4. Monitor resolution

The same image may appear very different when viewed on a dull CRT monitor or on a bright LCD monitor. The resolution can also affect the display of an image; a higher resolution may result in the image taking up less physical area on the screen, but this may be counteracted by a loss in the color depth: the monitor may only be able to display 24-bit color at low resolutions. If the monitor is bathed in bright light (sunlight, for example), the display of the image may be compromised. Furthermore, the individual's own visual system will affect the appearance of an image: the same image, viewed by two people, may appear to have different characteristics to each person. For our purpose, we shall assume that the computer setup is as optimal as possible, and the monitor is able to accurately reproduce the necessary gray values or colors in any image.

3.2   The `imshow` Function

Grayscale Images

MATLAB and Octave can display a grayscale image with `imshow` if the image matrix x is of type `uint8`, or of type `double` with all values in the range $0.0 - 1.0$. They can also display images of type `uint16`, but for Octave this requires that the underlying ImageMagick library is compiled to handle 16-bit images. Then

```
imshow(x)
```

will display x as an image.

This means that any image processing that produces an output of type `double` must either be scaled to the appropriate range or converted to type `uint8` before display.

If scaling is not done, then `imshow` will treat all values greater than 1 as white, and all values lower than 0 as black. Suppose we take an image and convert it to type `double`:

```
>> c = imread('caribou.png');
>> cd = double(c);
>> imshow(c),figure,imshow(cd)
```
MATLAB/Octave

The results are shown in Figure 3.1.

(a) The original image   (b) After conversion to type `double`

FIGURE 3.1: An attempt at data type conversion

As you can see, Figure 3.1(b) doesn't look much like the original picture at all! This is because any original values that were greater than 1 are now being displayed as white. In fact, the minimum value is 21, so that *every* pixel will be displayed as white. To display the matrix cd, we need to scale it to the range 0—1. This is easily done simply by dividing all values by 255:

```
>> imshow(cd/255)
```
MATLAB/Octave

and the result will be the caribou image as shown in Figure 3.1(a).

We can vary the display by changing the scaling of the matrix. Results of the commands:

```
>> imshow(cd/512)
>> imshow(cd/128)
```

are shown in Figure 3.2.

(a) The matrix **cd** divided by 512   (b) The matrix **cd** divided by 128

FIGURE 3.2: Scaling by dividing an image matrix by a scalar

Dividing by 512 darkens the image, as all matrix values are now between 0 and 0.5, so that the brightest pixel in the image is a mid-gray. Dividing by 128 means that the range is 0—2, and all pixels in the range 1—2 will be displayed as white. Thus, the image has an over-exposed, washed-out appearance.

The display of the result of a command whose output is a matrix of type **double** can be greatly affected by a judicious choice of a scaling factor.

We can convert the original image to **double** more properly using the function **im2double**. This applies correct scaling so that the output values are between 0 and 1. So the commands

```
>> cd = im2double(c);
>> imshow(cd)
```

will produce a correct image. It is important to make the distinction between the two functions **double** and **im2double**: **double** changes the data type but does not change the numeric values; **im2double** changes both the numeric data type *and* the values. The exception of course is if the original image is of type **double**, in which case **im2double** does nothing. Although the command **double** is not of much use for direct image display, it can be very useful for image arithmetic. We have seen examples of this above with scaling.

Corresponding to the functions **double** and **im2double** are the functions **uint8** and **im2uint8**. If we take our image **cd** of type **double**, properly scaled so that all elements are between 0 and 1, we can convert it back to an image of type **uint8** in two ways:

```
>> c2 = uint8(255*cd);
>> c3 = im2uint8(cd);
```

Use of `im2uint8` is to be preferred; it takes other data types as input, and always returns a correct result.

Python is less prescriptive about values in an image. The `io.imshow` method will automatically scale the image for display. So, for example:

```
In :  c = io.imread('caribou.png')
In :  cd1 = c.astype(float)
In :  cd2 = util.img_as_float(c)
```
Python

all produce images that are equally displayable with `io.imshow`. Note that the data types and values are all different:

```
In :  c.dtype, cd1.dtype, cd2.dtype
Out:  (dtype('uint8'), dtype('float64'), dtype('float64'))
```
Python

To see the numeric values, just check the minima and maxima of each array:

```
In :  c.min(), c.max()
Out:  (21, 254)
In :  cd1.min(), cd1.max()
Out:  (21.0, 254.0)
In :  cd2.min(), cd2.max()
Out:  (0.082352941176470587, 0.99607843137254903)
```
Python

This means that multiplying and dividing an image by a fixed value will not affect the display, as the result will be scaled. In order to obtain the effects of darkening and lightening, the default scaling, which uses the maximum and minimum values of the image matrix, can be adjusted by the use of two parameters: `vmin`, which gives the minimum gray level to be viewed, and `vmax`, which gives the maximum gray level.

For example, to obtain a dark caribou image:

```
In :  io.imshow(cd1/2,vmin=0,vmax=255)
```
Python

and to obtain a light, washed out image:

```
In :  io.imshow(cd1*2,vmin=0,vmax=255)
```
Python

Without the `vmin` and `vmax` values given, the image would be automatically scaled to look like the original.

Note that MATLAB/Octave and Python differ in their handling of arithmetic on uint8 numbers. For example:

```
>> a = uint8([0 80;160 240]);
>> (a-10)*2
ans =

     0    80
   160   255
```
MATLAB/Octave

MATLAB and Octave *clip* the output, so that values greater than 255 are set equal to 255, and values less than 0 are set equal to zero. But in Python the result is different:

```
In :   a = array([[0,80],[160,240]]).astype(uint8)
In :   (a-10)*2
Out:
array([[236, 140],
       [ 44, 204]], dtype=uint8)
```
Python

Python works *modulo* 256: numbers outside the range 0—255 are divided by 256 and the remainder given.

Binary images

Recall that a binary image will have only two values: 0 and 1. MATLAB and Octave do not have a `binary` data type as such, but they do have a `logical` flag, where `uint8` values as 0 and 1 can be interpreted as logical data. The logical flag will be set by the use of relational operations such as `==`, `<`, or `>` or any other operations that provide a yes/no answer. For example, suppose we take the caribou matrix and create a new matrix with

```
>> cl = c>120;
```
MATLAB/Octave

(we will see more of this type of operation in Chapter 4.) If we now check all of our variables with `whos`, the output will include the line:

```
>> whos c cl
  Name          Size           Bytes  Class      Attributes

  c          256x256          65536  uint8
  cl         256x256          65536  logical
```
MATLAB

The Octave output is slightly different:

```
  Attr Name       Size                    Bytes  Class
  ==== ====       ====                    =====  =====
       c          256x256                 65536  uint8
       cl         256x256                 65536  logical
```
Octave

This means that the command

```
>> imshow(cl)
```
MATLAB/Octave

will display the matrix as a binary image; the result is shown in Figure 3.3.

Suppose we remove the logical flag from `cl`; this can be done by a simple command:

```
>> clu = uint8(cl);
```
MATLAB/Octave

Now the output of `whos` will include the line:

(a) The caribou image turned binary   (b) After conversion to type `uint8`

FIGURE 3.3: Making the image binary

```
    clu      256x256        65536  uint8
```
MATLAB/Octave

If we now try to display this matrix with `imshow`, we obtain the result shown in Figure 3.3(b). A very disappointing image! But this is to be expected; in a matrix of type `uint8`, white is 255, 0 is black, and 1 is a very dark gray which is indistinguishable from black.

To get back to a viewable image, we can either turn the logical flag back on, and the view the result:

```
>> imshow(logical(clu))
```
MATLAB/Octave

or simply convert to type **double**:

```
>> imshow(double(clu))
```
MATLAB/Octave

Both these commands will produce the image seen in Figure 3.3.

Python   Python does have a boolean type:

```
In :  cl = c>120
In :  cl.dtype
Out:  dtype('bool')
```
Python

and the elements of the matrix `cl` are either **True** or **False**. This is still quite displayable with `io.imshow`. This matrix can be turned into a numeric matrix with any of:

```
In :  sk.util.img_as_ubyte(cl)
In :  uint8(cl*255)
In :  float16(cl*1)
```
Python

A numeric matrix x can be made boolean by

```
In :   x.astype(bool)
In :   util.img_as_bool(x)
```

`Python`

3.3 Bit Planes

Grayscale images can be transformed into a sequence of binary images by breaking them up into their *bit-planes*. If we consider the gray value of each pixel of an 8-bit image as an 8-bit binary word, then the 0th bit plane consists of the last bit of each gray value. Since this bit has the least effect in terms of the magnitude of the value, it is called the *least significant bit*, and the plane consisting of those bits the *least significant bit plane*. Similarly the 7th bit plane consists of the first bit in each value. This bit has the greatest effect in terms of the magnitude of the value, so it is called the *most significant bit*, and the plane consisting of those bits the *most significant bit plane*.

If we have a grayscale image of type `uint8`:

```
>> c = imread('cameraman.png');
```

`MATLAB/Octave`

then its bit planes can be accessed by the `bitget` function. In general, the function call

```
>> bitget(x,n)
```

`MATLAB/Octave`

will isolate the n-th rightmost bit from every element in x

All bitplanes can be simultaneously displayed using `subplot`, which places graphic objects such as images or plots on a rectangular grid in a single figure, but in order to minimize the distance between the images we can use the `position` parameter, starting off by creating the x and y positions for each subplot:

```
>> posx = [0 1 2 0 1 2 0 1]/3
>> posy = [2 2 2 1 1 1 0 0]/3
```

`MATLAB/Octave`

Then they can be plotted as:

```
>> for i = 1:8
>     subplot("position",[posx(i),posy(i),0.3,0.3]),
>     imshow(logical(bitget(c,i))),axis image
> end
```

`MATLAB/Octave`

The result is shown in Figure 3.4. Note that the least significant bit plane, at top left, is to all intents and purposes a random array, and that as the index value of the bit plane increases, more of the image appears. The most significant bit plane, which is at the bottom center, is actually a *threshold* of the image at level 128:

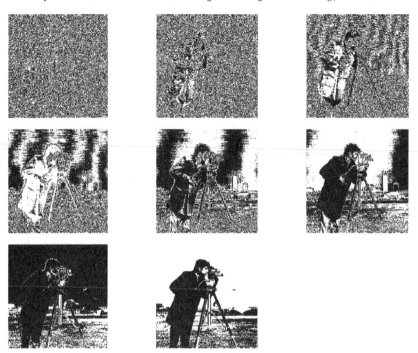

FIGURE 3.4: The bit planes of an 8-bit grayscale image

```
>> ct = c>127;
>> b8 = logical(bitget(c,8));
>> isequal(ct,b8)

ans =

    1
```
`MATLAB/Octave`

We shall discuss thresholding in Chapter 9.

We can recover and display the original image by multiplying each bitplane by an appropriate power of two and adding them; each scaled bitplane is added to an empty array which is initiated with the useful **zeros** function:

```
>> y = zeros(size(c));
>> for i = 1:8
>    y = y + 2^(i-1)*double(bitget(c,i));
>>   end
```
`MATLAB/Octave`

This new image **y** is the cameraman image:

```
>> isequal(c,uint8(y))
ans = 1
```
`MATLAB/Octave`

Python does not have a `bitget` function, however the equivalent result can be obtained by `bitshifting`: taking the binary string corresponding to a grayscale, and shifting it successively to the right, so the rightmost bits successively "drop off." Python uses the "»" notation for bitshifting:

```
In :   n = uint8(175)
In :   for i in range(8):
...:       print bin(n>>i)
...:
0b10101111
0b1010111
0b101011
0b10101
0b1010
0b101
0b10
0b1
```
Python

The rightmost bits can be obtained by using the modulo operator:

```
In :   for i in range(8):
...:       print (n>>i)%2
...:
1
1
1
1
0
1
0
1
```
Python

So the following generates and displays all bit planes using the `pyplot` module of the `matplotlib` library:

```
In :   import matplotlib.pyplot as plt
In :   c = io.imread('cameraman.png')
In :   bps = [(c>>i)%2 for i in range(8)]
In :   for i in range(8):
...:       plt.subplot(3,3,i+1)
...:       io.imshow(bps[i])
...:       plt.axis('off')
```
Python

and the bit planes can be reassembled with

```
In :   z = sum(bps[i]*2**i for i in range(8))
In :   (c==z).all()
Out:   True
```
Python

3.4   Spatial Resolution

Spatial resolution is the density of pixels over the image: the greater the spatial resolution, the more pixels are used to display the image. We can experiment with spatial resolution with MATLAB's `imresize` function. Suppose we have a 256×256 8-bit grayscale image saved to the matrix `x`. Then the command

```
imresize(x,1/2);
```

will halve the size of the image. It does this by taking out every other row and every other column, thus leaving only those matrix elements whose row and column indices are even:

$$
\begin{array}{cccccc}
x_{11} & x_{12} & x_{13} & x_{14} & x_{15} & x_{16} \quad \cdots \\
x_{21} & \boxed{x_{22}} & x_{23} & \boxed{x_{24}} & x_{25} & \boxed{x_{26}} \quad \cdots \\
x_{31} & x_{32} & x_{33} & x_{34} & x_{35} & x_{36} \quad \cdots \\
x_{41} & \boxed{x_{42}} & x_{43} & \boxed{x_{44}} & x_{45} & \boxed{x_{46}} \quad \cdots \\
x_{51} & x_{52} & x_{53} & x_{54} & x_{55} & x_{56} \quad \cdots \\
x_{61} & \boxed{x_{62}} & x_{63} & \boxed{x_{64}} & x_{65} & \boxed{x_{66}} \quad \cdots \\
\vdots & \vdots & \vdots & \vdots & \vdots & \vdots \quad \ddots
\end{array}
\longrightarrow \text{imresize(x,1/2)} \longrightarrow
\begin{array}{cccc}
x_{22} & x_{24} & x_{26} & \cdots \\
x_{42} & x_{44} & x_{46} & \cdots \\
x_{62} & x_{64} & x_{66} & \cdots \\
\vdots & \vdots & \vdots & \ddots
\end{array}
$$

If we apply `imresize` to the result with the parameter `2` and the method "`nearest`," rather than `1/2`, all the pixels are repeated to produce an image with the same size as the original, but with half the resolution in each direction:

$$
\begin{array}{ccc}
\boxed{\begin{array}{cc} x_{22} & x_{22} \\ x_{22} & x_{22} \end{array}} &
\boxed{\begin{array}{cc} x_{24} & x_{24} \\ x_{24} & x_{24} \end{array}} &
\boxed{\begin{array}{cc} x_{26} & x_{26} \\ x_{26} & x_{26} \end{array}} \cdots \\[2ex]
\boxed{\begin{array}{cc} x_{42} & x_{42} \\ x_{42} & x_{42} \end{array}} &
\boxed{\begin{array}{cc} x_{44} & x_{44} \\ x_{44} & x_{44} \end{array}} &
\boxed{\begin{array}{cc} x_{46} & x_{46} \\ x_{46} & x_{46} \end{array}} \cdots \\[2ex]
\boxed{\begin{array}{cc} x_{62} & x_{62} \\ x_{62} & x_{62} \end{array}} &
\boxed{\begin{array}{cc} x_{64} & x_{64} \\ x_{64} & x_{64} \end{array}} &
\boxed{\begin{array}{cc} x_{66} & x_{66} \\ x_{66} & x_{66} \end{array}} \cdots \\[2ex]
\vdots & \vdots & \vdots \quad \ddots
\end{array}
$$

The effective resolution of this new image is only 128×128. We can do all this in one line:

```
x2=imresize(imresize(x,1/2),2,'nearest');
```

By changing the parameters of `imresize`, we can change the effective resolution of the image to smaller amounts:

Command	Effective resolution
`imresize(imresize(x,1/4),4,'nearest');`	64×64
`imresize(imresize(x,1/8),8,'nearest');`	32×32
`imresize(imresize(x,1/16),16,'nearest');`	16×16
`imresize(imresize(x,1/32),32,'nearest');`	8×8

To see the effects of these commands, suppose we apply them to the image `thylacine.png`:[1]

---

[1] This is a famous image showing the last known *thylacine*, or Tasmanian Tiger, a now extinct carnivorous marsupial, in the Hobart Zoo in 1933.

```
>>  x = imread('thylacine.png;
>>_for_i_=_1:4
>____subplot(2,2,i),imshow(imresize(imresize(x,1/(2^(i+1))),2^(i+1),'
    nearest'))
>     end
```

Since this image has size 320×400, the effective resolutions will be these dimensions divided by powers of two.

(a) The original image   (b) at 160×200 resolution

FIGURE 3.5: Reducing resolution of an image

(a) At 80×100 resolution   (b) At 40×50 resolution

FIGURE 3.6: Further reducing the resolution of an image

The effects of increasing blockiness or *pixelization* become quite pronounced as the resolution decreases; even at 160×200 resolution fine detail, such as the wire mesh of the enclosure, are less clear, and at 80×100 all edges are now quite blocky. At 40×50, the image is barely recognizable, and at 20×25 and 10×12 the image becomes unrecognizable.

Python again is similar to MATLAB and Octave. To change the spatial resolution, use the **rescale** method from the **skimage.transform** module:

```
In :  import skimage.transform as tr
In :  x4 = tr.rescale(tr.rescale(x,0.25),4,order=0)
```

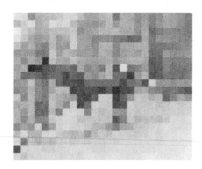
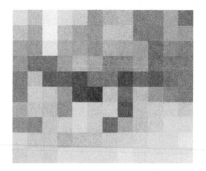

(a) At 20 × 25 resolution   (b) At 10 × 12 resolution

FIGURE 3.7: Reducing the resolution of an image even more

The "`order`" parameter indicates the method of interpolation used; 0 corresponds to nearest neighbor interpolation.

3.5   Quantization and Dithering

Quantization refers to the number of grayscales used to represent the image. As we have seen, most images will have 256 grayscales, which is more than enough for the needs of human vision. However, there are circumstances in which it may be more practical to represent the image with fewer grayscales. One simple way to do this is by *uniform quantization*: to represent an image with only n grayscales, we divide the range of grayscales into n equal (or nearly equal) ranges, and map the ranges to the values 0 to $n - 1$. For example, if $n = 4$, we map grayscales to output values as follows:

Original values	Output value
0–63	0
64–127	1
128–191	2
192–255	3

and the values 0, 1, 2, 3 may need to be scaled for display. This mapping can be shown graphically, as in Figure 3.8.

To perform such a mapping in MATLAB, we can perform the following operations, supposing x to be a matrix of type `uint8`:

```
>> q = (f/64)*64
```

MATLAB/Octave

Since `f` is of type `uint8`, dividing by 64 will also produce values of type `uint8`; so any fractional parts will be removed. For example:

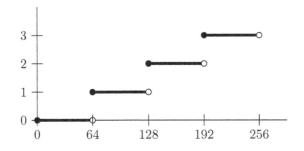

FIGURE 3.8: A mapping for uniform quantization

```
>> y = uint8(reshape(0:16:16*15,4,4))
y =

     0    64   128   192
    16    80   144   208
    32    96   160   224
    48   112   176   240
>> z = y/64
z =

     0     1     2     3
     0     1     2     3
     1     2     3     4
     1     2     3     4
>> z*64
ans =

     0    64   128   192
     0    64   128   192
    64   128   192   255
    64   128   192   255
```
MATLAB/Octave

The original array has now been quantized to only five different values. Given a `uint8` array x, and a value v, then in general the command

```
>> (x/v)*v
```
MATLAB/Octave

will produce a maximum of $256/v + 1$ different possible grayscales. So with $v = 64$ as above, we would expect $256/64 + 1 = 5$ possible output grayscales.

Python acts similarly:

```
In :  y = uint8(16*np.reshape(range(16),(4,4)))
In :  y
Out:
array([[  0,  16,  32,  48],
       [ 64,  80,  96, 112],
       [128, 144, 160, 176],
       [192, 208, 224, 240]], dtype=uint8)


In: z = y/64; z
Out:
array([[0, 0, 0, 0],
       [1, 1, 1, 1],
       [2, 2, 2, 2],
       [3, 3, 3, 3]], dtype=uint8)


In :  z*64
Out:
array([[  0,   0,   0,   0],
       [ 64,  64,  64,  64],
       [128, 128, 128, 128],
       [192, 192, 192, 192]], dtype=uint8)
```

Python

Note that dividing gives a different result in MATLAB/Octave and Python. In the former, the result is the closest integer (by rounding) to the value of $n/64$; in Python, the result is the floor.

If we wish to precisely specify the number of output grayscales, then we can use the **grayslice** function. Given an image matrix x and an integer n, the MATLAB command **grayslice(x,n)** produces a matrix whose values have been reduced to the values $0, 1, \ldots, n - 1$. So, for example

```
>> x4 = grayslice(x,4);
```

MATLAB

will produce a **uint8** version of our image with values 0, 1, 2 and 3. Note that Octave works slightly differently from MATLAB here; Octave requires that the input image be of type **double**:

```
>> x4 = grayslice(im2double(x),4);
```

Octave

We cannot view this directly, as it will appear completely black: the four values are too close to zero to be distinguishable. We need to treat this matrix as the indices to a color map, and the color map we shall use is **gray(4)**, which produces a color map of four evenly spaced gray values between 0 (black) and 1.0 (white). In general, given an image x and a number of grayscales n, the commands

```
>> y = grayslice(x,n);
>> imshow(x,gray(n))
```

MATLAB/Octave

will display the quantized image. Note that if you are using Octave, you will need to ensure that x is of type **double**. Quantized images can be displayed in one go using **subplot**:

```
>> qs = [256,64,32,16,4,2]
>> px = [1/6,1/2,1/6,1/2,1/6,1/2]
>> py = [2/3,2/3,1/3,1/3,0,0]
>> for i = 1:6
>     subplot("position",[px[i],py[i],1/3,1/3]),imshow(grayslice(x,qs[i]),
   gray(qs[i]))
>   end
```

MATLAB/Octave

If we apply these commands to the thylacine image, we obtain the results shown in Figures 3.9 to 3.12.

(a) The image quantized to 128 grayscales (b) The image quantized to 64 grayscales

FIGURE 3.9: Quantization (1)

(a) The image quantized to 32 grayscales (b) The image quantized to 16 grayscales

FIGURE 3.10: Quantization (2)

One immediate consequence of uniform quantization is that of "false contours," most noticeable with fewer grayscales. For example, in Figure 3.11, we can see on the ground and rear fence that the gray texture is no longer smooth; there are observable discontinuities between different gray values. We may expect that if fewer grayscales are used, and the jumps between consecutive grayscales becomes larger, that such false contours will occur.

(a) The image quantized to 8 grayscales   (b) The image quantized to 4 grayscales

FIGURE 3.11: Quantization (3)

FIGURE 3.12: The image quantized to 2 grayscales

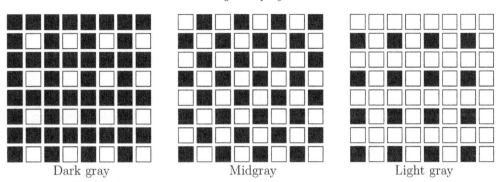

Dark gray   Midgray   Light gray

FIGURE 3.13: Patterns for dithering output

Dithering

Dithering, in general terms, refers to the process of reducing the number of colors in an image. For the moment, we shall be concerned only with grayscale dithering. Dithering is necessary sometimes for display, if the image must be displayed on equipment with a limited number of colors, or for printing. Newsprint, in particular, only has two scales of gray: black and white. Representing an image with only two tones is also known as *halftoning*.

One method of dealing with such false contours involves adding random values to the image before quantization. Equivalently, for quantization to 2 grayscales, we may compare the image to a random matrix r. The trick is to devise a suitable matrix so that grayscales are represented evenly in the result. For example, an area containing mid-way gray level (around 127) would have a checkerboard pattern; a darker area will have a pattern containing more black than white, and a light area will have a pattern containing more white than black. Figure 3.13 illustrates this. One standard matrix is

$$D = \begin{pmatrix} 0 & 128 \\ 192 & 64 \end{pmatrix}$$

which is repeated until it is as big as the image matrix, when the two are compared. Suppose $d(i,j)$ is the matrix obtained by replicating D. Thus, an output pixel $p(i,j)$ is defined by

$$p(i,j) = \begin{cases} 1 & \text{if } x(i,j) > d(i,j) \\ 0 & \text{if } x(i,j) \leq d(i,j) \end{cases}$$

This approach to quantization is called *dithering*, and the matrix D is an example of a *dither matrix*. Another dither matrix is given by

$$D_2 = \begin{pmatrix} 0 & 128 & 32 & 160 \\ 192 & 64 & 224 & 96 \\ 48 & 176 & 16 & 144 \\ 240 & 112 & 208 & 80 \end{pmatrix}$$

We can apply the matrices to our thylacine image matrix x by using the following commands in MATLAB/Octave:

```
>> D = [0 128;192 64]
>> r = repmat(D,160,200);
>> x2 = x>r; imshow(x2)
>> D2 = [0 128 32 160;192 64 224 96;48 176 16 144;240 112 208 80];
>> r2 = repmat(D2,80,100);
>> x4 = x>r2; imshow(x4)
```

MATLAB/Octave

or in Python:

```
In :  D = array([[0,128],[192,64]])
In :  r = np.tile(D,(160,200));
In :  x2 = (x>r).astype(uint8)
In :  io.imshow(x2)
In :  D2 = array([[0, 128, 32, 160],[192, 64, 224, 96],[48, 176, 16, 144],\
...:  [240, 112, 208, 80]])
In :  r2 = np.tile(D2,80,100);
In :  x4 = (x>r2).astype(uint8)
In :  io.imshow(x2)
```

Python

The results are shown in Figure 3.14. The dithered images are an improvement on the uniformly quantized image shown in Figure 3.12. General dither matrices are provided by Hawley [15].

(a) The thylacine image dithered using D

(b) The thylacine image dithered using D_2

FIGURE 3.14: Examples of dithering

Dithering can be easily extended to more than two output gray values. Suppose, for example, we wish to quantize to four output levels 0, 1, 2, and 3. Since $255/3 = 85$, we first quantize by dividing the gray value $x(i,j)$ by 85:

$$q(i,j) = [x(i,j)/85].$$

This will produce only the values 0, 1, and 2, except for when $x(i,j) = 255$. Suppose now that our replicated dither matrix $d(i,j)$ is scaled so that its values are in the range $0\ldots 85$. The final value $p(i,j)$ is then defined by

$$p(i,j) = q(i,j) + \begin{cases} 1 & \text{if } x(i,j) - 85q(i,j) > d(i,j) \\ 0 & \text{if } x(i,j) - 85q(i,j) \le d(i,j) \end{cases}$$

This can be easily implemented in a few commands, modifying D slightly from above and using the `repmat` function which simply tiles an array as many times as given:

```
>> D = [0 56;84 28]
>> r = repmat(D,128,128);
>> x = double(x);
>> q = floor(x/85);
>> x4 = q+(x-85*q>r);
>> imshow(uint8(85*x4))
```
MATLAB/Octave

The commands in Python are similar:

```
In :  D = array([[0, 56],[84, 28]])
In :  r = np.tile(D,(128,128))
In :  x = x.astype(float64)
In :  q = floor(x/85)
In :  x4 = q+(x-85*q>r)
In :  io.imshow(uint8(x4*85))
```
Python

and the result is shown in Figure 3.15(a). We can dither to eight gray levels by using $255/7 = 37$ (we round the result up to ensure that our output values stay within range) instead of 85 above; our starting dither matrix will be

$$D = \begin{pmatrix} 0 & 24 \\ 36 & 12 \end{pmatrix}$$

and the result is shown in Figure 3.15(b). Note how much better these images look than the corresponding uniformly quantized images in Figures 3.11(b) and 3.11(a). The eight-level result, in particular, is almost indistinguishable from the original, in spite of the quantization.

(a) Dithering to 4 output grayscales   (b) Dithering to 8 output grayscales

FIGURE 3.15: Dithering to more than two grayscales

Error Diffusion

A different approach to quantization from dithering is that of *error diffusion*. The image is quantized at two levels, but for each pixel we take into account the *error* between its gray value and its quantized value. Since we are quantizing to gray values 0 and 255, pixels close to these values will have little error. However, pixels close to the center of the range: 128, will have a large error. The idea is to spread this error over neighboring pixels. A popular method, developed by Floyd and Steinberg, works by moving through the image pixel by pixel, starting at the top left, and working across each row in turn. For each pixel $p(i,j)$ in the image we perform the following sequence of steps:

1. Perform the quantization.

2. Calculate the quantization error. This is defined as:

$$E = \begin{cases} p(i,j) & \text{if } p(i,j) < 128 \\ p(i,j) - 255 & \text{if } p(i,j) >= 128 \end{cases}$$

3. Spread this error E over pixels to the right and below according to this table:

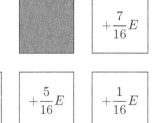

There are several points to note about this algorithm:

- The error is spread to pixels *before* quantization is performed on them. Thus, the error diffusion will affect the quantization level of those pixels.

- Once a pixel has been quantized, its value will never be affected. This is because the error diffusion only affects pixels to the right and below, and we are working from the left and above.

- To implement this algorithm, we need to embed the image in a larger array of zeros, so that the indices do not go outside the bounds of our array.

The `dither` function, when applied to a grayscale image, actually implements Floyd-Steinberg error diffusion. However, it is instructive to write a simple MATLAB function to implement it ourselves; one possibility is given at the end of the chapter.

The result of applying this function to the thylacine image is shown in Figure 3.16. Note that the result is very pleasing; so much so that it is hard to believe that every pixel in the image is either back or white, and so that we have a binary image.

FIGURE 3.16: The thylacine image after Floyd-Steinberg error diffusion

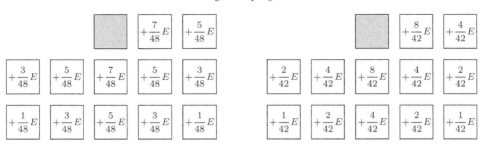

FIGURE 3.17: Different error diffusion schemes

Other error diffusion schemes are possible; two such schemes are Jarvis-Judice-Ninke and Stucki, which have error diffusion schemes shown in Figure 3.17.

They can be applied by modifying the Floyd-Steinberg function given at the end of the chapter. The results of applying these two error diffusion methods are shown in Figure 3.18.

(a) Result of Jarvis-Judice-Ninke error diffusion   (b) Result of Stucki error diffusion

FIGURE 3.18: Using other error-diffusion schemes

A good introduction to error diffusion (and dithering) is given by Schumacher [44].

3.6   Programs

Here are programs for error diffusion, first Floyd-Steinberg (with arbitrary levels), in MATLAB/Octave:

```MATLAB/Octave
function y = fs(x,k)
height = size(x,1);
width = size(x,2);
ed = [0 0 0 7 0;0 3 5 1 0;0 0 0 0 0]/16;
y = uint8(zeros(height,width));
z = zeros(height+4,width+4);
z(3:height+2,3:width+2) = x;
for i = 3:height+2,
  for j = 3:width+2,
    quant = floor(255/(k-1))*floor(z(i,j)*k/256);
    y(i-2,j-2) = quant;
    e = z(i,j)-quant;
    z(i:i+2,j-2:j+2) = z(i:i+2,j-2:j+2)+e*ed;
  endfor
endfor
endfunction
```

and now Jarvis-Judice-Ninke with two levels:

```MATLAB/Octave
function out = jjn(im)
height = size(im,1);
width = size(im,2);
out = zeros(size(im));
ed = [0 0 0 7 5;3 5 7 5 3;1 3 5 3 1]/48;
z = zeros(size(im)+4);
z(3:height+2,3:width+2) = double(im);
for i = 3:height+2,
  for j = 3:width+2,
    quant = 255*(z(i,j)>=128);
    out(i-2,j-2) = quant;
    e = z(i,j)-quant;
    z(i:i+2,j-2:j+2) = z(i:i+2,j-2:j+2)+e*ed;
  endfor
endfor
out = im2uint8(out);
endfunction
```

Here is Floyd-Steinberg in Python:

```Python
def fs(im,k):  #FS error diffusion at k levels
    rs,cs = im.shape
    ed = array([[0,0,7],[3,5,1]])/16.0
    z = zeros((rs+2,cs+2))
    z[1:rs+1,1:cs+1] = float64(im)
    for i in range(1,rs+1):
        for j in range(1,cs+1):
            old = z[i,j]
            new = (old//(255//k))*(255//(k-1))
            z[i,j] = new
            E = old - new
            z[i:i+2,j-1:j+2] = z[i:i+2,j-1:j+2]+E*ed;
    return uint8(z[1:rs+1,1:cs+1])
```

and Jarvis-Judice-Ninke:

```Python
def jjn(im): #JJN error diffusion at two levels
    rs,cs = im.shape
    ed = array([[0,0,0,7,5],[3,5,7,5,3],[1,3,5,3,1]])/48.0
    z = zeros((rs+4,cs+4))
    z[2:rs+2,2:cs+2] = float64(im)
    for i in range(2,rs+2):
        for j in range(2,cs+2):
            old = z[i,j]
            new = (old//128)*255
            z[i,j] = new
            E = old - new
            z[i:i+3,j-2:j+3] = z[i:i+3,j-2:j+3]+E*ed;
    return uint8(z[2:rs+2,2:cs+2]>0)
```

Exercises

1. Open the grayscale image `cameraman.png` and view it. What data type is it?

2. Enter the following commands (if you are using MATLAB or Octave):

```MATLAB/Octave
>> [em,map] = imread('emu.png');
>> e = ind2gray(em,map);
```

and if you are using Python:

```Python
In :  import skimage.color as co
In :  em = io.imread('emu.png')
In :  e = co.rgb2gray(em)
```

These will produce a grayscale image of type **double**. View this image.

3. Enter the command

```MATLAB/Octave
>> e2 = im2uint8(e);
```

or

```Python
In :  import skimage.util as ut
In :  e2 = ut.img_as_ubyte(e)
```

and view the output.

What does the function `im2uint8`/`img_as_ubyte` do? What affect does it have on

(a) The appearance of the image?

(b) The elements of the image matrix?

4. What happens if you apply that function to the cameraman image?

5. Experiment with reducing spatial resolution of the following images:

(a) `cameraman.png`

(b) The grayscale emu image

(c) `blocks.png`

(d) `buffalo.png`

In each case, note the point at which the image becomes unrecognizable.

6. Experiment with reducing the quantization levels of the images in the previous question. Note the point at which the image becomes seriously degraded. Is this the same for all images, or can some images stand lower levels of quantization than others?

Check your hypothesis with some other grayscale images.

7. Look at a grayscale photograph in a newspaper with a magnifying glass. Describe the colors you see.

8. Show that the 2×2 dither matrix D provides appropriate results on areas of unchanging gray. Find the results of $D > G$ when G is a 2×2 matrix of values (a) 50, (b) 100, (c) 150, (d) 200.

9. What are the necessary properties of D to obtain the appropriate patterns for the different input gray levels?

10. How do quantization levels effect the result of dithering? Use `gray2ind` to display a grayscale image with fewer grayscales, and apply dithering to the result.

11. Apply each of Floyd-Steinberg, Jarvis-Judice-Ninke, and Stucki error diffusion to the images in Question 5. Which of the images looks best? Which error-diffusion method seems to produce the best results?

Can you isolate what aspects of an image will render it most suitable for error-diffusion?

Chapter 4

Point Processing

4.1 Introduction

Any image processing operation transforms the values of the pixels. However, image processing operations may be divided into three classes based on the information required to perform the transformation. From the most complex to the simplest, they are:

1. **Transforms.** A "transform" represents the pixel values in some other, but equivalent form. Transforms allow for some very efficient and powerful algorithms, as we shall see later on. We may consider that in using a transform, the entire image is processed as a single large block. This may be illustrated by the diagram shown in Figure 4.1.

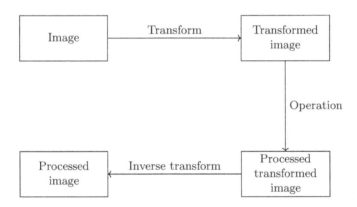

FIGURE 4.1: Schema for transform processing

2. **Neighborhood processing.** To change the gray level of a given pixel, we need only know the value of the gray levels in a small neighborhood of pixels around the given pixel.

3. **Point operations.** A pixel's gray value is changed without any knowledge of its surrounds.

Although point operations are the simplest, they contain some of the most powerful and widely used of all image processing operations. They are especially useful in image *pre-processing*, where an image is required to be modified before the main job is attempted.

4.2   Arithmetic Operations

These operations act by applying a simple function

$$y = f(x)$$

to each gray value in the image. Thus, $f(x)$ is a function which maps the range $0 \ldots 255$ onto itself. Simple functions include adding or subtracting a constant value to each pixel:

$$y = x \pm C$$

or multiplying each pixel by a constant:

$$y = Cx.$$

In each case, the output will need to be adjusted in order to ensure that the results are integers in the $0 \ldots 255$ range. We have seen that MATLAB and Octave work by "clipping" the values by setting:

$$y \leftarrow \begin{cases} 255 & \text{if } y > 255, \\ 0 & \text{if } y < 0. \end{cases}$$

and Python works by dividing an overflow by 256 and returning the remainder. We can see this by considering a list of `uint8` values in each system, first in MATLAB/Octave

```
>> x = uint8(0:32:255)
ans =
    0   32   64   96  128  160  192  224

>> x + 100
ans =
  100  132  164  196  228  255  255  255

>> x - 100
ans =
    0    0    0    0   28   60   92  124
```
MATLAB/Octave

and then in Python with `arange`, which produces a list of values as an array:

```
In : x = uint8(array(arange(0,255,32)));x
Out: array([  0,  32,  64,  96, 128, 160, 192, 224], dtype=uint8)

In : x + 100
Out: array([100, 132, 164, 196, 228,   4,  36,  68], dtype=uint8)

In : x - 100
Out: array([156, 188, 220, 252,  28,  60,  92, 124], dtype=uint8)
```
Python

We can obtain an understanding of how these operations affect an image by plotting $y = f(x)$. Figure 4.2 shows the result of adding or subtracting 64 from each pixel in the image. Notice that when we add 128, all gray values of 127 or greater will be mapped to 255. And

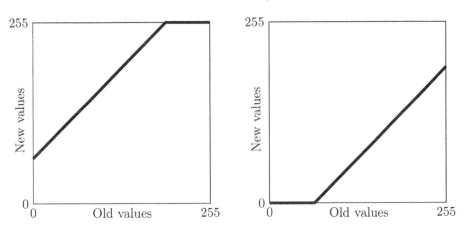

FIGURE 4.2: Adding and subtracting a constant in MATLAB and Octave

when we subtract 128, all gray values of 128 or less will be mapped to 0. By looking at these graphs, we observe that in general adding a constant will lighten an image, and subtracting a constant will darken it.

Figure 4.3 shows the results in Python.

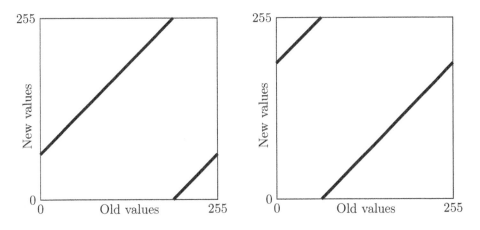

FIGURE 4.3: Adding and subtracting a constant in Python

We can test this on the "blocks" image **blocks.png**, which we can see in Figure 1.4. We start by reading the image in:

```
>> b = imread('blocks.png');
>> class(b)
ans =
    uint8
```
MATLAB/Octave

The point of the second command **class** is to find the numeric data type of b; it is **uint8**. This tells us that adding and subtracting constants will result in automatic clipping:

Adding 128   Subtracting 128

FIGURE 4.4: Arithmetic operations on an image: adding or subtracting a constant

```
>> imshow(b+128)
>> figure, imshow(b-128)
```
MATLAB/Octave

and the results are seen in Figure 4.4. Because of Python's arithmetic, the results of

```
In : b = io.imread('blocks.png');
In : io.imshow(b+128)
```
Python

will not be a brightened image: it is shown in Figure 4.5.

FIGURE 4.5: Adding 128 to an image in Python

In order to obtain the same results as given by MATLAB and Octave, first we need to use floating point arithmetic, and clip the output with `np.clip`:

```
In :  bf = float64(b)
In :  b1 = uint8(np.clip(bf+128,0,255))
In :  b2 = uint8(np.clip(bf-128,0,255))
```
Python

Then **b1** and **b2** will be the same as the images shown in Figure 4.4.

We can also perform lightening or darkening of an image by multiplication; Figure 4.6 shows some examples of functions that will have these effects. Again, these functions assume clipping. All these images can be viewed with `imshow`; they are shown in Figure 4.7.

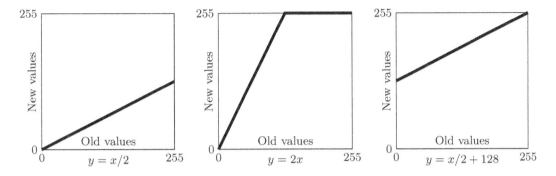

FIGURE 4.6: Using multiplication and division

| imshow(b/2) | imshow(b*2) | b/2+128 |

FIGURE 4.7: Arithmetic operations on an image: multiplication and division

Compare the results of darkening with division by two and subtraction by 128. Note that the division result, although darker than the original, is still quite clear, whereas a lot of information was lost by the subtraction process. This is because in image **b2** all pixels with gray values 128 or less have become zero.

A similar loss of information has occurred by adding 128, and so a better result is obtained by **b/2+128**.

Complements

The *complement* of a grayscale image is its photographic negative. If an image matrix m is of type `uint8` and so its gray values are in the range 0 to 255, we can obtain its negative with the command

```
>> 255 - m
```

MATLAB/Octave

(and the same in Python). If the image is binary, we can use

```
>> ~m
```
MATLAB/Octave

which also works for boolean arrays in Python.

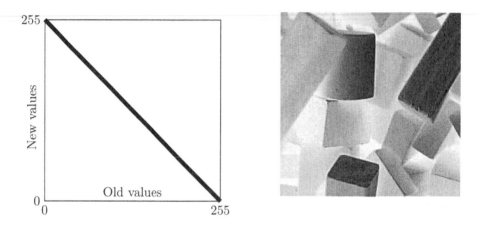

FIGURE 4.8: Image complementation: `255-b`

Interesting special effects can be obtained by complementing only *part* of the image; for example by taking the complement of pixels of gray value 128 or less, and leaving other pixels untouched. Or, we could take the complement of pixels that are 128 or greater and leave other pixels untouched. Figure 4.9 shows these functions.

The effect of these functions is called *solarization*, and will be discussed further in Chapter 16.

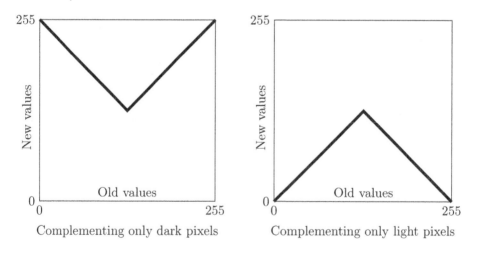

FIGURE 4.9: Part complementation

4.3 Histograms

Given a grayscale image, its *histogram* consists of the histogram of its gray levels; that is, a graph indicating the number of times each gray level occurs in the image. We can infer a great deal about the appearance of an image from its histogram, as the following examples indicate:

- In a dark image, the gray levels (and hence the histogram) would be clustered at the lower end.

- In a uniformly bright image, the gray levels would be clustered at the upper end.

- In a well-contrasted image, the gray levels would be well spread out over much of the range.

We can view the histogram of an image in MATLAB by using the imhist function:

```
>> c = imread('chickens.png');
>> imshow(c),figure,imhist(c),axis tight
```

MATLAB/Octave

(the axis tight command ensures the axes of the histogram are automatically scaled to fit all the values in). In Python, the commands are:

```
In : c = io.imread('chickens.png')
In : io.imshow(c)
In : f = figure(); f.show(plt.hist(c.flatten(),bins=256))
```

Python

The result is shown in Figure 4.10. Since most of the gray values are all clustered

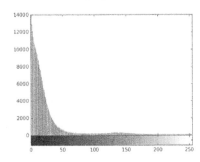

FIGURE 4.10: The image chickens.png and its histogram

together at the left of the histogram, we would expect the image to be dark and poorly contrasted, as indeed it is.

Given a poorly contrasted image, we would like to enhance its contrast by spreading out its histogram. There are two ways of doing this.

Histogram Stretching (Contrast Stretching)

Suppose we have an image with the histogram shown in Figure 4.11, associated with a

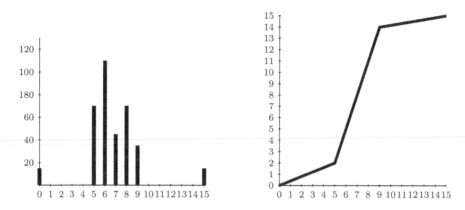

FIGURE 4.11: A histogram of a poorly contrasted image and a stretching function

table of the numbers n_i of gray values:

Gray level i	0	1	2	3	4	5	6	7	8	9	10	11	12	13	14	15
n_i	15	0	0	0	0	70	110	45	70	35	0	0	0	0	0	15

(with $n = 360$, as before.) We can stretch the gray levels in the center of the range out by applying the piecewise linear function shown at the right in Figure 4.11. This function has the effect of stretching the gray levels 5–9 to gray levels 2–14 according to the equation:

$$j = \frac{14 - 2}{9 - 5}(i - 5) + 2$$

where i is the original gray level and j its result after the transformation. Gray levels outside this range are either left alone (as in this case) or transformed according to the linear functions at the ends of the graph above. This yields:

i	5	6	7	8	9
j	2	5	8	11	14

and the corresponding histogram given in Figure 4.12:

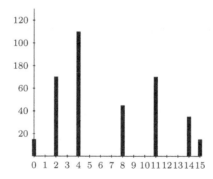

FIGURE 4.12: Histogram after stretching

which indicates an image with greater contrast than the original.

MATLAB/Octave: Use of `imadjust`.

To perform histogram stretching in MATLAB or Octave the `imadjust` function may be used. In its simplest incarnation, the command

```
imadjust(im,[a,b],[c,d])
```

stretches the image according to the function shown in Figure 4.13. Since `imadjust` is

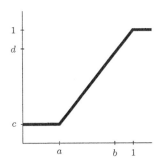

FIGURE 4.13: The stretching function given by `imadjust`

designed to work equally well on images of type `double`, `uint8`, or `uint16`, the values of a, b, c, and d must be between 0 and 1; the function automatically converts the image (if needed) to be of type `double`.

Note that `imadjust` does not work quite in the same way as shown in Figure 4.11. Pixel values less than a are all converted to c, and pixel values greater than b are all converted to d. If either of `[a,b]` or `[c,d]` are chosen to be `[0,1]`, the abbreviation `[]` may be used. Thus, for example, the command

```
>> imadjust(im,[],[])
```

MATLAB/Octave

does nothing, and the command

```
>> imadjust(im,[],[1,0])
```

MATLAB/Octave

inverts the gray values of the image, to produce a result similar to a photographic negative.

The `imadjust` function has one other optional parameter: the *gamma* value, which describes the shape of the function between the coordinates (a, c) and (b, d). If `gamma` is equal to 1, which is the default, then a linear mapping is used, as shown in Figure 4.13. However, values less than one produce a function that is concave downward, as shown on the left in Figure 4.14, and values greater than one produce a figure that is concave upward, as shown on the right in Figure 4.14.

The function used is a slight variation on the standard line between two points:

$$y = \left(\frac{x-a}{b-a}\right)^{\gamma}(d-c)+c.$$

Use of the gamma value alone can be enough to substantially change the appearance of the image. For example, with the chickens image:

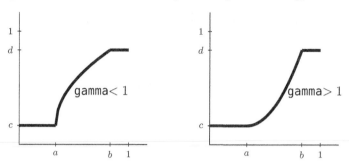

FIGURE 4.14: The `imadjust` function with `gamma` not equal to 1

```
>> ca1 = imadjust(t,[],[],0.5);
>> ca2 = imadjust(t,[],[],0.25);
>> imshow(ca1),figure,imshow(ca2)
```
MATLAB/Octave

produces the result shown in Figure 4.15.

FIGURE 4.15: The chickens image with different adjustments with the `gamma` value

Both results show background details that are hard to determine in the original image. Finding the correct gamma value for a particular image will require some trial and error.

The `imadjust` stretching function can be viewed with the `plot` function. For example,

```
>> plot(c,ca1,'.'),axis tight
```
MATLAB/Octave

produces the plot shown in Figure 4.16. Since `p` and `ph` are matrices that contain the original values and the values after the `imadjust` function, the `plot` function simply plots them, using dots to do it.

Adjustment in Python

Adjustment methods are held in the **exposure** module of **skimage**. Adjustment with gamma can be achieved with:

```
In :   import skimage.exposure as ex
In :   c = io.imread('chickens.png')
In :   ca1 = ex.adjust_gamma(c,0.5)
```
Python

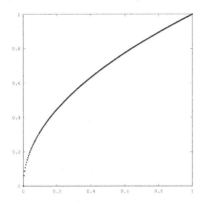

FIGURE 4.16: The function used in Figure 4.15

A Piecewise Linear Stretching Function

We can easily write our own function to perform piecewise linear stretching as shown in Figure 4.17. This is easily implemented by first specifying the points (a_i, b_i) involved and

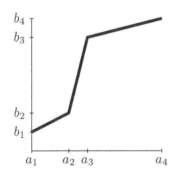

FIGURE 4.17: A piecewise linear stretching function

then using the one-dimensional interpolation function `interp1` to join them. In this case, we might have:

```
>> a = [0 100 150 255]
>> b = [40 100 220 255]
>> lin = interp1(a,b,0:255,'linear');
```
MATLAB/Octave

Then it can be applied to the image with

```
>> cl = uint8(lin(c+1));
```
MATLAB/Octave

Python also provides easy interpolation:

```
In :  a = array([0, 100, 150, 255])
In :  b = array([40, 100, 220, 255])
In :  lin = np.interp(range(256),a,b)
```
Python

Then it can be applied to the image with

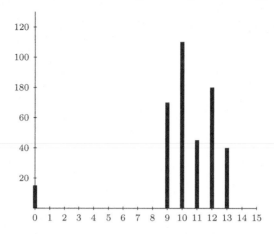

FIGURE 4.18: Another histogram indicating poor contrast

```
>> cl = uint8(lin[c])
```

Python

Histogram Equalization

The trouble with any of the above methods of histogram stretching is that they require user input. Sometimes a better approach is provided by *histogram equalization*, which is an entirely automatic procedure. The idea is to change the histogram to one that is uniform so that every bar on the histogram is of the same height, or in other words each gray level in the image occurs with the same frequency. In practice, this is generally not possible, although as we shall see the result of histogram equalization provides very good results.

Suppose our image has L different gray levels $0, 1, 2, \ldots L - 1$, and that gray level i occurs n_i times in the image. Suppose also that the total number of pixels in the image is n so that $n_0 + n_1 + n_2 + \cdots + n_{L-1} = n$. To transform the gray levels to obtain a better contrasted image, we change gray level i to

$$\left(\frac{n_0 + n_1 + \cdots + n_i}{n} \right) (L - 1).$$

and this number is rounded to the nearest integer.

An Example

Suppose a 4-bit grayscale image has the histogram shown in Figure 4.18 associated with a table of the numbers n_i of gray values:

Gray level i	0	1	2	3	4	5	6	7	8	9	10	11	12	13	14	15
n_i	15	0	0	0	0	0	0	0	0	70	110	45	80	40	0	0

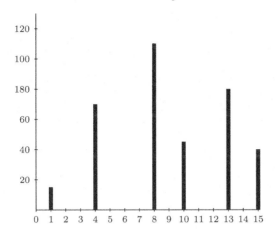

FIGURE 4.19: The histogram of Figure 4.18 after equalization

(with $n = 360$.) We would expect this image to be uniformly bright, with a few dark dots on it. To equalize this histogram, we form running totals of the n_i, and multiply each by $15/360 = 1/24$:

Gray level i	n_i	Σn_i	$(1/24)\Sigma n_i$	Rounded value
0	15	15	0.63	1
1	0	15	0.63	1
2	0	15	0.63	1
3	0	15	0.63	1
4	0	15	0.63	1
5	0	15	0.63	1
6	0	15	0.63	1
7	0	15	0.63	1
8	0	15	0.63	1
9	70	85	3.65	4
10	110	195	8.13	8
11	45	240	10	10
12	80	320	13.33	13
13	40	360	15	15
14	0	360	15	15
15	0	360	15	15

We now have the following transformation of gray values, obtained by reading off the first and last columns in the above table:

Original gray level i	0	1	2	3	4	5	6	7	8	9	10	11	12	13	14	15
Final gray level j	1	1	1	1	1	1	1	1	1	4	8	10	13	15	15	15

and the histogram of the j values is shown in Figure 4.19. This is far more spread out than the original histogram, and so the resulting image should exhibit greater contrast.

To apply histogram equalization in MATLAB or Octave, use the `histeq` function; for example:

```
>> c = imread('chickens.png');
>> ch = histeq(c);
>> imshow(ch),figure,imhist(ch),axis tight
```
MATLAB/Octave

Python supplies the **equalize_hist** method in the **exposure** module:

```
In :   c = io.imread('chickens.png')
In :   import sk.exposure as ex
In :   ch = ex.equalize_hist(c)
In :   f = figure(); f.show(plt.hist(ch.flatten(),bins=256))
```
Python

applies histogram equalization to the chickens image, and produces the resulting histogram. These results are shown in Figure 4.20. Notice the far greater spread of the histogram. This

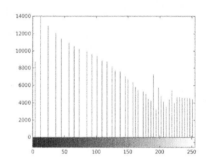

FIGURE 4.20: The histogram of Figure 4.10 after equalization

corresponds to the greater increase of contrast in the image.

We give one more example, that of a very dark image. For example, consider an under-exposed image of a sunset:

```
>> s = imread('sunset.png');
>> imshow(s)
```
MATLAB/Octave

It can be seen both from the image and the histogram shown in Figure 4.21 that the matrix e contains mainly low values, and even without seeing the image first we can infer from its histogram that it will appear very dark when displayed. We can display this matrix and its histogram with the usual commands:

```
>> sh = imhist(s);
>> imhow(sh),figure,imhist(sh),axis auto
```
MATLAB/Octave

or in Python as

```
In :   s = io.imread('sunset.png')
In :   io.imshow(s)
In :   f = figure(); f.show(plt.hist(s.flatten(),bins=256))
```
Python

and improve the contrast of the image with histogram equalization, as well as displaying the resulting histogram. The results are shown in the top row of Figure 4.21.

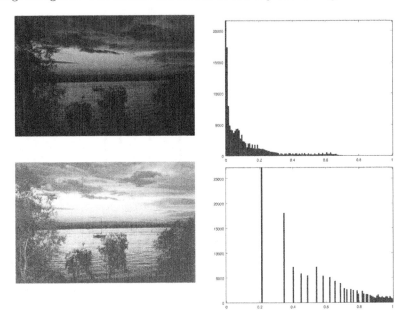

FIGURE 4.21: The sunset image and its histogram, with equalization

As you see, the very dark image has a corresponding histogram heavily clustered at the lower end of the scale.

But we can apply histogram equalization to this image, and display the results:

```
>> sh = histeq(s);
>> imshow(sh),figure,imhist(sh),axis tight
```

MATLAB/Octave

or

```
In :  sh = ex.equalize_hist(s)
In :  f = figure(); f.show(plt.hist(sh.flatten(),bins=256))
```

Python

and the results are shown in the bottom row of Figure 4.21. Note that many details that are obscured in the original image are now quite clear: the trunks of the foreground trees, the ripples on the water, the clouds at the very top.

Why It Works

Consider the histogram in Figure 4.18. To apply histogram stretching, we would need to stretch out the values between gray levels 9 and 13. Thus, we would need to apply a piecewise function similar to that shown in Figure 4.11.

Let's consider the cumulative histogram, which is shown in Figure 4.22. The dashed line is simply joining the top of the histogram bars. However, it can be interpreted as an appropriate histogram stretching function. To do this, we need to scale the y values so that they are between 0 and 15, rather than 0 and 360. But this is precisely the method described in Section 4.3.

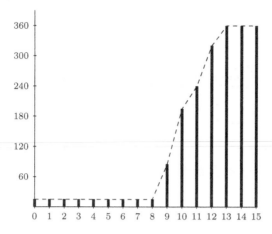

FIGURE 4.22: The cumulative histogram

As we have seen, none of the example histograms, after equalization, are uniform. This is a result of the discrete nature of the image. If we were to treat the image as a continuous function $f(x, y)$, and the histogram as the area between different contours (see, for example, Castleman [7]), then we can treat the histogram as a probability density function. But the corresponding cumulative density function will always have a uniform histogram; see, for example, Hogg and Craig [17].

4.4 Lookup Tables

Point operations can be performed very effectively by the use of a *lookup table*, known more simply as an LUT. For operating on images of type `uint8`, such a table consists of a single array of 256 values, each value of which is an integer in the range $0 \ldots 255$. Then our operation can be implemented by replacing each pixel value p by the corresponding value t_p in the table.

For example, the LUT corresponding to division by 2 looks like:

Index:	0	1	2	3	4	5	...	250	251	252	253	254	255
LUT:	0	0	1	1	2	2	...	125	125	126	126	127	127

This means, for example, that a pixel with value 4 will be replaced with 2; a pixel with value 253 will be replaced with value 126.

If `T` is a lookup table, and `im` is an image, then the lookup table can be applied by the simple command

    T(im) or T[im]

depending on whether we are working in MATLAB/Octave or Python.

For example, suppose we wish to apply the above lookup table to the blocks image. We can create the table with

```
>> T = uint8(floor(0:255)/2);
```

MATLAB/Octave

or

```
In :   T = uint8(arange(256))/2
```
Python

apply it to the blocks image **b** with

```
>> b2 = T(b);
```
MATLAB/Octave

or

```
In :   b2 = T[b]
```
Python

The image **b2** is of type **uint8**, and so can be viewed directly with the viewing command.

The piecewise stretching function discussed above was in fact an example of the use of a lookup table.

Exercises

Image Arithmetic

1. Describe lookup tables for

   (a) Multiplication by 2

   (b) Image complements

2. Enter the following command on the blocks image **b**:

   ```
   b2 = (b/64)*64
   ```

   Comment on the result. Why is the result not equivalent to the original image?

3. Replace the value 64 in the previous question with 32 and 16.

Histograms

4. Write informal code to calculate a histogram $h[f]$ of the gray values of an image $f[row][col]$.

5. The following table gives the number of pixels at each of the gray levels 0–7 in an image with those gray values only:

0	1	2	3	4	5	6	7
3244	3899	4559	2573	1428	530	101	50

   Draw the histogram corresponding to these gray levels, and then perform a histogram equalization and draw the resulting histogram.

6. The following tables give the number of pixels at each of the gray levels 0–15 in an image with those gray values only. In each case, draw the histogram corresponding to these gray levels, and then perform a histogram equalization and draw the resulting histogram.

(a)

0	1	2	3	4	5	6	7	8	9	10	11	12	13	14	15
20	40	60	75	80	75	65	55	50	45	40	35	30	25	20	30

(b)

0	1	2	3	4	5	6	7	8	9	10	11	12	13	14	15
0	0	40	80	45	110	70	0	0	0	0	0	0	0	0	15

7. The following small image has gray values in the range 0 to 19. Compute the gray level histogram and the mapping that will equalize this histogram. Produce an 8×8 grid containing the gray values for the new histogram-equalized image.

```
12   6   5  13  14  14  16  15
11  10   8   5   8  11  14  14
 9   8   3   4   7  12  18  19
10   7   4   2  10  12  13  17
16   9  13  13  16  19  19  17
12  10  14  15  18  18  16  14
11   8  10  12  14  13  14  15
 8   6   3   7   9  11  12  12
```

8. Is the histogram equalization operation idempotent? That is, is performing histogram equalization *twice* the same as doing it just once?

9. Apply histogram equalization to the indices of the image emu.png.

10. Create a dark image with

```
>> c = imread('cameraman.png');
>> [x,map] = gray2ind(c);
```
MATLAB/Octave

The matrix x, when viewed, will appear as a very dark version of the cameraman image. Apply histogram equalization to it and compare the result with the original image.

11. Using either c and ch or s and sh from Section 4.3, enter the command

```
>> figure,plot(c,ch,'.'),grid on
```
MATLAB/Octave

or

```
In :  plt.plot(c,ch,'.'),plt.grid('on'),plt.axis('image')
```
Python

What are you seeing here?

12. Experiment with some other grayscale images.

13. Using LUTs, and following the example given in Section 4.4, write a simpler function for performing piecewise stretching than the function described in Section 4.3.

Chapter 5

Neighborhood Processing

5.1 Introduction

We have seen in Chapter 4 that an image can be modified by applying a particular function to each pixel value. Neighborhood processing may be considered an extension of this, where a function is applied to a neighborhood of each pixel.

The idea is to move a "mask": a rectangle (usually with sides of odd length) or other shape over the given image. As we do this, we create a new image whose pixels have gray values calculated from the gray values under the mask, as shown in Figure 5.1. The

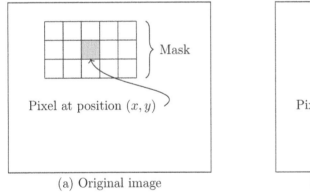

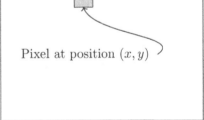

(a) Original image   (b) Image after filtering

FIGURE 5.1: Using a spatial mask on an image

combination of mask and function is called a *filter*. If the function by which the new gray value is calculated is a linear function of all the gray values in the mask, then the filter is called a *linear filter*.

A linear filter can be implemented by multiplying all elements in the mask by corresponding elements in the neighborhood, and adding up all these products. Suppose we have a 3×5 mask as illustrated in Figure 5.1. Suppose that the mask values are given by:

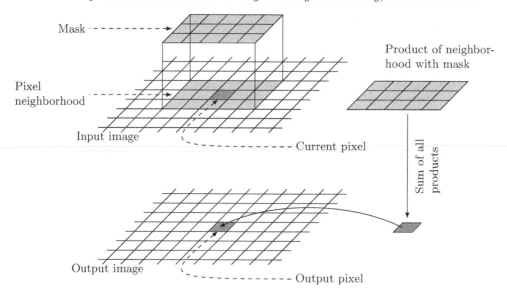

Mask

Product of neighbor-
hood with mask

Pixel
neighborhood

Input image

Current pixel

Sum of all
products

Output image

Output pixel

FIGURE 5.2: Performing linear spatial filtering

$m(-1,-2)$	$m(-1,-1)$	$m(-1,0)$	$m(-1,1)$	$m(-1,2)$
$m(0,-2)$	$m(0,-1)$	$m(0,0)$	$m(0,1)$	$m(0,2)$
$m(1,-2)$	$m(1,-1)$	$m(1,0)$	$m(1,1)$	$m(1,2)$

and that corresponding pixel values are

$p(i-1,j-2)$	$p(i-1,j-1)$	$p(i-1,j)$	$p(i-1,j+1)$	$p(i-1,j+2)$
$p(i,j-2)$	$p(i,j-1)$	$p(i,j)$	$p(i,j+1)$	$p(i,j+2)$
$p(i+1,j-2)$	$p(i+1,j-1)$	$p(i+1,j)$	$p(i+1,j+1)$	$p(i+1,j+2)$

We now multiply and add:

$$\sum_{s=-1}^{1} \sum_{t=-2}^{2} m(s,t)p(i+s,j+t).$$

A diagram illustrating the process for performing spatial filtering is given in Figure 5.2.
Spatial filtering thus requires three steps:

1. Position the mask over the current pixel

2. Form all products of filter elements with the corresponding elements of the neighborhood

3. Add up all the products

This must be repeated for every pixel in the image.

Allied to spatial filtering is spatial *convolution*. The method for performing a convolution is the same as that for filtering, except that the filter must be rotated by 180° before multiplying and adding. Using the $m(i,j)$ and $p(i,j)$ notation as before, the output of a convolution with a 3×5 mask for a single pixel is

$$\sum_{s=-1}^{1} \sum_{t=-2}^{2} m(-s,-t)p(i+s,j+t).$$

Note the negative signs on the indices of m. The same result can be achieved with

$$\sum_{s=-1}^{1} \sum_{t=-2}^{2} m(s,t)p(i-s+j-t).$$

Here we have rotated the *image* pixels by 180°; this does not of course affect the result. The importance of convolution will become apparent when we investigate the Fourier transform, and the convolution theorem. Note also that in practice, most filter masks are rotationally symmetric, so that spatial filtering and spatial convolution will produce the same output. Filtering as described above, where the filter is *not* rotated by 180°, is also called *correlation*.

An example: One important linear filter is to use a 3×3 mask and take the average of all nine values within the mask. This value becomes the gray value of the corresponding pixel in the new image. This operation may be described as follows:

a	b	c
d	e	f
g	h	i

$\longrightarrow \frac{1}{9}(a+b+c+d+e+f+g+h+i)$

where e is the gray value of the current pixel in the original image, and the average is the gray value of the corresponding pixel in the new image.

To apply this to an image, consider the 5×5 "image" obtained by multiplying a magic square by 10. In MATLAB/Octave:

```
>> x=uint8(10*magic(5))
x =

   170   240    10    80   150
   230    50    70   140   160
    40    60   130   200   220
   100   120   190   210    30
   110   180   250    20    90
```

and in Python:

```
In:  x = 10*uint8(array([[17,24,1,8,15],[23,5,7,14,16],\
...:  [4,6,13,20,22],[10,12,19,21,3],[11,18,25,2,9]]))
In: x
Out:
array([[170, 240,  10,  80, 150],
       [230,  50,  70, 140, 160],
       [ 40,  60, 130, 200, 220],
       [100, 120, 190, 210,  30],
       [110, 180, 250,  20,  90]], dtype=uint8)
```

Python

We may regard this array as being made of nine overlapping 3×3 neighborhoods. The output of our working will thus consist only of nine values. We shall see later how to obtain 25 values in the output.

Consider the top left 3×3 neighborhood of our image **x**:

170	240	10	80	150
230	50	70	140	160
40	60	130	200	220
100	120	190	210	30
110	180	250	20	90

Now we take the average of all these values in MATLAB/Octave as

```
>> mean2(x(1:3,1:3))

ans =

   111.1111
```

MATLAB/Octave

or in Python as

```
In : x[0:3,0:3].mean()
Out: 111.11111111111111
```

Python

which can be rounded to 111. Now we can move to the second neighborhood:

170	240	10	80	150
230	50	70	140	160
40	60	130	200	220
100	120	190	210	30
110	180	250	20	90

and take its average in MATLAB/Octave:

```
>> mean2(x(1:3,2:4))

ans =

   108.8889
```

and in Python:

```
In : x[0:3,1:4].mean()
Out: 108.88888888888889
```

and this can be rounded either down to 108, or to the nearest integer 109. If we continue in this manner, the following output is obtained:

```
111.1111   108.8889   128.8889
110.0000   130.0000   150.0000
131.1111   151.1111   148.8889
```

This array is the result of filtering $\mathbf{x}$ with the 3×3 averaging filter.

5.2 Notation

It is convenient to describe a linear filter simply in terms of the coefficients of all the gray values of pixels within the mask. This can be written as a matrix.

The averaging filter above, for example, could have its output written as

$$\frac{1}{9}a + \frac{1}{9}b + \frac{1}{9}c + \frac{1}{9}d + \frac{1}{9}e + \frac{1}{9}f + \frac{1}{9}g + \frac{1}{9}h + \frac{1}{9}i$$

and so this filter can be described by the matrix

$$\begin{bmatrix} \frac{1}{9} & \frac{1}{9} & \frac{1}{9} \\ \frac{1}{9} & \frac{1}{9} & \frac{1}{9} \\ \frac{1}{9} & \frac{1}{9} & \frac{1}{9} \end{bmatrix} = \frac{1}{9}\begin{bmatrix} 1 & 1 & 1 \\ 1 & 1 & 1 \\ 1 & 1 & 1 \end{bmatrix}$$

An example: The filter

$$\begin{bmatrix} 1 & -2 & 1 \\ -2 & 4 & -2 \\ 1 & -2 & 1 \end{bmatrix}$$

would operate on gray values as

a	b	c
d	e	f
g	h	i

$\longrightarrow a - 2b + c - 2d + 4e - 2f + g - 2h + i$

Borders of the Image

There is an obvious problem in applying a filter—what happens at the border of the image, where the mask partly falls outside the image? In such a case, as illustrated in Figure 5.3 there will be a lack of gray values to use in the filter function.

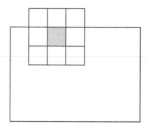

FIGURE 5.3: A mask at the edge of an image

There are a number of different approaches to dealing with this problem:

Ignore the edges. That is, the mask is only applied to those pixels in the image for which the mask will lie fully within the image. This means all pixels except for those along the borders, and results in an output image that is smaller than the original. If the mask is very large, a significant amount of information may be lost by this method.

We applied this method in our example above.

"Pad" with zeros. We assume that all necessary values outside the image are zero. This gives us all values to work with, and will return an output image of the same size as the original, but may have the effect of introducing unwanted artifacts (for example, dark borders) around the image.

Repeat the image. This means tiling the image in all directions, so that the mask always lies over image values. Since values might thus change across edges, this method may introduce unwanted artifacts in the result.

Reflect the image. This is similar to repeating, except that the image is reflected across all of its borders. This will ensure that there are no extra sudden changes introduced at the borders.

The last two methods are illustrated in Figure 5.4.

We can also see this by looking at the image which increases from top left to bottom right:

```
 20    40    60    80   100
 40    60    80   100   120
 60    80   100   120   140
 80   100   120   140   160
100   120   140   160   180
```

The results of filtering with a 3×3 averaging filter with zero padding, reflection, and repetition are shown in Figure 5.5.

Note that with repetition the upper left and bottom right values are "wrong"; this is because they have taken as extra values pixels that are very different. With zero-padding the upper left value has the correct size, but the bottom right values are far smaller than they should be, as they have been convolved with zeros. With reflection the values are all the correct size.

Repetition   Reflection

FIGURE 5.4: Repeating an image for filtering at its borders

17	33	46	59	44
33	60	80	100	73
46	80	100	120	86
60	100	120	140	100
44	73	86	99	71

Zero-padding

86	73	93	113	99
73	60	80	100	86
93	80	100	120	106
113	100	120	140	126
99	86	106	126	112

Repetition

32	46	66	86	99
46	60	80	100	113
66	80	100	120	133
86	100	120	140	153
99	113	133	153	166

Reflection

FIGURE 5.5: The result of filtering with different modes

5.3 Filtering in MATLAB and Octave

The `imfilter` function does the job of linear filtering for us; its use is

```
imfilter(image,filter,...)
```

and the result is a matrix of the same data type as the original image. There are other parameters for controlling the behavior of the filtering. Pixels at the boundary of the image can be managed by padding with zeros; this being the default method, but there are several other options:

Extra parameter	Implements
`'symmetric'`	Filtering with reflection
`'circular'`	Filtering with tiling repetition
`'replication'`	Filtering by repeating the border elements.
`'full'`	Padding with zero, and applying the filter at all places on and around the image where the mask intersects the image matrix.

The last option returns a result that is *larger* than the original:

```
>> imfilter(x,a,'full')

ans =

    19    46    47    37    27    26    17
    44    77    86    66    68    59    34
    49    88   111   109   129   106    59
    41    67   110   130   150   107    46
    28    68   131   151   149    86    38
    23    57   106   108    88    39    13
    12    32    60    50    40    12    10
```

MATLAB/Octave

The central 5×5 values of the last operation are the same values as provided by the command `imfilter(x,a)` and with no extra parameters.

There is no single "best" approach; the method must be dictated by the problem at hand, by the filter being used, and by the result required.

We can create our filters by hand or by using the `fspecial` function; this has many options which makes for easy creation of many different filters. We shall use the **average** option, which produces averaging filters of given size; thus,

```
>> fspecial('average',[5,7])
```

MATLAB/Octave

will return an averaging filter of size 5×7; more simply

```
>> fspecial('average',11)
```

MATLAB/Octave

will return an averaging filter of size 11×11. If we leave out the final number or vector, the 3×3 averaging filter is returned.

For example, suppose we apply the 3×3 averaging filter to an image as follows:

```
>> c = imread('cameraman.png');
>> f1 = fspecial('average');
>> cf1 = imfilter(c,f1);
```

MATLAB/Octave

The averaging filter blurs the image; the edges in particular are less distinct than in the original. The image can be further blurred by using an averaging filter of larger size. This is shown in Figure 5.6(c), where a 9×9 averaging filter has been used, and in Figure 5.6(d), where a 25×25 averaging filter has been used.

Notice how the default zero padding used at the edges has resulted in a dark border appearing around the image (c). This would be especially noticeable when a large filter is being used. Any of the above options can be used instead; image (d) was created with the **symmetric** option.

The resulting image after these filters may appear to be much "worse" than the original. However, applying a blurring filter to reduce detail in an image may the perfect operation for autonomous machine recognition, or if we are only concentrating on the "gross" aspects of the image: numbers of objects or amount of dark and light areas. In such cases, too much detail may obscure the outcome.

(a) Original image

(b) Average filtering

(c) Using a 9×9 filter

(d) Using a 25×25 filter

FIGURE 5.6: Average filtering

Separable Filters

Some filters can be implemented by the successive application of two simpler filters. For example, since

$$\frac{1}{9}\begin{bmatrix} 1 & 1 & 1 \\ 1 & 1 & 1 \\ 1 & 1 & 1 \end{bmatrix} = \frac{1}{3}\begin{bmatrix} 1 \\ 1 \\ 1 \end{bmatrix} \frac{1}{3}\begin{bmatrix} 1 & 1 & 1 \end{bmatrix}$$

the 3×3 averaging filter can be implemented by first applying a 3×1 averaging filter, and then applying a 1×3 averaging filter to the result. The 3×3 averaging filter is thus *separable* into two smaller filters. Separability can result in great time savings. Suppose an $n \times n$ filter is separable into two filters of size $n \times 1$ and $1 \times n$. The application of an $n \times n$ filter requires n^2 multiplications, and $n^2 - 1$ additions for each pixel in the image. But the application of an $n \times 1$ filter only requires n multiplications and $n - 1$ additions. Since this must be done twice, the total number of multiplications and additions are $2n$ and $2n - 2$, respectively. If n is large the savings in efficiency can be dramatic.

All averaging filters are separable; another separable filter is the Laplacian

$$\begin{bmatrix} 1 & -2 & 1 \\ -2 & 4 & -2 \\ 1 & -2 & 1 \end{bmatrix} = \begin{bmatrix} 1 \\ -2 \\ 1 \end{bmatrix} \begin{bmatrix} 1 & -2 & 1 \end{bmatrix}.$$

We will also consider other examples.

5.4   Filtering in Python

There is no Python equivalent to the **fspecial** and **imfilter** commands, but there are a number of commands for applying either a generic filter or a specific filter.

Linear filtering can be performed by using **convolve** or **correlate** from the **ndimage** module of the library, or **generic_filter**.

For example, we may apply 3×3 averaging to the 5×5 magic square array:

```
In:  import scipy.ndimage as ndi
In:  x = uint8(array([[17,24,1,8,15],[23,5,7,14,16],[4,6,13,20,22],\
     [10,12,19,21,3],[11,18,25,2,9]])*10)
In:  a = ones((3,3))/9
In:  ndi.convolve(x,a,mode='constant')
Out:
array([[ 76,  85,  65,  67,  58],
       [ 87, 111, 108, 128, 105],
       [ 66, 109, 130, 150, 106],
       [ 67, 131, 151, 148,  85],
       [ 56, 105, 107,  87,  38]], dtype=uint8)
```

Python

Note that this result is the same as that obtained with MATLAB's **imfilter(x,a)** command, and Python also automatically converts the result to the same data type as the input; here as an unsigned 8-bit array. The **mode** parameter, here set to **'constant'**, tells the function that the image is to be padded with constant values, the default value of which is zero, although other values can be specified with the **cval** parameter.

So to obtain, for example, the image in Figure 5.6(c), you could use the following Python commands:

```
In:  c = io.imread('cameraman.png')
In:  cf = ndi.convolve(c,ones((9,9))/81,mode='constant')
In:  io.imshow(cfs)
```
`Python`

Alternately, you could use the built in `uniform_filter` function:

```
In:  cf = ndi.uniform_filter(c,[9,9],mode='constant')
```
`Python`

Python differs from MATLAB and Octave here in that the default behavior of spatial convolution at the edges of the image is to reflect the image in all its edges. This will ameliorate the dark borders seen in Figure 5.6. So with leaving the `mode` parameter in its default setting, the commands

```
In:  cf = ndi.uniform_filter(c,25)
In:  cf2 = ndi.uniform_filter(c,50)
```
`Python`

will produce the images shown in Figure 5.7.

 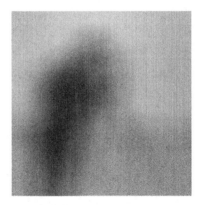

(a) Using a 25 × 25 filter   (b) Using a 50 × 50 filter

FIGURE 5.7: Average filtering in Python

5.5   Frequencies; Low and High Pass Filters

It will be convenient to have some standard terminology by which we can discuss the effects a filter will have on an image, and to be able to choose the most appropriate filter for a given image processing task. One important aspect of an image which enables us to do this is the notion of *frequencies*. Roughly speaking, the frequencies of an image are a measure of the amount by which gray values change with distance. This concept will be given a more formal setting in Chapter 7. *High frequency components* are characterized by large changes in gray values over small distances; examples of high frequency components are edges and noise. *Low frequency components*, on the other hand, are parts of the image characterized by little change in the gray values. These may include backgrounds or skin textures. We then say that a filter is a

high pass filter if it "passes over" the high frequency components, and reduces or eliminates low frequency components

low pass filter if it "passes over" the low frequency components, and reduces or eliminates high frequency components

For example, the 3 × 3 averaging filter is a low pass filter, as it tends to blur edges. The filter

$$\begin{bmatrix} 1 & -2 & 1 \\ -2 & 4 & -2 \\ 1 & -2 & 1 \end{bmatrix}$$

is a high pass filter.

We note that the sum of the coefficients (that is, the sum of all e elements in the matrix), in the high pass filter is zero. This means that in a low frequency part of an image, where the gray values are similar, the result of using this filter is that the corresponding gray values in the new image will be close to zero. To see this, consider a 4 × 4 block of similar values pixels, and apply the above high pass filter to the central four:

150	152	148	149
147	152	151	150
152	148	149	151
151	149	150	148

$\longrightarrow$

11	6
−13	−5

The resulting values are close to zero, which is the expected result of applying a high pass filter to a low frequency component. We shall see how to deal with negative values later.

High pass filters are of particular value in edge detection and edge enhancement (of which we shall see more in Chapter 9). But we can provide a sneak preview, using the cameraman image.

```
>> f=fspecial('laplacian')

f =

    0.1667    0.6667    0.1667
    0.6667   -3.3333    0.6667
    0.1667    0.6667    0.1667

>> cf=imfilter(c,f,'symmetric');
>> f1=fspecial('log')

f1 =

    0.0448    0.0468    0.0564    0.0468    0.0448
    0.0468    0.3167    0.7146    0.3167    0.0468
    0.0564    0.7146   -4.9048    0.7146    0.0564
    0.0468    0.3167    0.7146    0.3167    0.0468
    0.0448    0.0468    0.0564    0.0468    0.0448

>> cf1=imfilter(c,f1,'symmetric');
```

MATLAB/Octave

The images are shown in Figure 5.8. Image (a) is the result of the Laplacian filter; image (b) shows the result of the Laplacian of Gaussian ("log") filter. We discuss Gaussian filters in Section 5.6.

(a) Laplacian filter   (b) Laplacian of Gaussian ("log") filtering

FIGURE 5.8: High pass filtering

In each case, the sum of all the filter elements is zero.

Note that both MATLAB and Octave in fact do more than merely apply the Laplacian or LoG filters to the image; the results in Figure 5.8 have been "cleaned up" from the raw filtering. We can see this by applying the Laplacian filter in Python:

```
In:  f = array([[1,4,1],[4,-20,4],[1,4,1]])
In:  cf = ndi.convolve(c,f)
In:  io.imshow(cf)
```

Python

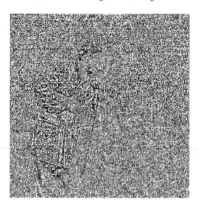

FIGURE 5.9: Laplacian filtering without any extra processing

of which the result is shown in Figure 5.9. We shall see in Chapter 9 how Python can be used to obtain results similar to those in Figure 5.8.

Values Outside the Range 0–255

We have seen that for image display, we would like the gray values of the pixels to lie between 0 and 255. However, the result of applying a linear filter may be values that lie outside this range. We may consider ways of dealing with values outside of this "displayable" range.

Make negative values positive. This will certainly deal with negative values, but not with values greater than 255. Hence, this can only be used in specific circumstances; for example, when there are only a few negative values, and when these values are themselves close to zero.

Clip values. We apply the following thresholding type operation to the gray values x produced by the filter to obtain a displayable value y:

$$y = \begin{cases} 0 & \text{if } x < 0 \\ x & \text{if } 0 \leq x \leq 255 \\ 255 & \text{if } x > 255 \end{cases}$$

This will produce an image with all pixel values in the required range, but is not suitable if there are many gray values outside the 0–255 range; in particular, if the gray values are equally spread over a larger range. In such a case, this operation will tend to destroy the results of the filter.

Scaling transformation. Suppose the lowest gray value produced by the filter if g_L and the highest value is g_H. We can transform all values in the range g_L–g_H to the range 0–255 by the linear transformation illustrated below:

Since the gradient of the line is $255/(g_H - g_L)$ we can write the equation of the line as

$$y = 255 \frac{x - g_L}{g_H - g_L}$$

and applying this transformation to all gray levels x produced by the filter will result (after any necessary rounding) in an image that can be displayed.

As an example, let's apply the high pass filter given in Section 5.5 to the cameraman image:

```
>> f2 = [1 -2 1;-2 4 -2;1 -2 1];
>> cf2 = imfilter(double(c),f2);
```

We have to convert the image to type "double" before the filtering, or else the result will be an automatically scaled image of type uint8. The maximum and minimum values of the matrix cf2 are 593 and −541, respectively. The mat2gray function automatically scales the matrix elements to displayable values; for any matrix M, it applies a linear transformation to its elements, with the lowest value mapping to 0.0, and the highest value mapping to 1.0. This means the output of mat2gray is always of type double. The function also requires that the input type is double.

```
>> figure,imshow(mat2gray(cf2));
```

To do this by hand, so to speak, applying the linear transformation above, we can use:

```
>> maxcf2 = max(cf2(:));
>> mincf2 = min(cf2(:));
>> cf2g = (cf2-mincf2)/(maxcf2-mncf2);
```

The result will be a matrix of type double, with entries in the range 0.0–1.0. This can be be viewed with imshow. We can make it a uint8 image by multiplying by 255 first. The result can be seen in Figure 5.10.

Sometimes a better result can be obtained by dividing an output of type double by a constant before displaying it:

```
>> figure,imshow(cf2/60)
```

and this is also shown in Figure 5.10.

Using mat2gray   Dividing by a constant

FIGURE 5.10: Using a high pass filter and displaying the result

High pass filters are often used for edge detection. These can be seen quite clearly in the right-hand image of Figure 5.10.

In Python, as we have seen, the `convolve` operation returns a `uint8` array as output if the image is of type `uint8`. To apply a linear transformation, we need to start with an output of type `float32`:

```
In:  cf2 = ndi.convolve(float32(c),f,mode='constant')
In:  maxcf2 = cf2.max()
In:  mincf2 = cf1.min()
In:  cf2f = (cf2-mincf2)/(maxcf2-mincf2)
In:  io.imshow(cf2f)
```
Python

Note that Python's `imshow` commands automatically scale the image to full brightness range. To switch off that behavior the parameters `vmax` and `vmin` will give the "true" range, as opposed to the displayed range:

```
In:  io.imshow(cf2/60,vmax=1.0,vmin=0.0)
```
Python

These last two `imshow` commands will produce the same images as shown in Figure 5.10.

5.6   Gaussian Filters

We have seen some examples of linear filters so far: the averaging filter and a high pass filter. The `fspecial` function can produce many different filters for use with the `imfilter` function; we shall look at a particularly important filter here.

Gaussian filters are a class of low pass filters, all based on the Gaussian probability distribution function

$$f(x) = e^{-\frac{x^2}{2\sigma^2}}$$

where σ is the standard deviation: a large value of σ produces to a flatter curve, and a small value leads to a "pointier" curve. Figure 5.11 shows examples of such one-dimensional Gaussians. Gaussian filters are important for a number of reasons:

1. They are mathematically very "well behaved"; in particular the Fourier transform (see Chapter 7) of a Gaussian filter is another Gaussian.

2. They are rotationally symmetric, and so are very good starting points for some edge detection algorithms (see Chapter 9),

3. They are *separable*; in that a Gaussian filter may be applied by first applying a one-dimension Gaussian in the x direction, followed by another in the y direction. This can lead to very fast implementations.

4. The convolution of two Gaussians is another Gaussian.

A two-dimensional Gaussian function is given by

$$f(x,y) = e^{-\frac{x^2+y^2}{2\sigma^2}}$$

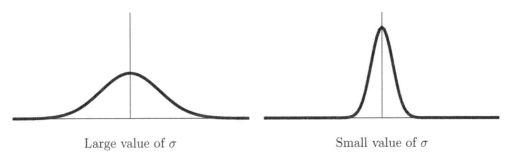

Large value of σ Small value of σ

FIGURE 5.11: One-dimensional Gaussians

The command `fspecial('gaussian')` produces a discrete version of this function. We can draw pictures of this with the `surf` function, and to ensure a nice smooth result, we shall create a large filter (size 50×50) with different standard deviations.

```
>> a = 50; s = 3;
>> g = fspecial('gaussian',[a a],s);
>> surf(1:a,1:a,g)
>> s = 9;
>> g2 = fspecial('gaussian',[a a],s);
>> figure,surf(1:a,1:a,g2)
```
MATLAB/Octave

The surfaces are shown in Figure 5.12.

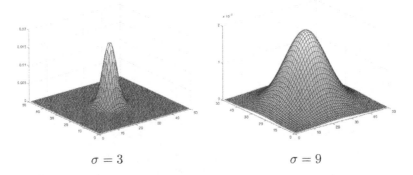

$\sigma = 3$ $\sigma = 9$

FIGURE 5.12: Two-dimensional Gaussians

Gaussian filters have a blurring effect which looks very similar to that produced by neighborhood averaging. Let's experiment with the cameraman image, and some different Gaussian filters.

```
>> g1 = fspecial('gaussian',[5,5]);
>> g1 = fspecial('gaussian',[5,5],2);
>> g1 = fspecial('gaussian',[11,11],1);
>> g1 = fspecial('gaussian',[11,11],5);
```
MATLAB/Octave

The final parameter is the standard deviation; which if not given, defaults to 0.5. The second parameter (which is also optional), gives the size of the filter; the default is 3×3. If the filter is to be square, as in all the previous examples, we can just give a single number in each case.

Now we can apply the filter to the cameraman image matrix c and view the result.

```
>> imshow(imfilter(c,g1))
>> figure,imshow(imfilter(c,g2))
>> figure,imshow(imfilter(c,g3,'symmetric'))
>> figure,imshow(imfilter(c,g4,'symmetric'))
```
MATLAB/Octave

As for the averaging filters earlier, the `symmetric` option is used to prevent the dark borders which would arise from low pass filtering using zero padding. The results are shown in Figure 5.13. Thus, to obtain a spread out blurring effect, we need a large standard deviation.

5×5, $\sigma = 0.5$ 5×5, $\sigma = 2$

11×11, $\sigma = 1$ 11×11, $\sigma = 5$

FIGURE 5.13: Effects of different Gaussian filters on an image

In fact, if we let the standard deviation grow large without bound, we obtain the averaging filters as limiting values. For example:

```
>> fspecial('gaussian',3,100)

ans =

    0.1111    0.1111    0.1111
    0.1111    0.1111    0.1111
    0.1111    0.1111    0.1111
```
MATLAB/Octave

and we have the 3×3 averaging filter.

Although the results of Gaussian blurring and averaging look similar, the Gaussian filter has some elegant mathematical properties that make it particularly suitable for blurring.

In Python, the command `gaussian_filter` can be used, which takes as parameters the standard variation, and also a `truncate` parameter, which gives the number of standard deviations at which the filter should be cut off. This means that the images in Figure 5.13 can be obtained with:

```
In:   cg1 = ndi.gaussian_filter(c,0.5,truncate=4.5)
In:   cg2 = ndi.gaussian_filter(c,2,truncate=1)
In:   cg3 = ndi.gaussian_filter(c,1,truncate=5)
In:   cg4 = ndi.gaussian_filter(c,5,truncate=1)
```
Python

As with the uniform filter, the default behavior is to reflect the image in its edges, which is the same result as the `symmetric` option used in MATLAB/Octave. To see the similarity between MATLAB/Octave and Python, we can apply a Gaussian filter in Python to an image consisting of all zeros except for a central one. This will produce the same output as MATLAB's `fspecial` function:

```
>> fspecial('gaussian',[5,5],1)
ans =

    0.0029690    0.0133062    0.0219382    0.0133062    0.0029690
    0.0133062    0.0596343    0.0983203    0.0596343    0.0133062
    0.0219382    0.0983203    0.1621028    0.0983203    0.0219382
    0.0133062    0.0596343    0.0983203    0.0596343    0.0133062
    0.0029690    0.0133062    0.0219382    0.0133062    0.0029690
```
MATLAB/Octave

and the Python equivalent:

```
In:   x = ones((5,5)); x(2,2)=1
In:   ndi.gaussian_filter(x,1,truncate=2).round(7)
Out:
array([[ 0.002969 ,  0.0133062,  0.0219382,  0.0133062,  0.002969 ],
       [ 0.0133062,  0.0596343,  0.0983203,  0.0596343,  0.0133062],
       [ 0.0219382,  0.0983203,  0.1621028,  0.0983203,  0.0219382],
       [ 0.0133062,  0.0596343,  0.0983203,  0.0596343,  0.0133062],
       [ 0.002969 ,  0.0133062,  0.0219382,  0.0133062,  0.002969 ]])
```
Python

Other filters will be discussed in future chapters; also check the documentation for `fspecial` for other filters.

5.7   Edge Sharpening

Spatial filtering can be used to make edges in an image slightly sharper and crisper, which generally results in an image more pleasing to the human eye. The operation is variously called "edge enhancement," "edge crispening," or "unsharp masking." This last term comes from the printing industry.

Unsharp Masking

The idea of unsharp masking is to subtract a scaled "unsharp" version of the image from the original. In practice, we can achieve this affect by subtracting a scaled blurred image from the original. The schema for unsharp masking is shown in Figure 5.14.

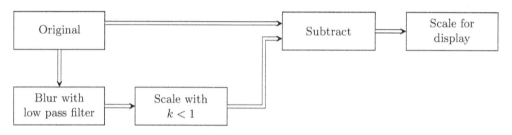

FIGURE 5.14: Schema for unsharp masking

In fact, all these steps can be built into the one filter. Starting with the filter

$$id = \begin{bmatrix} 0 & 0 & 0 \\ 0 & 1 & 0 \\ 0 & 0 & 0 \end{bmatrix}$$

which is an "identity" filter (in that applying it to an image leaves the image unchanged), all the steps in Figure 5.14 could be implemented by the filter

$$u = s(id - \frac{1}{k}a)$$

where s is the scaling factor. To ensure the output image has the same levels of brightness as the original, we need to ensure that the sum of all the elements of u is 1, that is, that

$$s(1 - \frac{1}{k}) = 1$$

or that

$$s = \frac{k}{k-1}$$

Suppose for example we let $k = 1.5$. Then $s = 3$ and so the filter is

$$3 \begin{bmatrix} 0 & 0 & 0 \\ 0 & 1 & 0 \\ 0 & 0 & 0 \end{bmatrix} - 2 \left(\frac{1}{9} \begin{bmatrix} 1 & 1 & 1 \\ 1 & 1 & 1 \\ 1 & 1 & 1 \end{bmatrix} \right) = \frac{1}{9} \begin{bmatrix} -2 & -2 & -2 \\ -2 & 25 & -2 \\ -2 & -2 & -2 \end{bmatrix}$$

An alternative derivation is to write the equation relating s and k as

$$s - \frac{s}{k} = 1.$$

If $s/k = t$, say, then $s = t + 1$. So if $s/k = 2/3$, then $s = 5/3$ and so the filter is

$$\frac{5}{3}\begin{bmatrix} 0 & 0 & 0 \\ 0 & 1 & 0 \\ 0 & 0 & 0 \end{bmatrix} - \frac{2}{3}\left(\frac{1}{9}\begin{bmatrix} 1 & 1 & 1 \\ 1 & 1 & 1 \\ 1 & 1 & 1 \end{bmatrix}\right) = \frac{1}{27}\begin{bmatrix} -2 & -2 & -2 \\ -2 & 43 & -2 \\ -2 & -2 & -2 \end{bmatrix}$$

For example, consider an image of an iconic Australian animal:

```
>> k = imread('koala.png');
```

The first filter above can be constructed as

```
>> id = [0 0 0;0 1 0;0 0 0];
>> f = fspecial('average');
>> u = 3*id - 2*f
u =

  -0.22222  -0.22222  -0.22222
  -0.22222   2.77778  -0.22222
  -0.22222  -0.22222  -0.22222
```

and applied to the image in the usual way:

```
>> ku = imfilter(k,u);
>> imshow(ku)
```

The image k is shown in Figure 5.15(a), then the result of unsharp masking is given in Figure 5.15(b). The result appears to be a better image than the original; the edges are crisper and more clearly defined.

(a) Original image   (b) The image after unsharp masking

FIGURE 5.15: An example of unsharp masking

The same effect can be seen in Python, where we convolve with the image converted to floats, so as not to lose any information in the arithmetic:

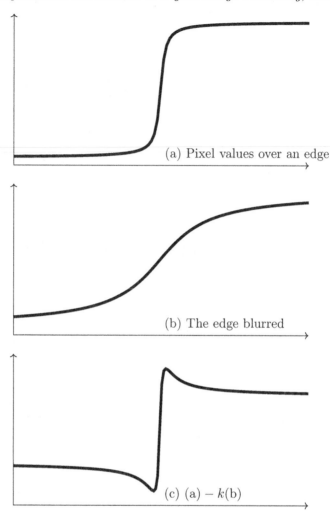

(a) Pixel values over an edge

(b) The edge blurred

(c) (a) $- k$(b)

FIGURE 5.16: Unsharp masking

```
In:   k = io.imread('koala.png')
In:   u = array([[-2,-2,-2],[-2,25,-2],[-2,-2,-2]])/9.0
In:   ku = ndi.convolve(k.astype(float),u)
In:   io.imshow(ku/255,vmax=1.0,vmin=0.0)
```

Python

To see why this works, we may consider the function of gray values as we travel across an edge, as shown in Figure 5.16.

As a scaled blur is subtracted from the original, the result is that the edge is enhanced, as shown in graph (c) of Figure 5.16.

We can in fact perform the filtering and subtracting operation in one command, using the linearity of the filter, and that the 3×3 filter

$$\begin{bmatrix} 0 & 0 & 0 \\ 0 & 1 & 0 \\ 0 & 0 & 0 \end{bmatrix}$$

is the "identity filter."

Hence, unsharp masking can be implemented by a filter of the form

$$f = \begin{bmatrix} 0 & 0 & 0 \\ 0 & 1 & 0 \\ 0 & 0 & 0 \end{bmatrix} - \frac{1}{k} \begin{bmatrix} 1/9 & 1/9 & 1/9 \\ 1/9 & 1/9 & 1/9 \\ 1/9 & 1/9 & 1/9 \end{bmatrix}$$

where k is a constant chosen to provide the best result. Alternatively, the unsharp masking filter may be defined as

$$f = k \begin{bmatrix} 0 & 0 & 0 \\ 0 & 1 & 0 \\ 0 & 0 & 0 \end{bmatrix} - \begin{bmatrix} 1/9 & 1/9 & 1/9 \\ 1/9 & 1/9 & 1/9 \\ 1/9 & 1/9 & 1/9 \end{bmatrix}$$

so that we are in effect subtracting a blur from a scaled version of the original; the scaling factor may also be split between the identity and blurring filters.

The **unsharp** option of **fspecial** produces such filters; the filter created has the form

$$\frac{1}{\alpha + 1} \begin{bmatrix} -\alpha & \alpha - 1 & -\alpha \\ \alpha - 1 & \alpha + 5 & \alpha - 1 \\ -\alpha & \alpha - 1 & -\alpha \end{bmatrix}$$

where α is an optional parameter which defaults to 0.2. If $\alpha = 0.5$, the filter is

$$\frac{1}{3} \begin{bmatrix} -1 & -1 & -1 \\ -1 & 11 & -1 \\ -1 & -1 & -1 \end{bmatrix} = 4 \begin{bmatrix} 0 & 0 & 0 \\ 0 & 1 & 0 \\ 0 & 0 & 0 \end{bmatrix} - 3 \begin{bmatrix} 1/9 & 1/9 & 1/9 \\ 1/9 & 1/9 & 1/9 \\ 1/9 & 1/9 & 1/9 \end{bmatrix}$$

Figure 5.17 was created using the MATLAB commands

```
>> p = imread('pelicans.png');
>> u = fspecial('unsharp',0.5);
>> pu = imfilter(p,u);
>> imshow(p),figure,imshow(pu)
```

MATLAB/Octave

Figure 5.17(b), appears much sharper and "cleaner" than the original. Notice in particular the rocks and trees in the background, and the ripples on the water.

Although we have used averaging filters above, we can in fact use any low pass filter for unsharp masking.

Exactly the same affect can be produced in Python; starting by creating an **unsharp** function to produce the filter:

```
In :  def unsharp(alpha=0.2):
...:      A1 = array([[-1,1,-1],[1,1,1],[-1,1,-1]])
...:      A2 = array([[0,-1,0],[-1,5,-1],[0,-1,0]])
...:      return (alpha*A1+A2)/(alpha+1)
```

Python

This can be applied with **convolve**:

```
In:  p = io.imread('pelicans.png')
In:  u = unsharp(0.5)
In:  pu = ndi.convolve(p.astype(float),u)
In:  io.imshow(pu/255,vmax=1.0,vmin=0.0)
```

Python

(a) The original   (b) After unsharp masking

FIGURE 5.17: Edge enhancement with unsharp masking

High Boost Filtering

Allied to unsharp masking filters are the *high boost* filters, which are obtained by

high boost = A(original) − (low pass).

where A is an "amplification factor." If $A = 1$, then the high boost filter becomes an ordinary high pass filter. If we take as the low pass filter the 3×3 averaging filter, then a high boost filter will have the form

$$\frac{1}{9}\begin{bmatrix} -1 & -1 & -1 \\ -1 & z & -1 \\ -1 & -1 & -1 \end{bmatrix}$$

where $z > 8$. If we put $z = 11$, we obtain a filtering very similar to the unsharp filter above, except for a scaling factor. Thus, the commands:

```
>> f = [-1 -1 -1;-1 11 -1;-1 -1 -1]/9;
>> xf = imfilter(f,x);
>> imshow(xf/80)
```

MATLAB/Octave

will produce an image similar to that in Figure 5.15. The value 80 was obtained by trial and error to produce an image with similar intensity to the original.

We can also write the high boost formula above as

$$\begin{aligned} \text{high boost} &= A(\text{original}) - (\text{low pass}) \\ &= A(\text{original}) - ((\text{original}) - (\text{high pass})) \\ &= (A-1)(\text{original}) + (\text{high pass}). \end{aligned}$$

Best results for high boost filtering are obtained if we multiply the equation by a factor w so that the filter values sum to 1; this requires

$$wA - w = 1$$

or

$$w = \frac{1}{A-1}.$$

So a general unsharp masking formula is

$$\frac{A}{A-1}(\text{original}) - \frac{1}{A-1}(\text{low pass}).$$

Another version of this formula is

$$\frac{A}{2A-1}(\text{original}) - \frac{1-A}{2A-1}(\text{low pass})$$

where for best results A is taken so that

$$\frac{3}{5} \leq A \leq \frac{5}{6}.$$

If we take $A = 3/5$, the formula becomes

$$\frac{3/5}{2(3/5)-1}(\text{original}) - \frac{1-(3/5)}{2(3/5)-1}(\text{low pass}) = 3(\text{original}) - 2(\text{low pass})$$

If we take $A = 5/6$ we obtain

$$\frac{5}{4}(\text{original}) - \frac{1}{4}(\text{low pass})$$

Using the identity and averaging filters, we can obtain high boost filters by:

```
>> id=[0 0 0;0 1 0;0 0 0];
>> f=fspecial('average');
>> hb1=3*id-2*f

hb1 =

   -0.2222   -0.2222   -0.2222
   -0.2222    2.7778   -0.2222
   -0.2222   -0.2222   -0.2222

>> hb2=1.25*id-0.25*f

hb2 =

   -0.0278   -0.0278   -0.0278
   -0.0278    1.2222   -0.0278
   -0.0278   -0.0278   -0.0278
```

MATLAB/Octave

If each of the filters `hb1` and `hb2` are applied to an image with `imfilter`, the result will have enhanced edges. The images in Figure 5.18 show these results; Figure 5.18(a) was obtained with

```
>> k1 = imfilter(k,hb1);
>> imshow(k1)
```

MATLAB/Octave

(a) High boost filtering with `hb1`   (b) High boost filtering with `hb2`

FIGURE 5.18: High boost filtering

and Figure 5.18(b) similarly.

With Python, these images could be obtained easily as:

```python
In:  k = im2fl(io.imread('koala.png'))
In:  kf = ndi.uniform_filter(k,3)
In:  hb1 = 3*k - 2*kf
In:  hb2 = 1.25*k - 0.25*kf
In:  subplot(121),io.imshow(hb1,vmax=1.0,vmin=0.0)
In:  subplot(122),io.imshow(hb2,vmax=1.0,vmin=0.0)
```

Python

Of the two filters, `hb1` appears to produce the best result; `hb2` produces an image not much crisper than the original.

5.8   Non-Linear Filters

Linear filters, as we have seen in the previous sections, are easy to describe, and can be applied very quickly and efficiently.

A *non-linear filter* is obtained by a non-linear function of the grayscale values in the mask. Simple examples are the *maximum filter*, which has as its output the maximum value under the mask, and the corresponding *minimum filter*, which has as its output the minimum value under the mask.

Both the maximum and minimum filters are examples of *rank-order filters*. In such a filter, the elements under the mask are ordered, and a particular value returned as output. So if the values are given in increasing order, the minimum filter is a rank-order filter for which the *first* element is returned, and the maximum filter is a rank-order filter for which the *last* element is returned

For implementing a general non-linear filter in MATLAB or Octave, the function to use is `nlfilter`, which applies a filter to an image according to a pre-defined function. If the function is not already defined, we have to create an m-file that defines it.

Here are some examples; first to implement a maximum filter over a 3×3 neighborhood (note that the commands are slightly different for MATLAB and for Octave):

```
>> cmax = nlfilter(c,[3,3],'max(x(:))');      # This is MATLAB
>> cmax = nlfilter(c,[3,3],@(x) max(x(:)));   # This is Octave
```

MATLAB/Octave

Python has a different syntax, made easier for maxima and minima in that the `max` function is applied to the entire array, rather than just its columns:

```
In:  cmax = ndi.generic_filter(c,max,[3,3])
```

Python

The `nlfilter` function and the `generic_filter` function each require three arguments: the image matrix, the size of the filter, and the function to be applied. The function must be a matrix function that returns a scalar value. The result of these operations is shown in Figure 5.19(a). Replacing `max` with `min` in the above commands implements a minimum filter, and the result is shown in Figure 5.19(b).

    (a) Using a maximum filter   (b) Using a minimum filter

FIGURE 5.19: Using non-linear filters

Note that in each case the image has lost some sharpness, and has been brightened by the maximum filter, and darkened by the minimum filter. The `nlfilter` function is very slow; in general, there is little call for non-linear filters except for a few that are defined by their own commands. We shall investigate these in later chapters.

However, if a non-linear filter is needed in MATLAB or Octave, a faster alternative is to use the `colfilt` function, which rearranges the image into columns first. For example, to apply the maximum filter to the cameraman image, we can use

```
>> cmax = colfilt(c,[3,3],'sliding',@max);
```

MATLAB/Octave

The parameter `sliding` indicates that overlapping neighborhoods are being used (which of course is the case with filtering). This particular operation is almost instantaneous, as compared with the use of `nlfilter`.

To implement the maximum and minimum filters as rank-order filters, we may use the MATLAB/Octave function `ordfilt2`. This requires three inputs: the image, the index value of the ordered results to choose as output, and the definition of the mask. So to apply the maximum filter on a 3×3 mask, we use

```
>> cmax = ordfilt2(c,9,ones(3,3));
```
MATLAB/Octave

and the minimum filter can be applied with

```
>> cmin = ordfilt2(c,1,ones(3,3));
```
MATLAB/Octave

A very important rank-order filter is the *median filter*, which takes the *central* value of the ordered list. We could apply the median filter with

```
>> cmed = ordfilt2(c,5,ones(3,3));
```
MATLAB/Octave

However, the median filter has its own command, medfilt2, which we discuss in more detail in Chapter 8.

Python has maximum and minimum filters in the scipy.ndimage module and also in the skimage.filter.rank module. As an example of each:

```
In:  cmin = ndi.minimum_filter(c, size=(3,3))
In:  cmax = rk.minimum(c,ones((3,3)))
```
Python

Rank order filters are implemented by the rank_filter function in the ndimage module, and one median filter is also provided by ndimage as median_filter. Thus the two commands

```
In:  cm = ndi.median_filter(c,size=(3,3))
In:  cm2 = ndi.rank_filter(c,4,size=(3,3))
```
Python

produce the same results. (Recall that in Python arrays and lists are indexed starting with zero, so the central value in a 3×3 array has index value 4.) User-defined filters can be applied using generic_filter. For example, suppose we consider the *root-mean-square filter*, which returns the value

$$\sqrt{\frac{1}{N} \sum_{x \in M} x^2}$$

where M is the mask, and N is the number of its elements. First, this must be defined as a Python function:

```
In :  def rms(x):
...:      return sqrt(mean(x**2))
```
Python

Then it can be applied, for example, to the cameraman image c:

```
In :  cr = ndi.generic_filter(c,rms,size=(3,3))
```
Python

Other non-linear filters are the *geometric mean filter*, which is defined as

$$\left(\prod_{(i,j)\in M} x(i,j) \right)^{1/N}$$

where as for the root-mean-square filter M is the filter mask, and N its size; and the *alpha-trimmed mean filter*, which first orders the values under the mask, trims off elements at either end of the ordered list, and takes the mean of the remainder. So, for example, if we have a 3×3 mask, and we order the elements as

$$x_1 \le x_2 \le x_3 \le \cdots \le x_9$$

and trim off two elements at either end, the result of the filter will be

$$(x_3 + x_4 + x_5 + x_6 + x_7)/5.$$

Both of these filters have uses for image restoration; again, see Chapter 8.

Non-linear filters are used extensively in image restoration, especially for cleaning noise; and these will be discussed in Chapter 8.

5.9 Edge-Preserving Blurring Filters

With the uniform (average) and Gaussian filters, blurring occurs across the entire image, and edges are blurred as well as backgrounds and other low frequency components. However, there is a large class of filters that blur low frequency components but keep the edges fairly sharp.

The median filter mentioned above is one such; we shall explore it in greater detail in Chapter 8. But just for a quick preview of its effects, consider median filters of size 3×3 and 5×5 applied to the cameraman image. These are shown in Figure 5.20. Even though

(a) Using 3×3 filter

(b) Using a 5×5 filter

FIGURE 5.20: Using a median filter

much of the fine detail has gone, especially when using the larger filter, the edges are still sharp.

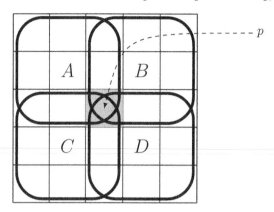

FIGURE 5.21: The neighborhoods used in the Kuwahara filter

Kuwahara Filters

These are a family of filters in which the variance of several regions in the neighborhood are computed, and the output is the mean of the region with the lowest variance. The simplest version takes a 5×5 neighborhood of a given pixel p looks at the four overlapping 3×3 neighborhoods of which it is a corner as shown in Figure 5.21.

The filter works by first computing the variances of the four neighborhoods around p, and then the output is the mean of the neighborhood with the lowest variance. So for example, consider the following neighborhood:

```
169  140  105  126  110
140   65  175  247   79
 40  178  240  171   37
 56   28  203   55   53
208  193   75  165  212
```

Then the variances of each of the neighborhoods can be found as

```
In:  x[0:3,0:3].var(), x[0:3,2:5].var(), x[2:5,0:3].var(), x[2:5,2:5].var()
```
Python

or as

```
>>> var(x(1:3,1:3)(:),1),var(x(1:3,3:5)(:),1),var(x(3:5,1:3)(:),1),...
    var(x(3:5,3:5)(:),1)
```
MATLAB/Octave

The variances of the neighborhoods A, B, C, and D will be found to be 3372.54, 4465.11, 6278.0, 5566.69, respectively. Of these four values, the variance of A is the smallest, so the output is the mean of A, which (when rounded to an integer) is 139.

None of MATLAB, Octave, or Python support the Kuwahara filter directly, but it is very easy to program it. Recall that for a random variable X, its variance can be computed as

$$\overline{X^2} - \left(\overline{X}\right)^2$$

where $\overline{(\cdot)}$ is the mean. This means that the variances in all 3×3 neighborhoods of an image x can be found by first filtering x^2 with the averaging filter, then squaring the result of filtering x with the averaging filter, and subtracting them:

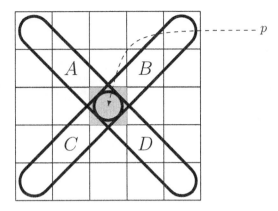

FIGURE 5.22: Alternative neighborhoods for a Kuwahara filter

```
>> cd = float(c);
>> cdm = imfilter(cd,ones(3)/9,'symmetric');
>> cd2f = imfilter(cd.^2,ones(3)/9,'symmetric');
>> cdv = cd2f - cdm.^2;
```
MATLAB/Octave

```
In:   cd = float32(c)
In:   cdm = ndi.uniform_filter(cd,(3,3))
In:   cd2f = ndi.uniform_filter(cd**2,(3,3))
In:   cdv = cd2f - cdm**2
```
Python

At this stage, the array `cdm` contains all the mean values, and `cdv` contains all the variances. At every point (i, j) in the image then:

1. Compute the list
$$\text{vars} = [\texttt{cdv}(i-1, j-1), \texttt{cdv}(i-1, j+1), \texttt{cdv}(i+1, j-1), \texttt{cdv}(i+1, j+1)].$$

2. Also compute the list
$$\text{means} = [\texttt{cdm}(i-1, j-1), \texttt{cdm}(i-1, j+1), \texttt{cdm}(i+1, j-1), \texttt{cdm}(i+1, j+1)].$$

3. Output the value of "means" corresponding to the lowest value of "vars."

Programs are given at the end of the chapter. Note that the neighborhoods can of course be larger than 3×3, any odd square size can used, or indeed any other shape, such as shown in Figure 5.22. Figure 5.23 shows the cameraman image first with the Kuwahara filter using 3×3 neighborhoods and with 7×7 neighborhoods.

Note that even with the significant blurring with the larger filter, the edges are still remarkably sharp.

Bilateral Filters

Linear filters (and many non-linear filters) are *domain filters*; the weights attached to each neighboring pixel depend on only the *position* of those pixels. Another family of filters are the *range filters*, where the weights depend on the relative *difference* between pixel

(a) Using 3×3 neighborhoods   (b) Using 7×7 neighborhoods

FIGURE 5.23: Using Kuwahara filters

values. For example, consider the Gaussian filter $e^{(-x^2-y^2)/2}$ (so with variance $\sigma^2 = 1$) over the region $-1 \le x, y \le 1$:

$$G =$$
$$
\begin{array}{ccc}
0.36788 & 0.60653 & 0.36788 \\
0.60653 & 1.00000 & 0.60653 \\
0.36788 & 0.60653 & 0.36788
\end{array}
$$

Applied to the neighborhood

$$N =$$
$$
\begin{array}{ccc}
0.8 & 0.1 & 0.6 \\
0.3 & 0.5 & 0.7 \\
0.4 & 0.9 & 0.2
\end{array}
$$

the output is simply the sum of all the products of corresponding elements of G and N.

As a *range filter*, however, with variance σ_r^2, the output would be the Gaussian function applied to the difference of the elements of N with its central value:

$$N - 0.5 =$$
$$
\begin{array}{ccc}
0.3 & -0.4 & 0.1 \\
-0.2 & 0.0 & 0.2 \\
-0.1 & 0.4 & -0.3
\end{array}
$$

For each of these values v the filter consists $e^{-v^2/2\sigma_r^2}$.

Thus, a domain filter is based on the *closeness* of pixels, and a range filter is based on the *similarity* of pixels. Clearly, the larger the value of σ_r the flatter the filter, and so there will be less distinction of closeness.

The following three arrays show the results with $\sigma_r = 0.1, 1, 10$, respectively:

$\sigma_r = 0.1$:

$$
\begin{array}{ccc}
0.01111 & 0.00034 & 0.60653 \\
0.13534 & 1.00000 & 0.13534 \\
0.60653 & 0.00034 & 0.01111
\end{array}
$$

$\sigma_r = 1$:

$$
\begin{array}{ccc}
0.95600 & 0.92312 & 0.99501 \\
0.98020 & 1.00000 & 0.98020 \\
0.99501 & 0.92312 & 0.95600
\end{array}
$$

$\sigma_r = 10$:

$$
\begin{array}{ccc}
0.99955 & 0.99920 & 0.99995 \\
0.99980 & 1.00000 & 0.99980 \\
0.99995 & 0.99920 & 0.99955
\end{array}
$$

In the last example, the values are very nearly equal, so this range filter treats all values as being (roughly) equally similar.

The idea of the *bilateral filter* is to use *both* domain and range filtering, both with Gaussian filters, and with variances σ_d^2 and σ_r^2. For each pixel in the image, a range filter is created using σ_r^2, which maps the similarity of pixels in that neighborhood. That filter is then convolved with the domain filter which uses σ_d.

By adjusting the size of the filter mask, and of the variances, it is possible to obtain varying amounts of blurring, as well as keeping the edges sharp.

At the time of writing, MATLAB's Image Processing Toolbox does not include bilateral filtering, but both Octave and Python do: the former with the `bilateral` parameter of the `imsmooth` function; the latter with the `denoise_bilateral` method in the `restoration` module of `skimage`.

However, a simple implementation can be written, and one is given at the chapter end.

Using any of these functions and applied to the cameraman image produces the results shown in Figure 5.24, where in each case w gives the size of the filter. Thus, the image in Figure 5.24(a) can be created with our function by:

```
>> cb = bilateral(c,2,2,0,2);
```

MATLAB/Octave

where the first parameter w (in this case 2) produces filters of size $2w + 1 \times 2w + 1$.

Notice that even when a large flat Gaussian is used as the domain filter, as in Figure 5.24(d), the corresponding use of an appropriate range filter keeps the edges sharp.

In Python, a bilateral filter can be applied to produce, for example, the image in Figure 5.24(b) with

```
In :  import skimage.restoration as re
In :  cb = re.denoise_bilateral(c,win_size=7,sigma_range=0.2,\
...:  sigma_spatial=10)
```

Python

The following is an example of the use of Octave's `imsmooth` function:

```
>> cb = imsmooth(c,'bilateral',sigma_d=2,sigma_r=0.1);
```

Octave

(a) $w = 5$, $\sigma_d = 2$, $\sigma_r = 0.2$ (b) $w = 7$, $\sigma_d = 10$, $\sigma_r = 0.2$

(c) $w = 11$, $\sigma_d = 3$, $\sigma_r = 0.1$ (d) $w = 11$, $\sigma_d = 5$, $\sigma_r = 0.5$

FIGURE 5.24: Bilateral filtering

and the window size of the filter is automatically generated from the σ values.

5.10   Region of Interest Processing

Often we may not want to apply a filter to an entire image, but only to a small region within it. A non-linear filter, for example, may be too computationally expensive to apply to the entire image, or we may only be interested in the small region. Such small regions within an image are called *regions of interest* or *ROIs*, and their processing is called *region of interest processing*.

Regions of Interest in MATLAB

Before we can process a ROI, we have to define it. There are two ways: by listing the coordinates of a polygonal region; or interactively, with the mouse. For example, suppose we take part of a monkey image:

```
>> m2 = imread('monkey.png');
>> m = m2(56:281,221:412);
```
MATLAB/Octave

and attempt to isolate its head. If the image is viewed with `impixelinfo`, then the co-ordinates of a hexagon that enclose the head can be determined to be $(60, 14)$, $(27, 38)$, $(14, 127)$, $(78, 177)$, $(130, 160)$ and $(139, 69)$, as shown in Figure 5.25. We can then define a region of interest using the **roipoly** function:

```
>> xi = [60 27 14 78 130 139]
>> yi = [14 38 127 177 160 69]
>> roi=roipoly(m,yi,xi);
```
MATLAB/Octave

Note that the ROI is defined by two sets of coordinates: first the columns and then the rows, taken in order as we traverse the ROI from vertex to vertex. In general, a ROI mask will be a binary image the same size as the original image, with 1s for the ROI, and 0s elsewhere. The function **roipoly** can also be used interactively:

```
>> roi=roipoly(m);
```
MATLAB/Octave

This will bring up the monkey image (if it isn't shown already). Vertices of the ROI can be selected with the mouse: a left click selects a new vertex, backspace or delete removes the most recently chosen vertex, and a right click finishes the selection.

Region of Interest Filtering

One of the simplest operations on a ROI is spatial filtering; this is implemented with the function **roifilt2**. With the monkey image and the ROI found above, we can experiment:

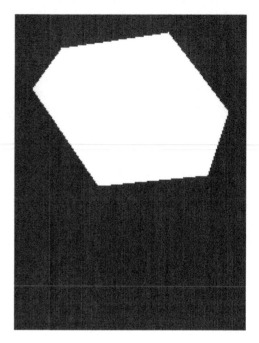

FIGURE 5.25: An image with an ROI, and the ROI mask

```
>> a = fspecial('average',15);
>> ma = roifilt2(a,m,roi);
>> u = fspecial('unsharp');
>> mu = roifilt2(u,m,roi);
>> l = fspecial('log');
>> ml = roifilt2(l,m,roi);
>> imshow(ma),figure,imshow(mu),figure,imshow(ml)
```
MATLAB

The images are shown in Figure 5.26.

Average filtering   Unsharp masking   Laplacian of Gaussian

FIGURE 5.26: Examples of the use of `roifilt2`

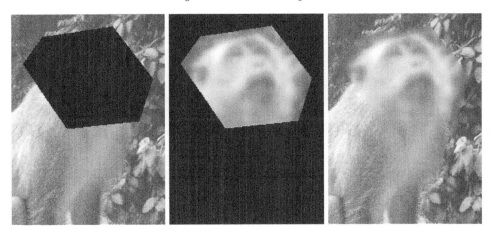

FIGURE 5.27: ROI filtering with a polygonal mask

Regions of Interest in Octave and Python

Neither Octave nor Python have `roipoly` or `roifilt2` commands. However, in each it is quite straightforward to create a mask defining the region of interest, and use that mask to restrict the act of filtering to the region.

With the monkey above, and the head region, suppose that the matrices for the image and the ROI are M and R, respectively. If M_f is the result of the image matrix after filtering, then

$$M_f R + M(1 - R)$$

will provide the result we want. This is shown in Figure 5.27 where the rightmost image is the sum of the other two.

In Octave, this can be achieved with:

```octave
>  m2 = imread('monkey.png'); m = m2(56:281,221:412);
>  [r,c] = size(m);
>  xi = [60 27 14 78 130 139]
>  yi = [14 38 127 177 160 69]
>  roi = poly2mask(yi,xi,r,c);
>  f = fspecial('gaussian',9,3);
>  mg = imfilter(m,f));
>  mr = imadd(mg.*roi,m.*~roi)
```

And in Python, using `zeros_like` which creates an array of zeros the same size as its input, as well as `polygon` from the `draw` module of `skimage`:

```
In:   m2 = io.imread('monkey.png'); m = m2[55:281,220:412]
In:   r,c = m.shape

In:   xi = np.array([60,27,14,78,130,139])
In:   yi = np.array([14,38,127,177,160,69])
In:   roi = np.zeros_like(m)
In:   r,c = polygon(yi,xi)
In:   roi[c,r] = 1

In:   mg = ut.img_as_ubyte(fl.gaussian_filter(m,g))
In:   mr = mg*roi + m*(1-roi)
```

Python

5.11   Programs

This is a simple program (which can be run in MATLAB or Octave) for bilateral filtering.

```
function out = bilateral(im,w,sigma_d,sigma_r)
   im = im2double(im);
   [r,c] = size(im);
   out = zeros(r,c);
   A = padarray(im,[w,w],'symmetric');
   G = fspecial('gaussian',2*w+1,sd);        # the domain filter
   for i = 1+w:r+w-1
       for j = 1+w:c+w-1
           R = A(i-w:i+w,j-w:j+w);           # region to be computed
           H = exp(-(R-A(i,j)).^2/(2*sr^2)); # the range filter
           F = H.*G;
           out(i-w,j-w) = sum(F(:).*R(:))/sum(F(:));
       end;
   end;
   close(h);
end
```

MATLAB/Octave

Exercises

1. The array below represents a small grayscale image. Compute the images that result when the image is convolved with each of the masks (a) to (h) shown. At the edge of the image use a restricted mask. (In other words, pad the image with zeros.)

```
20   20   20   10   10   10   10   10   10
20   20   20   20   20   20   20   20   10
20   20   20   10   10   10   10   20   10
20   20   10   10   10   10   10   20   10
```

```
20  10  10  10  10  10  10  20  10
10  10  10  10  20  10  10  20  10
10  10  10  10  10  10  10  10  10
20  10  20  20  10  10  10  20  20
20  10  10  20  10  10  20  10  20
```

	-1	-1	0		0	-1	-1		-1	-1	-1		-1	2	-1

(a)
```
-1  -1   0
-1   0   1
 0   1   1
```

(b)
```
 0  -1  -1
 1   0  -1
 1   1   0
```

(c)
```
-1  -1  -1
 2   2   2
-1  -1  -1
```

(d)
```
-1   2  -1
-1   2  -1
-1   2  -1
```

(e)
```
-1  -1  -1
-1   8  -1
-1  -1  -1
```

(f)
```
 1   1   1
 1   1   1
 1   1   1
```

(g)
```
-1   0   1
-1   0   1
-1   0   1
```

(h)
```
 0  -1   0
-1   4  -1
 0  -1   0
```

2. Check your answers to the previous question with using `imfilter` (if you are using MATLAB or Octave), or `ndi.correlate` if you are using Python.

3. Describe what each of the masks in the previous question might be used for. If you can't do this, wait until Question 5 below.

4. Devise a 3 × 3 mask for an "identity filter," which causes no change in the image.

5. Choose an image that has a lot of fine detail, and load it.

   Apply all the filters listed in Question 1 to this image. Can you now see what each filter does?

6. Apply larger and larger averaging filters to this image. What is the smallest sized filter for which the fine detail cannot be seen?

7. Repeat the previous question with Gaussian filters with the following parameters:

Size	Standard deviation		
[3,3]	0.5	1	2
[7,7]	1	3	6
[11,11]	1	4	8
[21,21]	1	5	10

   At what values do the fine details disappear?

8. Can you see any observable difference in the results of average filtering and of using a Gaussian filter?

9. If you are using MATLAB or Octave, read through the help page of the `fspecial` function, and apply some of the other filters to the cameraman image and to the mandrill image.

10. Apply different Laplacian filters to an image of your choice at and to the cameraman images. Which produces the best edge image?

11. Is the 3 × 3 median filter separable? That is, can this filter be implemented by a 3 × 1 filter followed by a 1 × 3 filter?

12. Repeat the above question for the maximum and minimum filters.

13. Apply a 3×3 averaging filter to the middle 9 values of the matrix

$$\begin{bmatrix} a & b & c & d & e \\ f & g & h & i & j \\ k & l & m & n & o \\ p & q & r & s & t \\ u & v & w & x & y \end{bmatrix}$$

and then apply another 3×3 averaging filter to the result.

Using your answer, describe a 5×5 filter that has the effect of two averaging filters.

Is this filter separable?

14. Use the appropriate commands to produce the outputs shown in Figure 5.5, starting with the diagonally increasing image

```
 20   40   60   80  100
 40   60   80  100  120
 60   80  100  120  140
 80  100  120  140  160
100  120  140  160  180
```

15. Display the difference between the cmax and cmin images obtained in Section 5.8. You can do this with an image subtraction.

What are you seeing here? Can you account for the output of these commands?

16. If you are using MATLAB or Octave, then use the tic and toc timer functions to compare the use of nlfilter and colfilt functions. If you are using the ipython enhanced shell of Python, try the %time function

17. Use colfilt (MATLAB/Octave) or generic_filter (Python) to implement the geometric mean and alpha-trimmed mean filters.

18. If you are using MATLAB or Octave, show how to implement the root-mean-square filter.

19. Can unsharp masking be used to reverse the effects of blurring? Apply an unsharp masking filter after a 3×3 averaging filter, and describe the result.

20. Rewrite the Kuwahara filter as a single function that can be applied with either colfilt (MATLAB/Octave) or generic_filter (Python).

Chapter 6

Image Geometry

There are many situations in which we might want to change the shape, size, or orientation of an image. We may wish to enlarge an image, to fit into a particular space, or for printing; we may wish also to reduce its size, say for inclusion on a web page. We might also wish to rotate it: maybe to adjust for an incorrect camera angle, or simply for affect. Rotation and scaling are examples of *affine transformations*, where lines are transformed to lines, and in particular parallel lines remain parallel after the transformation. Non-affine geometrical transformations include warping, which we will not consider.

6.1 Interpolation of Data

We will start with a simple problem: suppose we have a collection of 4 values, which we wish to enlarge to 8. How do we do this? To start, we have our points x_1, x_2, x_3, and x_4, which we suppose to be evenly spaced, and we have the values at those points: $f(x_1)$, $f(x_2)$, $f(x_3)$, and $f(x_4)$. Along the line $x_1 \ldots x_4$ we wish to space eight points $x'_1, x'_2, \ldots, x'_8$. Figure 6.1 shows how this would be done.

FIGURE 6.1: Replacing four points with eight

Suppose that the distance between each of the x_i points is 1; thus, the length of the line is 3. Thus, since there are seven increments from x'_i to x'_8, the distance between each two will be $3/7 \approx 0.4286$. To obtain a relationship between x and x' we draw Figure 6.1 slightly differently as shown in Figure 6.2. Then

$$x' = \frac{1}{3}(7x - 4),$$

$$x = \frac{1}{7}(3x' + 4).$$

As you see from Figure 6.1, none of the x'_i coincides exactly with an original x_j, except for the first and last. Thus we are going to have to "guess" at possible function values $f(x'_i)$. This guessing at function values is called *interpolation*. Figure 6.3 shows one way of doing this: we assign $f(x'_i) = f(x_j)$, where x_j is the original point closest to x'_i. This is called *nearest neighbor interpolation*.

FIGURE 6.2: Figure 6.1 slightly redrawn

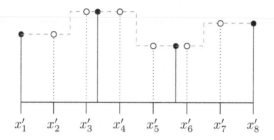

FIGURE 6.3: Nearest neighbor interpolation

The closed circles indicate the original function values $f(x_i)$; the open circles, the interpolated values $f(x_i')$.

Another way is to join the original function values by straight lines, and take our interpolated values as the values at those lines. Figure 6.4 shows this approach to interpolation; this is called *linear interpolation*.

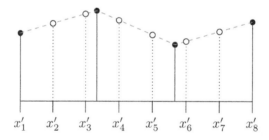

FIGURE 6.4: Linear interpolation

To calculate the values required for linear interpolation, consider the diagram shown in Figure 6.5.

In this figure we assume that $x_2 = x_1 + 1$, and that F is the value we require. By considering slopes:

$$\frac{F - f(x_1)}{\lambda} = \frac{f(x_2) - f(x_1)}{1}.$$

Solving this equation for F produces:

$$F = \lambda f(x_2) + (1 - \lambda)f(x_1). \tag{6.1}$$

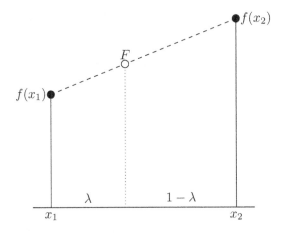

FIGURE 6.5: Calculating linearly interpolated values

As an example of how to use this, suppose we have the values $f(x_1) = 2$, $f(x_2) = 3$, $f(x_3) = 1.5$ and $f(x_4) = 2.5$. Consider the point x'_4. This is between x_2 and x_3, and the corresponding value for λ is $2/7$. Thus

$$f(x'_4) = \frac{2}{7}f(x_3) + \frac{5}{7}f(x_2)$$
$$= \frac{2}{7}(1.5) + \frac{5}{7}(3)$$
$$\approx 2.5714.$$

For x'_7, we are between x_3 and x_4 with $\lambda = 4/7$. So:

$$f(x'_7) = \frac{4}{7}f(x_4) + \frac{3}{7}f(x_3)$$
$$= \frac{4}{7}(2.5) + \frac{3}{7}(1.5)$$
$$\approx 2.0714.$$

6.2 Image Interpolation

The methods of the previous section can be applied to images. Figure 6.6 shows how a 4×4 image would be interpolated to produce an 8×8 image. Here the large open circles are the original points, and the smaller closed circles are the new points.

To obtain function values for the interpolated points, consider the diagram shown in Figure 6.7.

We can give a value to $f(x', y')$ by either of the methods above: by setting it equal to the function values of the closest image point, or by using linear interpolation. We can apply linear interpolation first along the top row to obtain a value for $f(x, y')$, and then along the bottom row to obtain a value for $f(x + 1, y')$. Finally, we can interpolate along the y' column between these new values to obtain $f(x', y')$. Using the formula given by

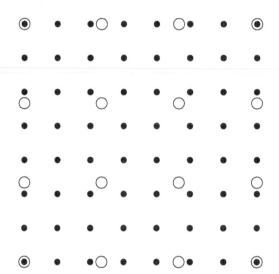

FIGURE 6.6: Interpolation on an image

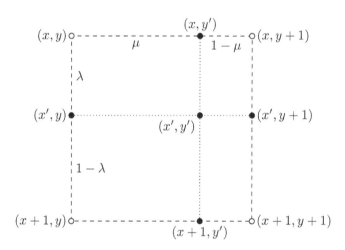

FIGURE 6.7: Interpolation between four image points

Equation 6.1, then

$$f(x, y') = \mu f(x, y+1) + (1-\mu)f(x, y)$$

and

$$f(x+1, y') = \mu f(x+1, y+1) + (1-\mu)f(x+1, y).$$

Along the y' column we have

$$f(x', y') = \lambda f(x+1, y') + (1-\lambda)f(x, y')$$

and substituting in the values just obtained produces

$$\begin{aligned}
f(x', y') &= \lambda(\mu f(x+1, y+1) + (1-\mu)f(x+1, y)) + (1-\lambda)(\mu f(x, y+1) \\
&\quad + (1-\mu)f(x, y)) \\
&= \lambda\mu f(x+1, y+1) + \lambda(1-\mu)f(x+1, y) + (1-\lambda)\mu f(x, y+1) \\
&\quad + (1-\lambda)(1-\mu)f(x, y)
\end{aligned}$$

This last equation is the formula for *bilinear interpolation*.

Now image scaling can be performed easily. Given our image, and either a scaling factor (or separate scaling factors for x and y directions), or a size to be scaled to, we first create an array of the required size. In our example above, we had a 4×4 image, given as an array (x, y), and a scale factor of two, resulting in an array (x', y') of size 8×8. Going right back to Figures 6.1 and 6.2, the relationship between (x, y) and (x', y') is

$$(x', y') = \left(\frac{1}{3}(7x - 4), \frac{1}{3}(7y - 4) \right),$$

$$(x, y) = \left(\frac{1}{7}(3x' + 4), \frac{1}{7}(3y' + 4) \right).$$

Given our (x', y') array, we can step through it point by point, and from the corresponding surrounding points from the (x, y) array calculate an interpolated value using either nearest neighbor or bilinear interpolation.

There's nothing in the above theory that requires the scaling factor to be greater than one. We can choose a scaling factor *less* than one, in which case the resulting image array will be *smaller* than the original. We can consider Figure 6.6 in this light: the small closed circles are the *original* image points, and the large open circles are the smaller array on which we are to find interpolated values.

MATLAB has the function `imresize` which does all this for us. It can be called with

```
imresize(A,k,'method')
```

where A is an image of any type, k is a scaling factor, and `'method'` is either `'nearest'` or `'bilinear'` (or another method to be described later). Another way of using `imresize` is

```
imresize(A,[m,n],'method')
```

where [m,n] provide the size of the scaled output. There is a further, optional parameter allowing you to choose either the size or type of low pass filter to be applied to the image before reducing its size—see the help file for details.

Let's try a few examples. We shall start by taking the head of the cameraman, and enlarging it by a factor of four:

```
>> c = imread('cameraman.png');
>> head = c(33:96,90:153);
>> imshow(head)
>> head4n = imresize(head,4,'nearest');imshow(head4n)
>> head4b = imresize(head,4,'bilinear');imshow(head4b)
```

MATLAB/Octave

Python has a `rescale` function in the `transform` module of `skimage`:

```
In :   c = io.imread('cameraman.png')
In :   head = c[32:96,89:153]
In :   io.imshow(head)
In :   head4n = tr.rescale(head,2,order=0)
In :   head4n = tr.rescale(head,2,order=1)
```

Python

The `order` parameter of the `rescale` method provides the order of the interpolating polynomial: 0 corresponds to nearest neighbor and 1 to bilinear interpolation.

The head is shown in Figure 6.8 and the results of the scaling are shown in Figure 6.9.

FIGURE 6.8: The head

Nearest neighbor interpolation gives an unacceptable blocky effect; edges in particular appear very jagged. Bilinear interpolation is much smoother, but the trade-off here is a certain blurriness to the result. This is unavoidable: interpolation can't predict values: we can't create data from nothing! All we can do is to guess at values that fit best with the original data.

6.3   General Interpolation

Although we have presented nearest neighbor and bilinear interpolation as two different methods, they are in fact two special cases of a more general approach. The idea is this: we wish to interpolate a value $f(x')$ for $x_1 \leq x' \leq x_2$, and suppose $x' - x_1 = \lambda$. We define an interpolation function $R(u)$, and set

$$f(x') = R(-\lambda)f(x_1) + R(1-\lambda)f(x_2). \tag{6.2}$$

(a) Nearest neighbor scaling   (b) Bilinear interpolation

FIGURE 6.9: Scaling by interpolation

Figure 6.10 shows how this works. The function $R(u)$ is centered at x', so x_1 corresponds

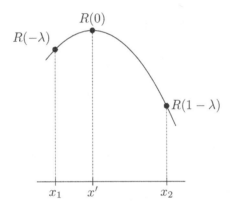

FIGURE 6.10: Using a general interpolation function

with $u = -\lambda$, and x_2 with $u = 1 - \lambda$. Now consider the two functions $R_0(u)$ and $R_1(u)$ shown in Figure 6.11. Both these functions are defined on the interval $-1 \le u \le 1$ only. Their formal definitions are:

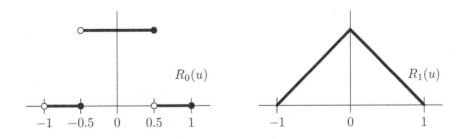

FIGURE 6.11: Two interpolation functions

$$R_0(u) = \begin{cases} 0 & \text{if } u \leq -0.5 \\ 1 & \text{if } -0.5 < u \leq 0.5 \\ 0 & \text{if } u > 0.5 \end{cases}$$

and

$$R_1(u) = \begin{cases} 1 + u & \text{if } u \leq 0 \\ 1 - u & \text{if } u \geq 0 \end{cases}$$

The function $R_1(u)$ can also be written as $1 - |x|$. Now substituting $R_0(u)$ for $R(u)$ in Equation 6.2 will produce nearest neighbor interpolation. To see this, consider the two cases $\lambda < 0.5$ and $\lambda \geq 0.5$ separately. If $\lambda < 0.5$, then $R_0(-\lambda) = 1$ and $R_0(1-\lambda) = 0$. Then

$$f(x') = (1)f(x_1) + (0)f(x_2) = f(x_1).$$

If $\lambda \geq 0.5$, then $R_0(-\lambda) = 0$ and $R_0(1 - \lambda) = 1$. Then

$$f(x') = (0)f(x_1) + (1)f(x_2) = f(x_2).$$

In each case $f(x')$ is set to the function value of the point closest to x'.

Similarly, substituting $R_1(u)$ for $R(u)$ in Equation 6.2 will produce linear interpolation. We have

$$f(x') = R_1(-\lambda)f(x_1) + R_1(1 - \lambda)f(x_2)$$
$$= (1 - \lambda)f(x_1) + \lambda f(x_2)$$

which is the correct equation.

The functions $R_0(u)$ and $R_1(u)$ are just two members of a family of possible interpolation functions. Another such function provides *cubic interpolation*; its definition is:

$$R_3(u) = \begin{cases} 1.5|u|^3 - 2.5|u|^2 + 1 & \text{if } |u| \leq 1, \\ -0.5|u|^3 + 2.5|u|^2 - 4|u| + 2 & \text{if } 1 < |u| \leq 2. \end{cases}$$

Its graph is shown in Figure 6.12. This function is defined over the interval $-2 \leq u \leq 2$,

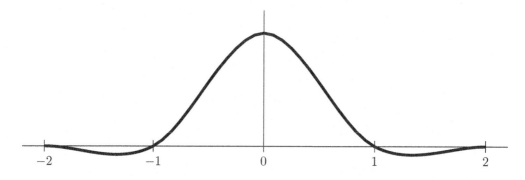

FIGURE 6.12: The cubic interpolation function $R_3(u)$

and its use is slightly different from that of $R_0(u)$ and $R_1(u)$, in that as well as using the function values $f(x_1)$ and $f(x_2)$ for x_1 and x_2 on either side of x', we use values of x further away. In fact the formula we use, which extends Equation 6.2, is:

$$f(x') = R_3(-1 - \lambda)f(x_1) + R_3(-\lambda)f(x_2) + R_3(1 - \lambda)f(x_3) + R_4(2 - \lambda)f(x_4)$$

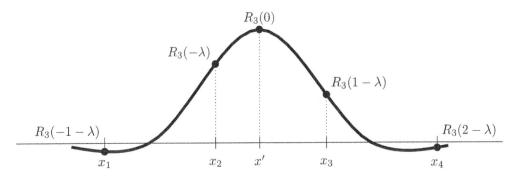

FIGURE 6.13: Using $R_3(u)$ for interpolation

where x' is between x_2 and x_3, and $x - x_2 = \lambda$. Figure 6.13 illustrates this. To apply this interpolation to images, we use the 16 known values around our point (x', y'). As for bilinear interpolation, we first interpolate along the rows, and then finally down the columns, as shown in Figure 6.14. Alternately, we could first interpolate down the columns, and then across the row. This means of image interpolation by applying cubic interpolation in both directions is called *bicubic interpolation*. To perform bicubic interpolation on an

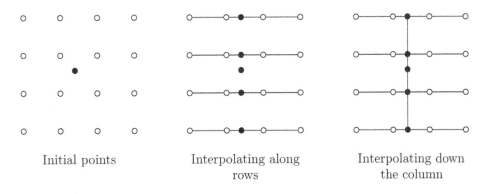

FIGURE 6.14: How to apply bicubic interpolation

image with MATLAB, we use the `'bicubic'` method of the `imresize` function. To enlarge the cameraman's head, we enter (for MATLAB or Octave):

```
>> head4c = imresize(head,4,'bicubic');imshow(head4c)
```
MATLAB/Octave

In Python the `order` parameter of `rescale` must be set to 3 for bicubic interpolation:

```
In :  head4c = tr.rescale(head,4,order=3)
```
Python

and the result is shown in Figure 6.15.

FIGURE 6.15: Enlargement using bicubic interpolation

6.4   Enlargement by Spatial Filtering

If we merely wish to enlarge an image by a power of 2, there is a quick and dirty method which uses linear filtering. We give an example. Suppose we take a simple 4×4 matrix:

```
>> m = magic(4)

m =

    16     2     3    13
     5    11    10     8
     9     7     6    12
     4    14    15     1
```
MATLAB/Octave

Our first step is to create a *zero-interleaved* version of this matrix. This is obtained by interleaving rows and columns of zeros between the rows and columns of the original matrix. Such a matrix will be double the size of the original, and will contain mostly zeros. If m_2 is the zero-interleaved version of m, then it is defined by:

$$m_2(i,j) = \begin{cases} m((i+1)/2, (j+1)/2) & \text{if } i \text{ and } j \text{ are both odd,} \\ 0 & \text{otherwise.} \end{cases}$$

This can be implemented very simply:

```
>> [r,c] = size(m);
>> m2 = zeros(2*r,2*c);
>> m2(1:2:1*r,1:2:2*c) = m

ans =

       16        0        2        0        3        0       13        0
        0        0        0        0        0        0        0        0
        5        0       11        0       10        0        8        0
        0        0        0        0        0        0        0        0
        9        0        7        0        6        0       12        0
        0        0        0        0        0        0        0        0
        4        0       14        0       15        0        1        0
        0        0        0        0        0        0        0        0
```

MATLAB/Octave

In Python, interleaving is also easily done:

```
In :   m = array([[16,2,3,13],[5,11,10,8],[9,7,6,12],[4,14,15,1]]);
In :   r,c = m.shape
In :   m2 = zeros((2*r,2*c))
In :   m2[::2,::2] = m
```

Python

We can now replace the zeros by applying a spatial filter to this matrix. The spatial filters

$$\begin{pmatrix} 1 & 1 & 0 \\ 1 & 1 & 0 \\ 0 & 0 & 0 \end{pmatrix} \qquad \frac{1}{4}\begin{pmatrix} 1 & 2 & 1 \\ 2 & 4 & 2 \\ 1 & 2 & 1 \end{pmatrix}$$

implement nearest neighbor interpolation and bilinear interpolation, respectively. We can test this with a few commands:

```
>> imfilter(m2,[1 1 0;1 1 0;0 0 0])

ans =

    16    16     2     2     3     3    13    13
    16    16     2     2     3     3    13    13
     5     5    11    11    10    10     8     8
     5     5    11    11    10    10     8     8
     9     9     7     7     6     6    12    12
     9     9     7     7     6     6    12    12
     4     4    14    14    15    15     1     1
     4     4    14    14    15    15     1     1

>> format bank
>> imfilter(m2,[1 2 1;2 4 2;1 2 1]/4)

ans =

    16.00     9.00     2.00     2.50     3.00     8.00    13.00     6.50
    10.50     8.50     6.50     6.50     6.50     8.50    10.50     5.25
     5.00     8.00    11.00    10.50    10.00     9.00     8.00     4.00
     7.00     8.00     9.00     8.50     8.00     9.00    10.00     5.00
     9.00     8.00     7.00     6.50     6.00     9.00    12.00     6.00
     6.50     8.50    10.50    10.50    10.50     8.50     6.50     3.25
     4.00     9.00    14.00    14.50    15.00     8.00     1.00     0.50
     2.00     4.50     7.00     7.25     7.50     4.00     0.50     0.25
```

MATLAB/Octave

(The `format bank` provides an output with only two decimal places.) In Python, the filters can be applied as:

```
In : ndi.convolve(m2,array([[0,0,0],[0,1,1],[0,1,1]]),mode='constant')
In : ndi.convolve(m2,array([[1,2,1],[2,4,2],[1,2,1]])/4.0,mode='constant')
```

Python

We can check these with the commands

```
>> m2b=imresize(m,[8,8],'nearest');m2b

>> m2b=imresize(m,[7,7],'bilinear');m2b
```

MATLAB/Octave

In the second command we only scaled up to 7×7, to ensure that the interpolation points lie exactly half-way between the original data values. The filter

$$\frac{1}{64} \begin{pmatrix} 1 & 4 & 6 & 4 & 1 \\ 4 & 16 & 24 & 16 & 4 \\ 6 & 24 & 36 & 24 & 6 \\ 4 & 16 & 24 & 16 & 4 \\ 1 & 4 & 6 & 4 & 1 \end{pmatrix}$$

can be used to approximate bicubic interpolation.

We can try all of these with the cameraman's head, doubling its size.

```
>> hz = uint8(zeros(size(head)*2));
>> hz(1:2:2nd,1:2:end) = head;
>> imshow(hz)
>> imshow(imfilter(hz, [1 1 0;1 1 0;0 0 0]))
>> imshow(imfilter(hz,[1 2 1;2 4 2;1 2 1]/4)
>> bfilt=[1 4 6 4 1;4 16 24 16 4;6 24 36 24 6;4 16 24 16 4;1 4 6 4 1]/64;
>> imshow(imfilter(hz,bfilt))
```

MATLAB/Octave

or in Python with

```
In :   r,c = head.shape
In :   hz = np.zeros((2*r,2*c)).astype('uint8')
In :   hz[::2,::2] = head
In :   ne = array([[0,0,0],[0,1,1],[1,1,1]])
In :   bi = array([[1,2,1],[2,4,2],[1,2,1]])/4.0
In :   bc = array
       ([[1,4,6,4,1],[4,16,24,16,4],[6,24,36,24,6],[4,16,24,16,4],[1,4,6,4,1]])
       /64.0
In :   io.imshow(ndi.correlate(hz,ne))
```

Python

and similarly for the others. The results are shown in Figure 6.16. We can enlarge more by simply taking the result of the filter, applying a zero-interleave to it, and then another filter.

| Zero interleaving | Nearest neighbor | Bilinear | Bicubic |

FIGURE 6.16: Enlargement by spatial filtering

6.5 Scaling Smaller

The business of making an image smaller is also called *image minimization*; one way is to take alternate pixels. If we wished to produce an image one-sixteenth the size of the original, we would take out only those pixels (i, j) for which i and j are both multiples of four. This method is called image *subsampling* and corresponds to the **nearest** option of **imresize**, and is very easy to implement.

However, it does not give very good results at high frequency components of an image. We shall give a simple example; we shall construct a large image consisting of a white square with a single circle on it. The **meshgrid** command provides row and column indices which span the array:

```
>> [x,y] = meshgrid(-255:256);
>> z = sqrt(x.^2+y.^2);
>> t = 1 - (z>254.5 & z<256);
>> imshow(t)
```

MATLAB/Octave

or

```
In :   r = range(-256,256)
In :   [x,y] = np.meshgrid(r,r)
In :   z = sqrt(x**2 + y**2)
In :   t = 1-((z>254.5) & (z<256))*1
In :   io.imshow(t)
```

Python

Now we can resize it by taking out most pixels:

```
>> t2 = imresize(t,0.25,'nearest');
```

MATLAB/Octave

or

```
In :   t2 = tr.rescale(t,0.25,order=0)
```

Python

and this is shown in Figure 6.17(a). Notice that because of the way that pixels were removed, the resulting circle contains gaps. If we were to use one of the other methods:

```
>> t3 = imresize(t,0.25,'bicubic');
```

MATLAB/Octave

or

```
In :   t3 = tr.rescale(t,0.25,order=3)
```

Python

a low pass filter is applied to the image first. The result is shown in Figure 6.17(b). The image in Figure 6.17(b) can be made binary by thresholding (which will be discussed in greater detail in Chapter 9). In this case

```
>> t4 = imresize(t,0.25,'bicubic')>0.9;
```

MATLAB/Octave

or

```
In :   t4 = tr.rescale(t,0.25,order=3)>0.9
```

Python

does the trick.

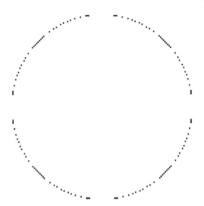 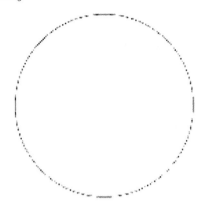

(a) Nearest neighbor minimization   (b) Bicubic interpolation for minimization

FIGURE 6.17: Minimization

6.6   Rotation

Having done the hard work of interpolation for scaling, we can easily apply the same theory to image rotation. First recall that the mapping of a point (x, y) to another (x', y') through a counter-clockwise rotation of θ as shown in Figure 6.18 is obtained by the matrix product

$$\begin{pmatrix} x' \\ y' \end{pmatrix} = \begin{pmatrix} \cos\theta & -\sin\theta \\ \sin\theta & \cos\theta \end{pmatrix} \begin{pmatrix} x \\ y \end{pmatrix}.$$

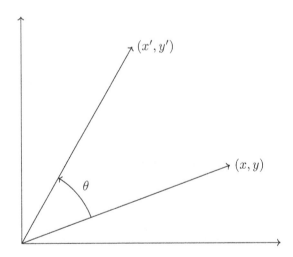

FIGURE 6.18: Rotating a point through angle θ

Similarly, since the matrix involved is orthogonal (its inverse is equal to its transpose), we have:

$$\begin{pmatrix} x \\ y \end{pmatrix} = \begin{pmatrix} \cos\theta & \sin\theta \\ -\sin\theta & \cos\theta \end{pmatrix} \begin{pmatrix} x' \\ y' \end{pmatrix}.$$

Now we can rotate an image by considering it as a large collection of points. Figure 6.19 illustrates the idea. In this figure, the dark circles indicate the original position; the light

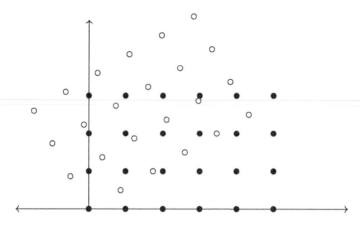

FIGURE 6.19: Rotating a rectangle

points their positions after rotation. However, this approach won't work for images. Since an image grid can be considered as pixels forming a subset of the Cartesian (integer valued) grid, we must ensure that even after rotation, the points remain in that grid. To do this we consider a rectangle that includes the rotated image, as shown in Figure 6.20. Now consider

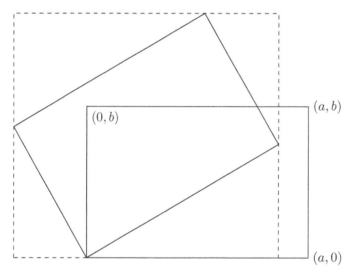

FIGURE 6.20: A rectangle surrounding a rotated image

all integer-valued points (x', y') in the dashed rectangle. A point will be in the image if, when rotated back, it lies within the original image limits. That is, if

$$
\begin{aligned}
0 &\leq x' \cos\theta + y' \sin\theta \leq a \\
0 &\leq -x' \sin\theta + y' \cos\theta \leq b
\end{aligned}
$$

If we consider the array of 6×4 points shown in Figure 6.19, then the points after rotation by $30°$ are shown in Figure 6.21 This gives us the position of the pixels in our rotated image,

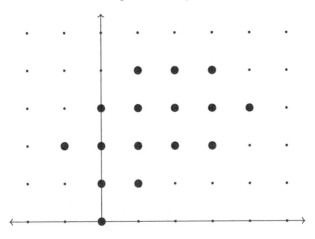

FIGURE 6.21: The points on a grid after rotation

but what about their value? Take a point (x', y') in the rotated image, and rotate it back into the original image to produce a point (x'', y''), as shown in Figure 6.22. Now the gray

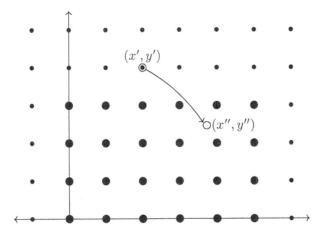

FIGURE 6.22: Rotating a point back into the original image

value at (x'', y'') can be found by interpolation using surrounding gray values. This value is then the gray value for the pixel at (x', y') in the rotated image.

Image rotation in MATLAB is obtained using the command `imrotate`; it has the syntax

```
imrotate(image,angle,'method')
```

where `method`, as with `imresize`, can be any one of `nearest`, `bilinear`, or `bicubic`. Also as with `imresize`, the `method` parameter may be omitted, in which case nearest neighbor interpolation is used. Python has the `rotate` method in the `transform` module of `skimage`, with syntax

```
transform.rotate(image,angle,[resize],[order],[mode])
```

Here the last three parameters are optional; `order` (default is 1) is the polynomial order of the interpolation.

For an example, let's take our old friend the cameraman, and rotate him by 60°. We shall do this twice; once with nearest neighbor interpolation, and once using bicubic interpolation. With MATLAB or Octave, the commands are:

```
>> cr = imrotate(c,60);
>> imshow(cr)
>> crc = imrotate(c,60,'bicubic');
>> figure,imshow(crc)
```

MATLAB/Octave

and in Python the commands are

```
In :   cr = tr.rotate(c,60,order=0)
In :   io.imshow(cr)
In :   crc = tr.rotate(c,60,order=3)
In :   f = figure(), f.show(io.imshow(crc))
```

Python

The results are shown in Figure 6.23. There's not a great deal of observable difference between the two images; however, nearest neighbor interpolation produces slightly more jagged edges.

(a) Nearest neighbor   (b) Bicubic interpolation

FIGURE 6.23: Rotation with interpolation

Notice that for angles that are integer multiples of 90° image rotation can be accomplished far more efficiently with simple matrix transposition and reversing of the order of rows and columns. These commands can be used:

MATLAB/Octave	Python	Result
flipud	np.flipud	Flips a matrix in the up/down direction
fliplr	np.fliplr	Flips a matrix in the left/right direction
rot90	np.rot90	Rotates a matrix by 90°

In fact, imrotate uses these simpler commands for these particular angles.

6.7 Correcting Image Distortion

This is in fact a huge topic: there are many different possible distortions caused by optical aberrations and effects. So this section will investigate just one: *perspective distortion,* which is exemplified by the image in Figure 6.24. Because of the position of the camera

FIGURE 6.24: Perspective distortion

lens relative to the building, the towers appear to be leaning inward. Fixing this requires a little algebra. First note that the building is contained within a symmetric trapezoid, the corners of which can be found by carefully moving a cursor over the image and checking its coordinates.

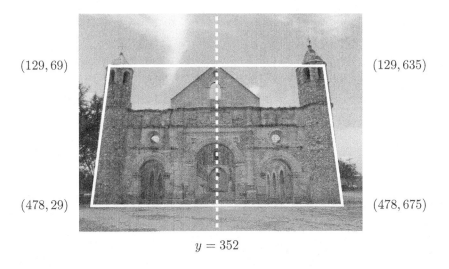

$y = 352$

FIGURE 6.25: The corners of the building

The trapezoid can be seen to be centered at $y = 352$. The problem now is to warp the trapezoid into a rectangle. This warping is built in to many current picture processing software tools; the user can implement such a warp by drawing lines over the edges and then dragging those lines in such a way to fix the distortion. Here we investigate the mathematics and the background processing behind such an operation.

In general, suppose a symmetric trapezoid of height r is centered on the y axis, with horizontal lengths of $2a$ and $2b$, as shown in Figure 6.26.

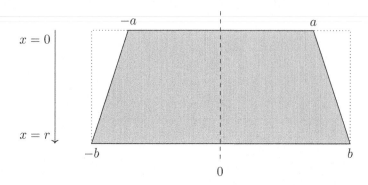

FIGURE 6.26: A general symmetric trapezoid

To warp this trapezoid to a rectangle, each horizontal line must be stretched by an amount $\mathrm{str}(x)$, where $\mathrm{str}(r) = 1$ and $\mathrm{str}(0) = b/a$. Since the stretching amount is a linear function of x, the values given mean that

$$\mathrm{str}(x) = \frac{\dfrac{b}{a} - 1}{-r}(x - r) + 1$$

$$= \frac{a - b}{ar}x + \frac{b}{a}.$$

If (x_1, y_1) and (x_2, y_2) are points on one of the slanted sides of the trapezoid, then finding the equation of line through them and substituting $x = 0$ and $x = r$ will produce the values of a and b:

$$a = y_1 - x_1\frac{y_2 - y_1}{x_2 - x_1}, \qquad b = y_1 + (r - x_1)\frac{y_2 - y_1}{x_2 - x_1}.$$

Consider the trapezoid in Figure 6.25, with a shift by subtracting 352 from all the y values, as shown in Figure 6.27.

The previous formulas can now be applied to $(x_1, y_1) = (478, 323)$ and $(x_2, y_2) = (129, 283)$ to obtain $a \approx 268.21$ and $b \approx 331.71$. To actually perform the stretching, the transformation from the old to new image is given by

$$(x, y) \to (x, (y - 352)\mathrm{str}(x) + 352).$$

In other words, a pixel at position (x, y) in the new image will take the value of the pixel at place

$$\left(x, \frac{y - 352}{\mathrm{str}(x)} + 352\right)$$

in the old image.

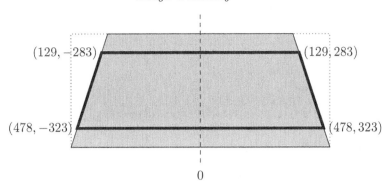

FIGURE 6.27: The trapezoid around the building

However, this involves a division, which is in general a slow operation. To rewrite as a multiplication, we would use replace pixel at (x, y) in the new image would take the place of a pixel at place

$$(x, \mathrm{sq}(x, y - 352) + 352)$$

where $\mathrm{sq}(x)$ ("sq" stands for "squash") is a linear function for which

$$\mathrm{sq}(0) = a/b$$
$$\mathrm{sq}(r) = 1.$$

This means that

$$\mathrm{sq}(x) = \frac{b - a}{br}(x) + \frac{a}{b}.$$

Here's how this can be implemented in MATLAB or Octave, assuming that the image has been read as `im`:

```
>> [r,c] = size(im);
>> z = zeros(r,c);
>> x1 = 129; y1=283; x2=478; y2=323;
>> a = y1-x1*(y2-y1)/(x2-x1)
>> b = y1+(r-x1)*(y2-y1)/(x2-x1)
>> [y,x] = meshgrid(1:c,1:r);
>> sq = floor((y-352).*((b-a)/(b*r)*x+a/b)+352);
>> for i = 1:r,...
>     for j = 1:c,...
>       z(i,j) = im(i,sq(i,j));
>     end;...
>   end;
>> im2 = uint8(z)
```

MATLAB/Octave

and in Python as

```
In :   r,c = im.shape
In :   x1, y1, x2, y2 = 129.0, 283.0, 478.0, 323.0
In :   a = y1-x1*(y2-y1)/(x2-x1)
In :   b = y1+(r-x1)*(y2-y1)/(x2-x1)
In :   z = np.zeros_like(im)
In :   x,y = np.mgrid[0:r,0:c]
In :   sq = np.floor((y-352)*((b-a)/(b*r)*x+a/b)+352)
In :   for i in range(r):
...:       for j in range(c):
...:           z[i,j] = im[i,sq[i,j]]
...:
In :   im2 = uint8(z)
```

`Python`

and the result `im2` is shown in Figure 6.28. Note that the Python commands used `mgrid` which can be used as an alternative to `meshgrid` and with a more concise syntax.

FIGURE 6.28: The image corrected for perspective distortion

Exercises

1. By hand, enlarge the list

    1 4 7 4 3 6

    to lengths
    (a) 9   (b) 11   (c) 2
    by

    (a) nearest neighbor interpolation
    (b) linear interpolation

Check your answers with your computer system.

2. By hand, enlarge the matrix

$$\begin{pmatrix} 8 & 6 & 13 & 9 \\ 1 & 13 & 1 & 15 \\ 5 & 4 & 7 & 7 \\ 5 & 10 & 3 & 7 \end{pmatrix}$$

to sizes

(i) 7×7 (ii) 8×8 (iii) 10×10

by

  (a) nearest neighbor interpolation,

  (b) bilinear interpolation.

Check your answers with your computer system.

3. Use zero-interleaving and spatial filtering to enlarge the cameraman's head by a factor of four in each dimension, using the three filters given. Use the following sequence of commands:

```
>> head2 = zeroint(head);
>> head2n = imfilter(head2,filt);
>> head4 = zeroint(head2n);
>> head4n = imfilter(head4,filt);
>> imshow(head4n/255)
```
MATLAB/Octave

where `filt` is a filter. Compare your results to those given by `imresize`. Are there any observable differences?

4. Take another small part of an image: say the head of the seagull in `seagull.png`. This can be obtained with:

```
>> g = imread('seagull.png');
>> head = g(110:173,272:367);
```
MATLAB/Octave

or with

```
In :  g = io.imread('seagull.png')
In :  head  = g[109:173,271:367]
```
Python

Enlarge the head to four times as big using both `imresize` with the different parameters, and the zero-interleave method with the different filters.

As above, compare the results.

5. Suppose an image is enlarged by some amount k, and the result decreased by the same amount. Should this result be exactly the same as the original? If not, why not?

6. What happens if the image is decreased first, and the result enlarged?

7. Create an image consisting of a white square with a black background. Rotate the image by 30° and 45°. Use (a) rotation with nearest neighbor interpolation, and (b) rotation with bilinear interpolation

   Compare the results.

8. For the rotated squares in the previous question, rotate back to the original orientation. How close is the result to the original square?

9. In general, suppose an image is rotated, and then the result rotated back. Should this result be exactly the same as the original? If not, why not?

10. Write a function to implement image enlargement using zero-interleaving and spatial filtering. The function should have the syntax

    ```
    imenlarge(image,n,filt)
    ```

    where n is the number of times the interleaving is to be done, and `filt` is the filter to use. So, for example, the command

    ```
    imenlarge(head,2,bfilt);
    ```

    would enlarge an image to four times its size, using the 5×5 filter described in Section 6.4.

Chapter 7

The Fourier Transform

7.1 Introduction

The Fourier Transform is of fundamental importance to image processing. It allows us to perform tasks that would be impossible to perform any other way; its efficiency allows us to perform other tasks more quickly. The Fourier Transform provides, among other things, a powerful alternative to linear spatial filtering; it is more efficient to use the Fourier Transform than a spatial filter for a large filter. The Fourier Transform also allows us to isolate and process particular image "frequencies," and so perform low pass and high pass filtering with a great degree of precision.

Before we discuss the Fourier Transform of images, we shall investigate the one-dimensional Fourier Transform, and a few of its properties.

7.2 Background

Our starting place is the observation that a periodic function may be written as the sum of sines and cosines of varying amplitudes and frequencies. For example, in Figure 7.1 we plot a function and its decomposition into sine functions.

Some functions will require only a finite number of functions in their decomposition; others will require an infinite number. For example, a "square wave," such as is shown in Figure 7.2, has the decomposition

$$f(x) = \sin x + \frac{1}{3}\sin 3x + \frac{1}{5}\sin 5x + \frac{1}{7}\sin 7x + \frac{1}{9}\sin 9x + \cdots \tag{7.1}$$

In Figure 7.2 we take the first five terms only to provide the approximation. The more terms of the series we take, the closer the sum will approach the original function.

This can be formalized; if $f(x)$ is a function of period $2T$, then we can write

$$f(x) = a_0 + \sum_{n=1}^{\infty}\left(a_n \cos \frac{n\pi x}{T} + b_n \sin \frac{n\pi x}{T}\right)$$

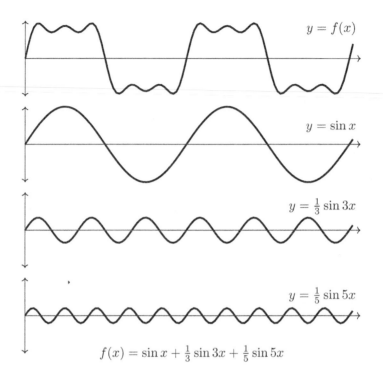

FIGURE 7.1: A function and its trigonometric decomposition

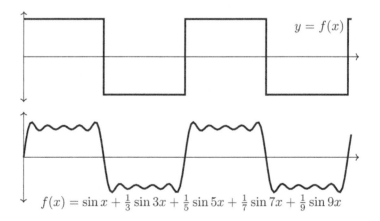

FIGURE 7.2: A square wave and its trigonometric approximation

where

$$a_0 = \frac{1}{2T} \int_{-T}^{T} f(x)\, dx$$

$$a_n = \frac{1}{T} \int_{-T}^{T} f(x) \cos \frac{n\pi x}{T}\, dx, \quad n = 1, 2, 3, \ldots$$

$$b_n = \frac{1}{T} \int_{-T}^{T} f(x) \sin \frac{n\pi x}{T}\, dx, \quad n = 1, 2, 3, \ldots$$

These are the equations for the *Fourier series expansion* of $f(x)$, and they can be expressed in complex form:

$$f(x) = \sum_{n=-\infty}^{\infty} c_n \exp\left(\frac{in\pi x}{T}\right) dx$$

where

$$c_n = \frac{1}{2T} \int_{-T}^{T} f(x) \exp\left(\frac{-in\pi x}{T}\right) dx.$$

If the function is non-periodic, we can obtain similar results by letting $T \to \infty$, in which case

$$f(x) = \int_{0}^{\infty} [a(\omega) \cos \omega x + b(\omega) \sin \omega x] d\omega$$

where

$$a(\omega) = \frac{1}{\pi} \int_{-\infty}^{\infty} f(x) \cos \omega x\, dx,$$

$$b(\omega) = \frac{1}{\pi} \int_{-\infty}^{\infty} f(x) \sin \omega x\, dx.$$

These equations can be written again in complex form:

$$f(x) = \int_{-\infty}^{\infty} F(\omega) e^{i\omega x} d\omega,$$

$$F(\omega) = \frac{1}{2\pi} \int_{-\infty}^{\infty} f(x) e^{i\omega x} dx.$$

In this last form, the functions $f(x)$ and $F(\omega)$ form a *Fourier transform pair*. Further details can be found, for example, in James [23].

7.3   The One-Dimensional Discrete Fourier Transform

When we deal with a *discrete* function, as we shall do for images, the situation from the previous section changes slightly. Since we only have to obtain a finite number of values, we only need a finite number of functions to do it.

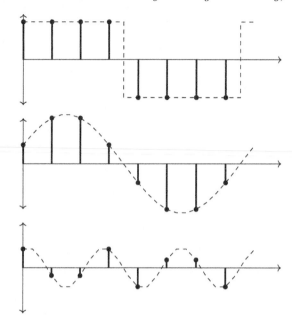

FIGURE 7.3: Expressing a discrete function as the sum of sines

Consider, for example the discrete sequence

$$1, \quad 1, \quad 1, \quad 1, \quad -1, \quad -1, \quad -1, \quad -1$$

which we may take as a discrete approximation to the square wave of Figure 7.2. This can be expressed as the sum of only *two* sine functions; this is shown in Figure 7.3. We shall see below how to obtain those sequences.

The Fourier Transform allows us to obtain those individual sine waves that compose a given function or sequence. Since we shall be concerned with discrete sequences, and of course images, we shall investigate only the *discrete Fourier Transform*, abbreviated DFT.

Definition of the One-Dimensional DFT

Suppose

$$\mathbf{f} = [f_0, f_1, f_2, \ldots, f_{N-1}]$$

is a sequence of length N. We define its *discrete Fourier Transform* to be the sequence

$$\mathbf{F} = [F_0, F_1, F_2, \ldots, F_{N-1}]$$

where

$$F_u = \frac{1}{N} \sum_{x=0}^{N-1} \exp\left[-2\pi i \frac{xu}{N}\right] f_x. \tag{7.2}$$

Note the similarity between this equation and the equations for the Fourier series expansion discussed in the previous section. Instead of an integral, we now have a finite sum. This definition can be expressed as a matrix multiplication:

$$F = \mathcal{F} f$$

where $\mathcal{F}$ is an $N \times N$ matrix defined by

$$\mathcal{F}_{m,n} = \frac{1}{N} \exp\left[-2\pi i \frac{mn}{N}\right].$$

Given N, we shall define

$$\omega = \exp\left[\frac{-2\pi i}{N}\right]$$

so that

$$\mathcal{F}_{m,n} = \frac{1}{N}\omega^{mn}.$$

Then we can write

$$\mathcal{F} = \frac{1}{N}\begin{bmatrix} 1 & 1 & 1 & 1 & 1 & \cdots & 1 \\ 1 & \omega^1 & \omega^2 & \omega^3 & \omega^4 & \cdots & \omega^{N-1} \\ 1 & \omega^2 & \omega^4 & \omega^6 & \omega^8 & \cdots & \omega^{2(N-1)} \\ 1 & \omega^3 & \omega^6 & \omega^9 & \omega^{12} & \cdots & \omega^{3(N-1)} \\ 1 & \omega^4 & \omega^8 & \omega^{12} & \omega^{16} & \cdots & \omega^{4(N-1)} \\ \vdots & \vdots & \vdots & \vdots & \vdots & \ddots & \vdots \\ 1 & \omega^{N-1} & \omega^{2(N-1)} & \omega^{3(N-1)} & \omega^{4(N-1)} & \cdots & \omega^{(N-1)^2} \end{bmatrix}$$

Example. Suppose $\mathbf{f} = [1, 2, 3, 4]$ so that $N = 4$. Then

$$\omega = \exp\left[\frac{-2\pi i}{4}\right]$$

$$= \exp\left[-\frac{\pi i}{2}\right]$$

$$= \cos\left(-\frac{\pi}{2}\right) + i \sin\left(-\frac{\pi}{2}\right)$$

$$= -i.$$

Then we have

$$\mathcal{F} = \begin{bmatrix} 1 & 1 & 1 & 1 \\ 1 & -i & (-i)^2 & (-i)^3 \\ 1 & (-i)^2 & (-i)^4 & (-i)^6 \\ 1 & (-i)^3 & (-i)^6 & (-i)^9 \end{bmatrix} = \begin{bmatrix} 1 & 1 & 1 & 1 \\ 1 & -i & -1 & i \\ 1 & -1 & 1 & -1 \\ 1 & i & -1 & -i \end{bmatrix}$$

and so

$$\mathbf{F} = \frac{1}{4}\begin{bmatrix} 1 & 1 & 1 & 1 \\ 1 & -i & -1 & i \\ 1 & -1 & 1 & -1 \\ 1 & i & -1 & -i \end{bmatrix}\begin{bmatrix} 1 \\ 2 \\ 3 \\ 4 \end{bmatrix} = \frac{1}{4}\begin{bmatrix} 10 \\ -2+2i \\ -2 \\ -2-2i \end{bmatrix}.$$

The Inverse DFT

The formula for the inverse DFT is very similar to the forward transform:

$$x_u = \sum_{x=0}^{N-1} \exp\left[2\pi i \frac{xu}{N}\right] F_u. \tag{7.3}$$

If you compare this equation with Equation 7.2, you will see that there are really only two differences:

1. There is no scaling factor $1/N$

2. The sign inside the exponential function has been changed to positive

As with the forward transform, we can express this as a matrix product:

$$f = \mathcal{F}^{-1} F$$

with

$$\mathcal{F}^{-1} = \begin{bmatrix} 1 & 1 & 1 & 1 & 1 & \cdots & 1 \\ 1 & \overline{\omega}^1 & \overline{\omega}^2 & \overline{\omega}^3 & \overline{\omega}^4 & \cdots & \overline{\omega}^{N-1} \\ 1 & \overline{\omega}^2 & \overline{\omega}^4 & \overline{\omega}^6 & \overline{\omega}^8 & \cdots & \overline{\omega}^{2(N-1)} \\ 1 & \overline{\omega}^3 & \overline{\omega}^6 & \overline{\omega}^9 & \overline{\omega}^{12} & \cdots & \overline{\omega}^{3(N-1)} \\ 1 & \overline{\omega}^4 & \overline{\omega}^8 & \overline{\omega}^{12} & \overline{\omega}^{16} & \cdots & \overline{\omega}^{4(N-1)} \\ \vdots & \vdots & \vdots & \vdots & \vdots & \ddots & \vdots \\ 1 & \overline{\omega}^{N-1} & \overline{\omega}^{2(N-1)} & \overline{\omega}^{3(N-1)} & \overline{\omega}^{4(N-1)} & \cdots & \overline{\omega}^{(N-1)^2} \end{bmatrix}$$

where

$$\overline{\omega} = \frac{1}{\omega} = \exp\left[\frac{2\pi i}{N}\right].$$

In MATLAB or Octave, we can calculate the forward and inverse transforms with `fft` and `ifft`. Here `fft` stands for *Fast Fourier Transform*, which is a fast and efficient method of performing the DFT (see below for details). For example:

```
a =

     1    2    3    4    5    6

>> fft(a')

ans =

   21.0000
   -3.0000 + 5.1962i
   -3.0000 + 1.7321i
   -3.0000
   -3.0000 - 1.7321i
   -3.0000 - 5.1962i
```
MATLAB/Octave

We note that to apply a DFT to a single vector in MATLAB or Octave, we should use a column vector.

In Python it is very similar; there are modules for the DFT in both the `scipy` and `numpy` libraries. For ease of use, we can first import all the functions we need from `numpy.fft`, and then apply them:

```
In:   from numpy.fft import *
In:   a = range(1,7)
In:   L = fft(a)
In:   for x in L: print '%0.4f%_+0.4fi' % (x.real, x.imag)


21.0000 +0.0000i
-3.0000 +5.1962i
-3.0000 +1.7320i
-3.0000 +0.0000i
-3.0000 -1.7321i
-3.0000 -5.1961i
```

`Python`

7.4 Properties of the One-Dimensional DFT

The one-dimensional DFT satisfies many useful and important properties. We will investigate some of them here. A more complete list is given, for example, by Jain [22].

Linearity. This is a direct consequence of the definition of the DFT as a matrix product. Suppose f and g are two vectors of equal length, and p and q are scalars, with $h = pf + qg$. If F, G, and H are the DFTs of f, g, and h, respectively, we have

$$H = pF + qG.$$

This follows from the definitions of

$$F = \mathcal{F}f, \quad G = \mathcal{F}g, \quad H = \mathcal{F}h$$

and the linearity of the matrix product.

Shifting. Suppose we multiply each element x_n of a vector x by $(-1)^n$. In other words, we change the sign of every second element. Let the resulting vector be denoted x'. The the DFT X' of x' is equal to the DFT X of x with the swapping of the left and right halves.

Let's do a quick example:

```
>> x = [2 3 4 5 6 7 8 1];

>> x1=(-1).^[0:7].*x

x1 =

    2    -3    4    -5    6    -7    8    -1

>> X=fft(x')

   36.0000
   -9.6569 + 4.0000i
   -4.0000 - 4.0000i
    1.6569 - 4.0000i
    4.0000
    1.6569 + 4.0000i
   -4.0000 + 4.0000i
   -9.6569 - 4.0000i

>> X1=fft(x1')

X1 =

    4.0000
    1.6569 + 4.0000i
   -4.0000 + 4.0000i
   -9.6569 - 4.0000i
   36.0000
   -9.6569 + 4.0000i
   -4.0000 - 4.0000i
    1.6569 - 4.0000i
```

MATLAB/Octave

Notice that the first four elements of **X** are the last four elements of **X1**, and vice versa. Note that in Python the above example could be obtained with:

```
In:   x = np.array([2, 3, 4, 5, 6, 7, 8, 1])
In:   x1 = copy(x)
In:   x1[1::2] = x1[1::2]*-1
In:   L = fft(x)
In:   L1 = fft(x1)
```

Python

and then the arrays **L** and **L1** can be displayed using the "print" example above.

Conjugate symmetry If **x** is real, and of length N, then its DFT **X** satisfies the condition that

$$X_k = \overline{X_{N-k}},$$

where $\overline{X_{N-k}}$ is the complex conjugate of X_{N-k}, for all $k = 1, 2, 3, \ldots, N-1$. So in our example of length 8, we have

$$X_1 = \overline{X_7}, \quad X_2 = \overline{X_6}, \quad X_3 = \overline{X_5}.$$

In this case, we also have $X_4 = \overline{X_4}$, which means X_4 must be real. In fact, if N is even, then $X_{N/2}$ will be real. Examples can be seen above.

Convolution. Suppose x and y are two vectors of the same length N. Then we define their *convolution* (or, more properly, their *circular convolution*) to be the vector

$$z = x * y$$

where

$$z_k = \frac{1}{n} \sum_{n=0}^{N-1} x_n y_{k-n}.$$

For example, if $N = 4$, then

$$z_0 = \frac{1}{4}(x_0 y_0 + x_1 y_{-1} + x_2 y_{-2} + x_3 y_{-3})$$

$$z_1 = \frac{1}{4}(x_0 y_1 + x_1 y_0 + x_2 y_{-1} + x_3 y_{-2})$$

$$z_2 = \frac{1}{4}(x_0 y_2 + x_1 y_1 + x_2 y_0 + x_3 y_{-1})$$

$$z_3 = \frac{1}{4}(x_0 y_3 + x_2 y_2 + x_2 y_1 + x_3 y_0)$$

The negative indices can be interpreted by imagining the y vector to be periodic, and can be indexed backward from 0 as well as forward:

$$
\begin{array}{ccccccccc}
\cdots & y_0 & y_1 & y_2 & y_3 & y_0 & y_1 & y_2 & y_3 & \cdots \\
= \cdots & y_0 & y_{-3} & y_{-2} & y_{-1} & y_0 & y_1 & y_2 & y_3 & \cdots
\end{array}
$$

Thus, $y_{-1} = y_3$, $y_{-2} = y_2$ and $y_{-3} = y_1$. It can be checked from the definition that convolution is *commutative* (the order of the operands is irrelevant):

$$x * y = y * x.$$

One way of thinking about circular convolution is by a "sliding" operation. Consider the array x against an array consisting of the array y backwards twice: Sliding the x array into different positions and multiplying the corresponding elements and adding will produce the result. This is shown in Figure 7.4.

Circular convolution as defined looks like a messy operation. However, it can be defined in terms of polynomial products. Suppose $p(u)$ is the polynomial in u whose coefficients are the elements of x. Let $q(u)$ be the polynomial whose coefficients are the elements of y. Form the product $p(u)q(u)(1+u^N)$, and extract the coefficients of u^N to u^{2N-1}. These will be our required circular convolution.

For example, suppose we have:

$$x = [1, \quad 2, \quad 3, \quad 4], \qquad y = [5, \quad 6, \quad 7, \quad 8].$$

Then we have

$$p(u) = 1 + 2u + 3u^2 + 4u^3$$

and

$$q(u) = 5 + 6u + 7u^2 + 8u^3.$$

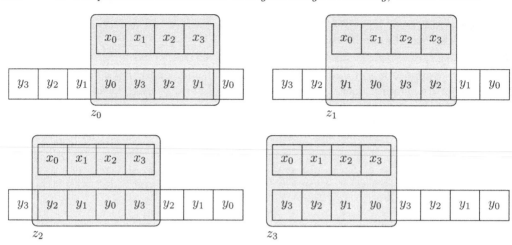

FIGURE 7.4: Visualizing circular convolution

Then we expand

$$p(u)q(u)(1+u^4) = 5+16x+34u^2+60u^3+66u^4+68u^5+66u^6+60u^7+61u^8+52u^9+32u^{10}.$$

Extracting the coefficients of $u^4, u^5, \ldots, u^7$ we obtain

$$x * y = [66, \quad 68, \quad 66, \quad 60].$$

MATLAB has a `conv` function, which produces the coefficients of the polynomial $p(u)q(u)$ as defined above:

```
>> a = [1 2 3 4]

a =

     1    2    3    4

>> b = [5 6 7 8]

b =

     5    6    7    8

>> conv(a,b)

ans =

     5   16   34   60   61   52   32
```
MATLAB/Octave

To perform circular convolution, we first notice that the polynomial $p(u)(1 + u^N)$ can be obtained by simply repeating the coefficients of x. So, given the above two arrays:

```
>> N = length(a);
>> C = conv(a,[b,b])
>> C(N:2*N-1)
ans =

   60   66   68   66
```

which is exactly what we obtained above. Python works very similarly, using the command convolve:

```
In:  a = [1,2,3,4]
In:  b = [5,6,7,8]
In:  convolve(a,b)
Out: array([ 5, 16, 34, 60, 61, 52, 32])
In:  convolve(a,b+b,'valid')[:-1]
Out: array([60, 66, 68, 66])
```

The importance of convolution is the *convolution theorem*, which states:

> Suppose x and y are vectors of equal length. Then the DFT of their circular convolution x ∗ y is equal to the element-by-element product of the DFTs of x and y.

So if Z, X, Y are the DFTs of z = x ∗ y, x and y, respectively, then

Z = X.Y.

We can check this with our vectors above:

```
>> C = conv(a,[b,b]);
>> fft(C')


ans =

   1.0e+02 *

   2.6000
        0 - 0.0800i
   0.0400
        0 + 0.0800i

>> fft(a').*fft(b');

ans =

   1.0e+02 *

   2.6000
        0 - 0.0800i
   0.0400
        0 + 0.0800i
```

The previous computations can be done in Python, but using numpy arrays instead of lists. The `convolve` function here from `numpy` acts only on one-dimensional arrays, and the `hstack` function concatenates two arrays horizontally:

```
In:  a = np.array([1,2,3,4])
In:  b = np.array([5,6,7,8])
In:  C = np.convolve(a,np.hstack([b,b]),'valid')[:-1]
In:  fft(C)
Out: array([ 260.+0.j,   -8.+0.j,   -4.+0.j,   -8.+0.j])
In:  fft(a)*fft(b)
Out: array([ 260.+0.j,   0.-8.j,   4.-0.j,   0.+8.j])
```

Python

Note that in each case the results are the same. The convolution theorem thus provides us with another way of performing convolution: multiply the DFTs of our two vectors and invert the result:

```
>> fft(a').*fft(b');
>> ifft(ans)'

ans =

    66    68    66    60
```

MATLAB/Octave

or

```
In:  ab = fft(a)*fft(b)
In:  real(ifft(ab))
Out: array([66., 68., 66., 60.])
```

Python

A formal proof of the convolution theorem for the DFT is given by Petrou [34].

The Fast Fourier Transform.   One of the many aspects that make the DFT so attractive for image processing is the existence of very fast algorithms to compute it. There are a number of extremely fast and efficient algorithms for computing a DFT; such an algorithm is called a *Fast Fourier Transform*, or FFT. The use of an FFT vastly reduces the time needed to compute a DFT.

One FFT method works recursively by dividing the original vector into two halves, computing the FFT of each half, and then putting the results together. This means that the FFT is most efficient when the vector length is a power of 2. This method is discussed in Appendix C.

Table 7.1 shows that advantage gained by using the FFT algorithm as opposed to the direct arithmetic definition of Equations 7.6 and 7.7 by comparing the number of multiplications required for each method. For a vector of length 2^n, the direct method takes $(2^n)^2 = 2^{2n}$ multiplications; the FFT only $n2^n$. The saving in time is thus of an order of $2^n/n$. Clearly the advantage of using an FFT algorithm becomes greater as the size of the vector increases.

Because of the computational advantage, any implementation of the DFT will use an FFT algorithm.

TABLE 7.1: Comparison of FFT and direct arithmetic

2^n	Direct arithmetic	FFT	Increase in speed
4	16	8	2.0
8	84	24	2.67
16	256	64	4.0
32	1024	160	6.4
64	4096	384	10.67
128	16384	896	18.3
256	65536	2048	32.0
512	262144	4608	56.9
1024	1048576	10240	102.4

7.5 The Two-Dimensional DFT

In two dimensions, the DFT takes a matrix as input, and returns another matrix, of the same size, as output. If the original matrix values are $f(x, y)$, where x and y are the indices, then the output matrix values are $F(u, v)$. We call the matrix F the *Fourier Transform of* f and write

$$F = \mathcal{F}(f).$$

Then the original matrix f is the *inverse Fourier Transform of* F, and we write

$$f = \mathcal{F}^{-1}(F).$$

We have seen that a (one-dimensional) function can be written as a sum of sines and cosines. Given that an image may be considered a two-dimensional function $f(x, y)$, it seems reasonable to assume that f can be expressed as sums of "corrugation" functions which have the general form

$$z = a \sin(bx + cy).$$

A sample such function is shown in Figure 7.5. And this is in fact exactly what the two-dimensional Fourier Transform does: it rewrites the original matrix in terms of sums of corrugations. The amplitude and period of each corrugation defines its position on the Fourier spectrum, as shown in Figure 7.6. From Figure 7.6 we note that the more spread out the corrugation, the closer to the center of the spectrum it will be positioned. This is because a spread out corrugation means large values of $1/x$ and $1/y$, hence small values of x and y. Similarly, "squashed" corrugations (with small values of $1/x$ and $1/y$) will be positioned further from the center.

The definition of the two-dimensional discrete Fourier Transform is very similar to that for one dimension. The forward and inverse transforms for an $M \times N$ matrix, where for notational convenience we assume that the x indices are from 0 to $M - 1$ and the y indices are from 0 to $N - 1$, are:

$$F(u, v) = \sum_{x=0}^{M-1} \sum_{y=0}^{N-1} f(x, y) \exp\left[-2\pi i \left(\frac{xu}{M} + \frac{yv}{N}\right)\right]. \tag{7.4}$$

$$f(x, y) = \frac{1}{MN} \sum_{u=0}^{M-1} \sum_{v=0}^{N-1} F(u, v) \exp\left[2\pi i \left(\frac{xu}{M} + \frac{yv}{N}\right)\right]. \tag{7.5}$$

These are horrendous looking formulas, but if we spend a bit of time pulling them apart, we shall see that they aren't as bad as they look.

Before we do this, we note that the formulas given in Equations 7.4 and 7.5 are not used by all authors. The main change is the position of the scaling factor $1/MN$. Some people put it in front of the sums in the forward formula. Others put a factor of $1/\sqrt{MN}$ in front of both sums. The point is the sums by themselves would produce a result (after both forward and inverse transforms) which is too large by a factor of MN. So somewhere in the forward-inverse formulas a corresponding $1/MN$ must exist; it doesn't really matter where.

Some Properties of the Two-Dimensional Fourier Transform

All the properties of the one-dimensional DFT transfer into two dimensions. But there are some further properties not previously mentioned, which are of particular use for image processing.

Similarity. First notice that the forward and inverse transforms are very similar, with the exception of the scale factor $1/MN$ in the inverse transform, and the negative sign in the exponent of the forward transform. This means that the same algorithm, only very slightly adjusted, can be used for both the forward and inverse transforms.

The DFT as a spatial filter. Note that the values

$$\exp\left[\pm 2\pi i \left(\frac{xu}{M} + \frac{yv}{N}\right)\right]$$

are independent of the values f or F. This means that they can be calculated in advance, and only then put into the formulas above. It also means that every value $F(u, v)$ is obtained by multiplying every value of $f(x, y)$ by a fixed value, and adding up all the results. But this is precisely what a linear spatial filter does: it multiplies all elements under a mask with fixed values, and adds them all up. Thus, we can consider the DFT as a linear spatial filter which is as big as the image. To deal with the problem of edges, we assume that the image is tiled in all directions, so that the mask always has image values to use.

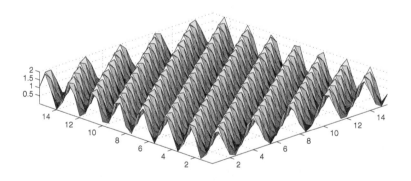

FIGURE 7.5: A "corrugation" function

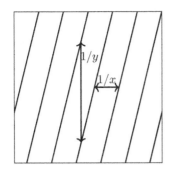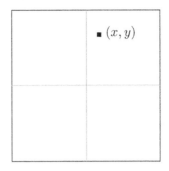

FIGURE 7.6: Where each corrugation is positioned on the spectrum

Separability. Notice that the Fourier Transform "filter elements" can be expressed as products:

$$\exp\left[2\pi i\left(\frac{xu}{M} + \frac{yv}{N}\right)\right] = \exp\left[2\pi i\frac{xu}{M}\right]\exp\left[2\pi i\frac{yv}{N}\right].$$

The first product value

$$\exp\left[2\pi i\frac{xu}{M}\right]$$

depends only on x and u, and is independent of y and v. Conversely, the second product value

$$\exp\left[2\pi i\frac{yv}{N}\right]$$

depends only on y and v, and is independent of x and u. This means that we can break down our formulas above to simpler formulas that work on single rows or columns:

$$F(u) = \sum_{x=0}^{M-1} f(x)\exp\left[-2\pi i\frac{xu}{M}\right], \tag{7.6}$$

$$f(x) = \frac{1}{M}\sum_{u=0}^{M-1} F(u)\exp\left[2\pi i\frac{xu}{M}\right]. \tag{7.7}$$

If we replace x and u with y and v, we obtain the corresponding formulas for the DFT of matrix columns. These formulas define the *one-dimensional DFT* of a vector, or simply the DFT.

The 2-D DFT can be calculated by using this property of "separability"; to obtain the 2-D DFT of a matrix, we first calculate the DFT of all the rows, and then calculate the DFT of all the columns of the result, as shown in Figure 7.7. Since a product is independent of the order, we can equally well calculate a 2-D DFT by calculating the DFT of all the columns first, and then calculating the DFT of all the rows of the result.

Linearity An important property of the DFT is its linearity; the DFT of a sum is equal to the sum of the individual DFTs, and the same goes for scalar multiplication:

$$\mathcal{F}(f + g) = \mathcal{F}(f) + \mathcal{F}(g)$$
$$\mathcal{F}(kf) = k\mathcal{F}(f)$$

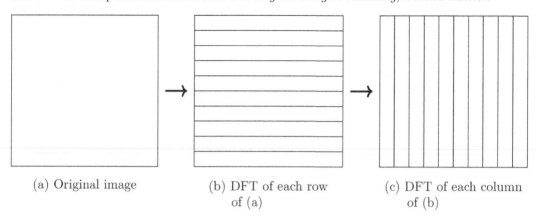

(a) Original image   (b) DFT of each row   (c) DFT of each column
of (a)   of (b)

FIGURE 7.7: Calculating a 2D DFT

where k is a scalar, and f and g are matrices. This follows directly from the definition given in Equation 7.4.

This property is of great use in dealing with image degradation such as noise, which can be modelled as a sum:

$$d = f + n$$

where f is the original image, n is the noise, and d is the degraded image. Since

$$\mathcal{F}(d) = \mathcal{F}(f) + \mathcal{F}(n)$$

we may be able to remove or reduce n by modifying the transform. As we shall see, some noise appears on the DFT in a way that makes it particularly easy to remove.

The convolution theorem.   This result provides one of the most powerful advantages of using the DFT. Suppose we wish to convolve an image M with a spatial filter S. Our method has been place S over each pixel of M in turn, calculate the product of all corresponding gray values of M and elements of S, and add the results. The result is called the *digital convolution* of M and S, and is denoted

$$M * S.$$

This method of convolution can be very slow, especially if S is large. The *convolution theorem* states that the result $M * S$ can be obtained by the following sequence of steps:

1. Pad S with zeros so that it is the same size as M; denote this padded result by S'.

2. Form the DFTs of both M and S, to obtain $\mathcal{F}(M)$ and $\mathcal{F}(S')$.

3. Form the element-by-element product of these two transforms:

$$\mathcal{F}(M) \cdot \mathcal{F}(S').$$

4. Take the inverse transform of the result:

$$\mathcal{F}^{-1}(\mathcal{F}(M) \cdot \mathcal{F}(S')).$$

Put simply, the convolution theorem states:

$$M * S = \mathcal{F}^{-1}(\mathcal{F}(M) \cdot \mathcal{F}(S'))$$

or equivalently that

$$\mathcal{F}(M * S) = \mathcal{F}(M) \cdot \mathcal{F}(S').$$

Although this might seem like an unnecessarily clumsy and roundabout way of computing something so simple as a convolution, it can have enormous speed advantages if S is large.

For example, suppose we wish to convolve a 512×512 image with a 32×32 filter. To do this directly would require $32^2 = 1024$ multiplications for each pixel, of which there are $512 \times 512 = 262,144$. Thus there will be a total of $1024 \times 262,144 = 268,435,456$ multiplications needed. Now look at applying the DFT (using an FFT algorithm). Each row requires 4608 multiplications by Table 7.1; there are 512 rows, so a total of $4608 \times 512 = 2,359,296$ multiplications; the same must be done again for the columns. Thus to obtain the DFT of the image requires 4,718,592 multiplications. We need the same amount to obtain the DFT of the filter, and for the inverse DFT. We also require 512×512 multiplications to perform the product of the two transforms.

Thus, the total number of multiplications needed to perform convolution using the DFT is

$$4,718,592 \times 3 + 262,144 = 14,417,920$$

which is an enormous saving compared to the direct method.

The DC coefficient. The value $F(0,0)$ of the DFT is called the *DC coefficient*. If we put $u = v = 0$ in the definition given in Equation 7.4, then

$$F(0,0) = \sum_{x=0}^{M-1} \sum_{y=0}^{N-1} f(x,y) \exp(0) = \sum_{x=0}^{M-1} \sum_{y=0}^{N-1} f(x,y).$$

That is, this term is equal to the sum of *all* terms in the original matrix.

Shifting. For purposes of display, it is convenient to have the DC coefficient in the center of the matrix. This will happen if all elements $f(x,y)$ in the matrix are multiplied by $(-1)^{x+y}$ before the transform. Figure 7.8 demonstrates how the matrix is shifted by this method. In each diagram, the DC coefficient is the top left-hand element of submatrix A, and is shown as a black square.

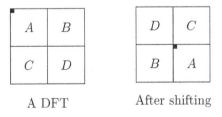

A DFT　　　　　After shifting

FIGURE 7.8: Shifting a DFT

Conjugate symmetry   An analysis of the Fourier Transform definition leads to a symmetry property; if we make the substitutions $u = -u$ and $v = -v$ in Equation 7.4, then

$$\mathcal{F}(u, v) = \mathcal{F}^*(-u + pM, -v + qN)$$

for any integers p and q. This means that half of the transform is a mirror image of the conjugate of the other half. We can think of the top and bottom halves, or the left and right halves, being mirror images of the conjugates of each other.

Figure 7.9 demonstrates this symmetry in a shifted DFT. As with Figure 7.8, the black square shows the position of the DC coefficient. The symmetry means that its information

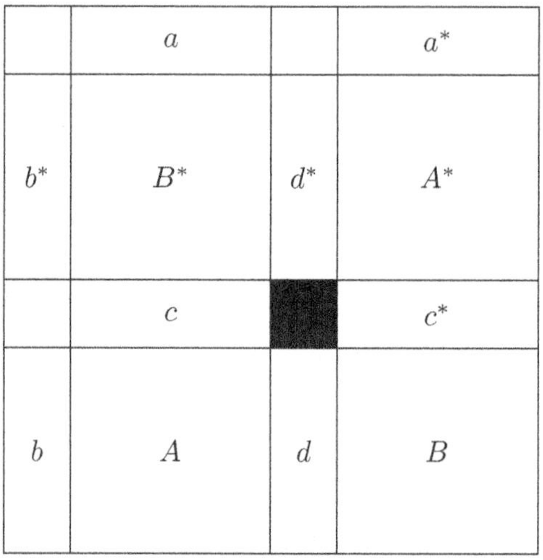

	a		a^*
b^*	B^*	d^*	A^*
	c	▮	c^*
b	A	d	B

FIGURE 7.9: Conjugate symmetry in the DFT

is given in just half of a transform, and the other half is redundant.

Displaying transforms.   Having obtained the Fourier Transform $F(u, v)$ of an image $f(x, y)$, we would like to see what it looks like. As the elements $F(u, v)$ are complex numbers, we can't view them directly, but we can view their magnitude $|F(u, v)|$. Since these will be numbers of type `double`, generally with large range, we would need to scale these absolute values so that they can be displayed.

One trouble is that the DC coefficient is generally very much larger than all other values. This has the effect of showing a transform as a single white dot surrounded by black. One way of stretching out the values is to take the logarithm of $|F(u, v)|$ and to display

$$\log(1 + |F(u, v)|).$$

The display of the magnitude of a Fourier Transform is called the *spectrum* of the transform. We shall see some examples later.

7.6   Experimenting with Fourier Transforms

The relevant functions for us are:

- fft, which takes the DFT of a vector

- ifft, which takes the inverse DFT of a vector

- fft2, which takes the DFT of a matrix

- ifft2, which takes the inverse DFT of a matrix

- fftshift, which shifts a transform as shown in Figure 7.8

of which we have seen the first two above. These functions are available in MATLAB and Octave, and can be brought into the top namespace of Python with the command

```
from numpy.fft import *
```

Before attacking a few images, let's take the Fourier Transform of a few small matrices to get more of an idea what the DFT "does."

Example 1. Suppose we take a constant matrix $f(x, y) = 1$. Going back to the idea of a sum of corrugations, then *no* corrugations are required to form a constant. Thus, we would hope that the DFT consists of a DC coefficient and zeros everywhere else. We will use the ones function, which produces an $n \times n$ matrix consisting of 1's, where n is an input to the function.

```
>> a1 = ones(8);
>> fft2(a1)
```
                                                                    MATLAB/Octave

The result is indeed as we expected:

```
ans =
      64     0     0     0     0     0     0     0
       0     0     0     0     0     0     0     0
       0     0     0     0     0     0     0     0
       0     0     0     0     0     0     0     0
       0     0     0     0     0     0     0     0
       0     0     0     0     0     0     0     0
       0     0     0     0     0     0     0     0
       0     0     0     0     0     0     0     0
```
                                                                    MATLAB/Octave

Note that the DC coefficient is indeed the sum of all the matrix values.

Example 2. Now we will take a matrix consisting of a single corrugation:

```
>> a2 = [100 200; 100 200];
>> a2 = repmat(a2,4,4)

ans =
       100    200    100    200    100    200    100    200
       100    200    100    200    100    200    100    200
       100    200    100    200    100    200    100    200
       100    200    100    200    100    200    100    200
       100    200    100    200    100    200    100    200
       100    200    100    200    100    200    100    200
       100    200    100    200    100    200    100    200
       100    200    100    200    100    200    100    200

>> af2 = fft2(a2)

ans =
      9600      0      0      0  -3200      0      0      0
         0      0      0      0      0      0      0      0
         0      0      0      0      0      0      0      0
         0      0      0      0      0      0      0      0
         0      0      0      0      0      0      0      0
         0      0      0      0      0      0      0      0
         0      0      0      0      0      0      0      0
         0      0      0      0      0      0      0      0
```
MATLAB/Octave

What we have here is really the sum of two matrices: a constant matrix each element of which is 150, and a corrugation which alternates -50 and 50 from left to right. The constant matrix alone would produce (as in example 1) a DC coefficient alone of value $64 \times 150 = 9600$; the corrugation a single value. By linearity, the DFT will consist of just the two values.

Example 3. We will take here a single step edge:

```
>> a3 = [zeros(8,4) ones(8,4)]
   a3 =

       0    0    0    0    1    1    1    1
       0    0    0    0    1    1    1    1
       0    0    0    0    1    1    1    1
       0    0    0    0    1    1    1    1
       0    0    0    0    1    1    1    1
       0    0    0    0    1    1    1    1
       0    0    0    0    1    1    1    1
       0    0    0    0    1    1    1    1
```
MATLAB/Octave

Now we shall perform the Fourier Transform with a shift, to place the DC coefficient in the center, and since it contains some complex values, for simplicity we shall just show the rounded absolute values:

```
>> af3 = fftshift(fft2(a3));
>> round(abs(af3))

ans =

        0     0     0     0     0     0     0     0
        0     0     0     0     0     0     0     0
        0     0     0     0     0     0     0     0
        0     0     0     0     0     0     0     0
        0     9     0    21    32    21     0     9
        0     0     0     0     0     0     0     0
        0     0     0     0     0     0     0     0
        0     0     0     0     0     0     0     0
```

MATLAB/Octave

The DC coefficient is of course the sum of all values of **a**; the other values may be considered to be the coefficients of the necessary sine functions required to from an edge, as given in Equation 7.1. The mirroring of values about the DC coefficient is a consequence of the symmetry of the DFT.

All these can also be easily performed in Python, with the three images constructed as

```
In:  a1 = ones((8,8))
In:  a2 = np.array([[100, 200],[100,200]])
In:  a2 = np.tile(a2,(4,4))
In:  a3 = np.hstack([zeros((8,4)),ones((8,4))])
```

Python

Then the Fourier spectra can be obtained by very similar commands to those above, but converting to integers for easy viewing:

```
In:  int16(fft2(a1))
In:  int16(fft2(a1))
In:  af3 = fftshift(fft2(a3))
In:  int16(np.round(abs(af3),0))
```

Python

7.7   Fourier Transforms of Synthetic Images

Before experimenting with images, it will be helpful to have a program that will enable the viewing of Fourier spectra, either with absolute values, or with log-scaling. Such a program, called fftshow, is given in all our languages at the end of the chapter.

Now we can create a few simple images, and see what the Fourier Transform produces.

Example 1. We shall produce a simple image consisting of a single edge, first in MATLAB/Octave

```
>> a = [zeros(256,128) ones(256,128)];
>> af = fftshift(fft2(a));
```
MATLAB/Octave

The image is displayed on the left in Figure 7.10. The spectrum can be displayed either directly or with log-scaling:

```
>> afl = abs(af);
>> imshow(mat2gray(afl))
>> afl2 = log(1+abs(af));
>> imshow(mat2gray(afl2));
```
MATLAB/Octave

Alternatively, the largest value can be isolated: this will be the DC coefficient that, because of shifting, will be at position $(129, 129)$, and divide the spectrum by this value for display. Either of the first two commands will work:

```
>> mx = max(afl(:))
>> mx = afl(129,129)
>> imshow(afl/mx)
```
MATLAB/Octave

The image and the two displays are shown in Figure 7.10.

In Python, the commands for constructing the image and viewing its Fourier spectrum are:

```
In:  a = np.hstack([zeros((256,128)),ones((256,128))])
In:  af = fftshift(fft2(a))
In:  afl = abs(af)
In:
```
Python

The image and its Fourier spectrum shown first as absolute values and then as log-scaled are shown in Figure 7.10. We observe immediately that the final (right-most) result is similar (although larger) to Example 3 in the previous section.

Example 2. Now we will create a box, and then its Fourier Transform:

FIGURE 7.10: A single edge and its DFT

```
>> a = zeros(256,256);
>> a(78:178,78:178) = 1;
>> imshow(a)
>> af = fftshift(fft2(a));
>> figure, imshow(mat2gray(log(1+abs(af))))
```
MATLAB/Octave

In Python, we can use the `rescale_intensity` function to scale the output to fit a given range:

```
In:  a = zeros((256,256))
In:  a[77:177,77:177] = 1
In:  af = fftshift(fft2(a))
In:  afl = ex.rescale_intensity(np.log(1+abs(af)),outrange=(0.0,1.0))
In:  io.imshow(afl)
```
Python

The box is shown on the left in Figure 7.11, and its Fourier transform is shown on the right.

 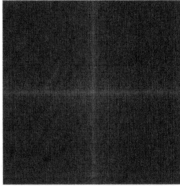

FIGURE 7.11: A box and its DFT

Example 3. Now we shall look at a box rotated 45°, first in MATLAB/Octave.

```
>> [x,y] = meshgrid(1:256,1:256);
>> b = (x+y<329)&(x+y>182)&(x-y>-67)&(x-y<73);
>> imshow(b)
>> bf = fftshift(fft2(b));
>> figure, imshow(mat2gray(log(1+abs(af))))
```
MATLAB/Octave

and next in Python:

```
In:   x,y = meshgrid(range(256),range(256))
In:   b = (x+y<329)&(x+y>182)&(x-y>-67)&(x-y<73);
In:   io.imshow(b)
In:   bf = fftshift(fft2(b))
In:   bfl = log(1+abs(bf))
In:   io.imshow(ex.rescale_intensity(bfl,out_range=(0.0,1.0)))
```
Python

The results are shown in Figure 7.12. Note that the transform of the rotated box is the

FIGURE 7.12: A rotated box and its DFT

rotated transform of the original box.

Example 4. We will create a small circle, and then transform it:

```
>> [x,y] = meshgrid(-128:217,-128:127);
>> z = sqrt(x.^2+y.^2);
>> c = (z<15);
```
MATLAB/Octave

or

```
In:   ar = np.arange(-128,128)
In:   x,y = meshgrid(ar,ar)
In:   z = sqrt(x**2+y**2)
In:   c = (z<15)*1
```
Python

The result is shown on the left in Figure 7.13. Now we will create its Fourier Transform and display it:

```
>> cf = fftshift(fft2(c));
>> imshow(mat2gray(log(1+abs(cf))))
```
MATLAB/Octave

or

```
In:  cf = fftshift(fft2(c))
In:  cfl = log(1+abs(cf))
In:  io.imshow(ex.rescale_intensity(cfl,out_range=(0.0,1.0)))
```
`Python`

and this is shown on the right in Figure 7.13. Note the "ringing" in the Fourier transform.

FIGURE 7.13: A circle and its DFT

This is an artifact associated with the sharp cutoff of the circle. As we have seen from both the edge and box images in the previous examples, an edge appears in the transform as a line of values at right angles to the edge. We may consider the values on the line as being the coefficients of the appropriate corrugation functions which sum to the edge. With the circle, we have lines of values radiating out from the circle; these values appear as circles in the transform.

A circle with a gentle cutoff, so that its edge appears blurred, will have a transform with no ringing. Such a circle can be made with the commands (given z above):

```
>> c2 = 1./(1+(z./15).^2);
```
`MATLAB/Octave`

or

```
In:  c2 = 1/(1+(z/15)**2)
```
`Python`

This image appears as a blurred circle, and its transform is very similar—check them out!

7.8 Filtering in the Frequency Domain

We have seen in Section 7.5 that one of the reasons for the use of the Fourier Transform in image processing is due to the convolution theorem: a spatial convolution can be performed by element-wise multiplication of the Fourier Transform by a suitable "filter matrix." In this section, we shall explore some filtering by this method.

Ideal Filtering

Low Pass Filtering

Suppose we have a Fourier Transform matrix F, shifted so that the DC coefficient is in the center. Recall from Figure 7.6 that the low frequency components are toward the center. This means we can perform low pass filtering by multiplying the transform by a matrix in such a way that center values are maintained, and values away from the center are either removed or minimized. One way to do this is to multiply by an *ideal low pass matrix*, which is a binary matrix m defined by:

$$m(x, y) = \begin{cases} 1 & \text{if } (x, y) \text{ is closer to the center than some value } D, \\ 0 & \text{if } (x, y) \text{ is further from the center than } D. \end{cases}$$

The circle c displayed in Figure 7.13 is just such a matrix, with $D = 15$. Then the inverse Fourier Transform of the element-wise product of F and m is the result we require:

$$\mathcal{F}^{-1}(F \cdot m).$$

Let's see what happens if we apply this filter to an image. First we obtain an image and its DFT.

```
>> cm = imread('cameraman.png');
>> cf = fftshift(fft2(cm));
>> imshow(mat2gray(log(1+abs(cf))))
```
MATLAB/Octave

The cameraman image and its DFT are shown in Figure 7.14. Now we can perform a low

FIGURE 7.14: The "cameraman" image and its DFT

pass filter by multiplying the transform matrix by the circle matrix we created earlier (recall that "dot asterisk" is the MATLAB syntax for element-wise multiplication of two matrices):

```
>> cfl = cf.*c;
>> imshow(mat2gray(log(1+abs(cfl))))
```
MATLAB/Octave

and this is shown in Figure 7.15(a). Now we can take the inverse transform and display the result:

```
>> cfli = ifft2(cfl);
>> imshow(mat2gray(abs(cfli)))
```

MATLAB/Octave

and this is shown in Figure 7.15(b). Note that even though `cfli` is supposedly a matrix of real numbers, we are still using `abs` to obtain displayable values. This is because the `fft2` and `fft2` functions, being numeric, will not produce mathematically perfect results, but rather very close numeric approximations. So, taking the absolute values of the result rounds out any errors obtained during the transform and its inverse. Note the "ringing"

(a) Ideal filtering on the DFT   (b) After inversion

FIGURE 7.15: Applying ideal low pass filtering

about the edges in this image. This is a direct result of the sharp cutoff of the circle. The ringing as shown in Figure 7.13 is transferred to the image.

We would expect that the smaller the circle, the more blurred the image, and the larger the circle; the less blurred. Figure 7.16 demonstrates this, using cutoffs of 5 and 30. Notice

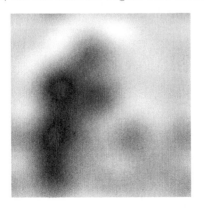

(a) Cutoff of 5   (b) Cutoff of 30

FIGURE 7.16: Ideal low pass filtering with different cutoffs

that ringing is still present, and clearly visible in Figure 7.16(b).

All these operations can be performed with almost the same commands in Python; first reading the image and producing its Fourier Transform:

```
In: cm = io.imread('cameraman.png')
In: cf = fftshift(fft2(cm))
```
Python

and then the low pass filter with a circle of cutoff 15:

```
In: d = 15
In: ar = range(-128,128)
In: x,y = meshgrid(ar,ar)
In: c = (x**2+y**2<d**2)*1
```
Python

Finally, the filter can be applied to the Fourier Transform and the inverse obtained and viewed:

```
In: cfl = cf*c
In: io.imshow(abs(ifft2(cfl)))
```
Python

This will produce the result shown in Figure 7.15. The other images can be produced by varying the value of d in the definition of the ideal filter.

High Pass Filtering

Just as we can perform low pass filtering by keeping the center values of the DFT and eliminating the others, so high pass filtering can be performed by the opposite: eliminating center values and keeping the others. This can be done with a minor modification of the preceding method of low pass filtering. First we create the circle:

```
>> [x,y] = meshgrid(-128:127,-128:127);
>> z = sqrt(x.^2+y.^2);
>> c = z>15;
```
MATLAB/Octave

and then multiply it by the DFT of the image:

```
>> cfh = cf.*c;
>> imshow(mat2gray(log(1+abs(cfh))))
```
MATLAB/Octave

This is shown in Figure 7.17(a). The inverse DFT can be easily produced and displayed:

```
>> cfhi = ifft2(cfh);
>> figure,imshow(abs(cfhi))
```
MATLAB/Octave

and this is shown in Figure 7.17(b). As with low pass filtering, the size of the circle influences the information available to the inverse DFT, and hence the final result. Figure 7.18 shows some results of ideal high pass filtering with different cutoffs. If the cutoff is large, then more information is removed from the transform, leaving only the highest frequencies. This can be observed in Figure 7.18(c) and (d); only the edges of the image remain. If we have small cutoff, such as in Figure 7.18(a), we are only removing a small amount of the transform.

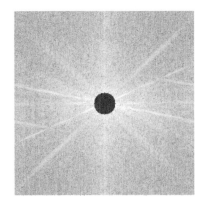

(a) The DFT after high pass filtering   (b) The resulting image

FIGURE 7.17: Applying an ideal high pass filter to an image

(a) Cutoff of 5   (b) The resulting image

(a) Cutoff of 30   (b) The resulting image

FIGURE 7.18: Ideal high pass filtering with different cutoffs

We would thus expect that only the lowest frequencies of the image would be removed. And this is indeed true, as seen in Figure 7.18(b); there is some grayscale detail in the final image, but large areas of low frequency are close to zero.

In Python, the commands are almost the same as before, except that with a cutoff of d the ideal filter will be produced with

```
In: c = (x**2+y**2>d**2)*1
```

Butterworth Filtering

Ideal filtering simply cuts off the Fourier Transform at some distance from the center. This is very easy to implement, as we have seen, but has the disadvantage of introducing unwanted artifacts: ringing, into the result. One way of avoiding this is to use as a filter matrix a circle with a less sharp cutoff. A popular choice is to use *Butterworth filters*.

Before we describe these filters, we shall look again at the ideal filters. As these are radially symmetric about the center of the transform, they can be simply described in terms of their cross-sections. That is, we can describe the filter as a function of the distance x from the center. For an ideal low pass filter, this function can be expressed as

$$f(x) = \begin{cases} 1 & \text{if } x < D, \\ 0 & \text{if } x \geq D \end{cases}$$

where D is the cutoff radius. Then the ideal high pass filters can be described similarly:

$$f(x) = \begin{cases} 1 & \text{if } x > D, \\ 0 & \text{if } x \leq D \end{cases}$$

These functions are illustrated in Figure 7.19. Butterworth filter functions are based on the

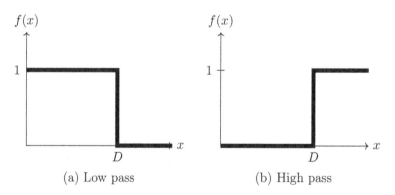

(a) Low pass   (b) High pass

FIGURE 7.19: Ideal filter functions

following functions for low pass filters:

$$f(x) = \frac{1}{1 + (x/D)^{2n}}$$

and for high pass filters:

$$f(x) = \frac{1}{1 + (D/x)^{2n}}$$

where in each case the parameter n is called the *order* of the filter. The size of n dictates the sharpness of the cutoff. These functions are illustrated in figures 7.20 and 7.21.

Note that on a two-dimensional grid, where x in the above formulas may be replaced with $\sqrt{x^2 + y^2}$ as distance from the grid's center, assuming the grid to be centered at $(0,0)$, then the formulas for the low pass and high pass filters may be written

$$f(x,y) = \frac{1}{1 + \left(\dfrac{x^2 + y^2}{D^2}\right)^n}$$

and

$$f(x,y) = \frac{1}{1 + \left(\dfrac{D^2}{x^2 + y^2}\right)^n}$$

respectively.

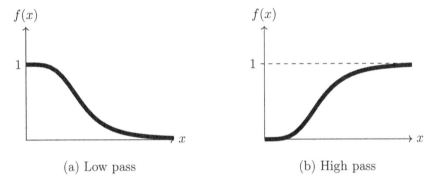

(a) Low pass (b) High pass

FIGURE 7.20: Butterworth filter functions with $n = 2$

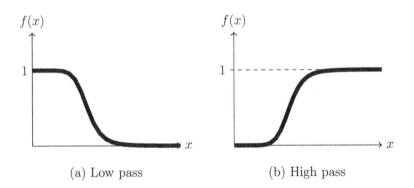

(a) Low pass (b) High pass

FIGURE 7.21: Butterworth filter functions with $n = 4$

It is easy to implement these filters; here are the commands to produce a Butterworth low pass filter of size 256×256 with $D = 15$ and order $n = 2$:

```
>> [x,y] = meshgrid(-128:217,-128:127));
>> bl = 1./(1+((x.^2+y.^2)/15).^2);
```

MATLAB/Octave

and in Python:

```
In:  ar = arange(-128,128,1.0)
In:  x,y = meshgrid(ar,ar)
In:  bl = 1/(1+((x**2+y**2)/15**2)**2)
```

Python

Note that in Python it is essential that the elements of the array have floating point values, so that all arithmetic will be done as floating point, rather than integers. Hence, we start by making the indexing array consist of floats by using **arange** instead of **range**.

Since a Butterworth high pass filter can be obtained by subtracting a low pass filter from 1, the above commands can be used to create high pass filters.

To apply a Butterworth low pass filter in Python, then, with $n = 2$ and $D = 15.0$, we can use the commands above, and then:

```
>> ar = range(-128,127)
>> x,y = meshgrid(ar,ar)
>> D = 15.0
>> bl = 1.0/(1.0+((x**2+y**2)/D**2)**2)
>> cfbl = cf*bl
>> io.imshow(ex.rescale_intensity(abs(ifft2(cfbl)),out_range=(0.0,1.0)))
```

Python

And to apply a Butterworth low pass filter to the DFT of the cameraman image in MATLAB/Octave, create the filter array as above and then apply it to the image transform:

```
>> cfbl = cf.*bl;
>> figure,fftshow(cfbl,'log')
```

MATLAB/Octave

and this is shown in Figure 7.22(a). Note that there is no sharp cutoff as seen in Figure 7.15; also that the outer parts of the transform are not equal to zero, although they are dimmed considerably. Performing the inverse transform and displaying it as we have done previously produces Figure 7.22(b). This is certainly a blurred image, but the ringing seen in Figure 7.15 is completely absent. Compare the transform after multiplying with a Butter-

(a) The DFT after Butterworth low pass filtering   (b) The resulting image

FIGURE 7.22: Butterworth low pass filtering

worth filter (Figure 7.22(a)) with the original transform (in Figure 7.14). The Butterworth filter does cause an attenuation of values away from the center, even if they don't become suddenly zero, as with the ideal low pass filter in Figure 7.15.

We can apply a Butterworth high pass filter similarly, first by creating the filter and applying it to the image transform:

```
In: bh = 1- 1/(1+((x**2+y**2)/D**2)**2)
In: cfbh = cf*bh;
```
Python

and then inverting and displaying the result:

```
In: io.imshow(abs(ifft2(cfbh)))
```
Python

In MATLAB/Octave the commands are similar:

```
>> D = 15
>> bh = 1 - 1./(1+((x.^2+y.^2)/15).^2);
>> cfbh = cf.*bh;
>> imshow(mat2gray(log(1+abs(cfbh))))
>> figure,imshow(mat2gray(abs(ifft2(cfbh))))
```
MATLAB/Octave

The images are shown in Figure 7.23

(a) The DFT after Butterworth high pass filtering   (b) The resulting image

FIGURE 7.23: Butterworth high pass filtering

Gaussian Filtering

We have met Gaussian filters in Chapter 5, and we saw that they could be used for low pass filtering. However, we can also use Gaussian filters in the frequency domain. As with ideal and Butterworth filters, the implementation is very simple: create a Gaussian filter, multiply it by the image transform, and invert the result. Since Gaussian filters have the very nice mathematical property that a Fourier Transform of a Gaussian is a Gaussian, we should get exactly the same results as when using a linear Gaussian spatial filter.

Gaussian filters may be considered to be the most "smooth" of all the filters we have discussed so far, with ideal filters the least smooth and Butterworth filters in the middle.

In Python, Gaussian filters are provided in the `ndimage` of `scipy`, and there is a wrapper in `skimage.filter`. However, we can also create a filter for ourselves. Using the cameraman image, and a standard deviation $\sigma = 10$, define:

```
In: sigma = 10
In: g = exp(-(x**2+y**2)/sigma**2)
In: g1 = g1/g1.sum()
```

where the last command is to ensure that the sum of all filter elements is one. This filter can then be applied to the image in the same way as the ideal and Butterworth filters above:

```
In: cg1 = cf*g1
In: io.imshow(abs(ifft2(cg1)))
```

and the results are shown in Figure 7.24(b) and (d). A high pass filter can be defined by

(a) $\sigma = 10$ (b) Resulting image

(c) $\sigma = 30$ (d) Resulting image

FIGURE 7.24: Applying a Gaussian low pass filter in the frequency domain

first scaling a low pass filter so that its maximum value is one, and then subtracting it from 1:

```python
In: h1 = 1-g/g.max()
```

Python

and it can then be applied to the image:

```python
In: ch1 = cf*h1
In: io.imshow(abs(ifft2(ch1)))
```

Python

The same sequence of commands can be produced starting with $\sigma = 30$, and the images are shown in Figure 7.25.

(a) Using $\sigma = 10$ (b) Using $\sigma = 30$

FIGURE 7.25: Applying a Gaussian high pass filter in the frequency domain

With MATLAB or Octave, Gaussian filters can be created using the `fspecial` function, and applied to our transform.

```matlab
>> g1=mat2gray(fspecial('gaussian',256,10));
>> cg1=cf.*g1;
>> imshow(mat2gray(log(1+abs(cg1))))
>> g2=mat2gray(fspecial('gaussian',256,30));
>> cg2=cf.*g2;
>> imshow(mat2gray(log(1+abs(cg1))))
```

MATLAB/Octave

Note the use of the `mat2gray` function. The `fspecial` function on its own produces a low pass Gaussian filter with a very small maximum:

```matlab
>> g=fspecial('gaussian',256,10);
>> format long, max(g(:)), format

ans =

   0.00158757552679
```

MATLAB/Octave

The reason is that `fspecial` adjusts its output to keep the volume under the Gaussian function always 1. This means that a wider function, with a large standard deviation, will have a low maximum. So we need to scale the result so that the central value will be 1; and `mat2gray` does that automatically.

The transforms are shown in Figure 7.24(a) and (c). In each case, the final parameter of the `fspecial` function is the standard deviation; it controls the width of the filter. Clearly, the larger the standard deviation, the wider the function, and so the greater amount of the transform is preserved.

The results of the transform on the original image can be produced using the usual sequence of commands:

```
>> cgi1 = ifft2(cg1);
>> cgi2 = ifft2(cg2);
>> fftshow(cgi1,'abs');
>> fftshow(cgi2,'abs');
```

MATLAB/Octave

We can apply a high pass Gaussian filter easily; we create a high pass filter by subtracting a low pass filter from 1.

```
>> h1 = 1-g1;
>> h2 = 1-g2;
>> ch1 = cf.*h1;
>> ch2 = cf.*h2;
>> chi1 = ifft2(ch1);
>> chi2 = ifft2(ch2);
>> imshow(abs(ifft2(chi1)))
>> figure, imshow(abs(ifft2(chi2)))
```

MATLAB/Octave

As with ideal and Butterworth filters, the wider the high pass filter, the more of the transform we are reducing, and the less of the original image will appear in the result.

7.9   Homomorphic Filtering

If we have an image that suffers from variable illumination (dark in some sections, light in others), we may wish to enhance the contrast locally, and in particular to enhance the dark regions. Such an image may be obtained if we are recording a scene with high intensity range (say, an outdoor scene on a sunny day which includes shadows) onto a medium with a smaller intensity range. The resulting image will contain very bright regions (those well illuminated); on the other hand, regions in shadow may appear very dark indeed.

In such a case histogram equalization won't be of much help, as the image already has high contrast. What we need to do is reduce the intensity range and at the same time increase the local contrast. We first note that the intensity of an object in an image may be considered to be a combination of two factors: the amount of light falling on it, and how much light is reflected by the object. In fact, if $f(x, y)$ is the intensity of a pixel at position (x, y) in our image, we may write:

$$f(x, y) = i(x, y)r(x, y)$$

where $i(x,y)$ is the *illumination* and $r(x,y)$ is the *reflectance*. These satisfy

$$0 < i(x,y) < \infty$$

and

$$0 < r(x,y) < 1.$$

There is no (theoretical) limit as to the amount of light that can fall on an object, but reflectance is strictly bounded.

To reduce the intensity range, we need to reduce the illumination, and to increase the local contrast, we need to increase the reflectance. However, this means we need to separate $i(x,y)$ and $r(x,y)$. We can't do this directly, as the image is formed from their product. However, if we take the logarithm of the image:

$$\log f(x,y) = \log i(x,y) + \log r(x,y)$$

we can then separate the logarithms of $i(x,y)$ and $r(x,y)$. The basis of homomorphic filtering is this working with the logarithm of the image, rather than with the image directly. A schema for homomorphic filtering is given in Figure 7.26.

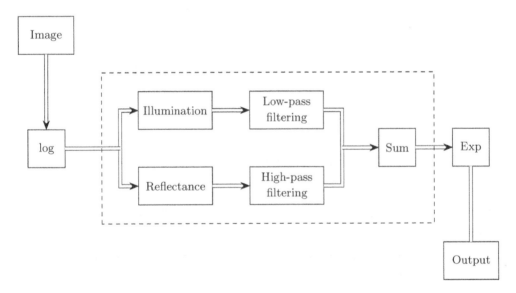

FIGURE 7.26: A schema for homomorphic filtering

The "exp" in the second to last box simply reverses the effect of the original logarithm. We assume that the logarithm of illumination will vary slowly, and the logarithm of reflectance will vary quickly, so that the filtering processes given in Figure 7.26 will have the desired effects.

Clearly this schema will be unworkable in practice: given a value $\log f(x,y)$ we can't determine the values of its summands $\log i(x,y)$ and $\log r(x,y)$. A simpler way is to replace the dashed box in Figure 7.26 with a high boost filter of the Fourier transform. This gives the simpler schema shown in Figure 7.27.

A simple function `homfilt` to apply homomorphic filtering to an image (using a Butterworth high-boost filter) is given at the end of the chapter.

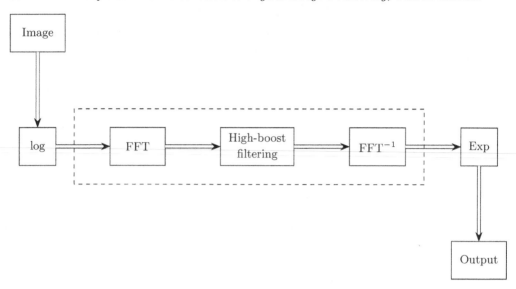

FIGURE 7.27: A simpler schema for homomorphic filtering

To see this in action, suppose we take an image of type **double** (so that its pixel values are between 0.0 and 1.0), and multiply it by a trigonometric function, scaled to between 0.1 and 1. If, for example, we use $\sin x$, then the function

$$y = 0.5 + 0.4 \sin x$$

satisfies $0.1 \leq y \leq 1.0$. The result will be an image with varying illumination.

Suppose we take an image of type **double** and of size 256×256. Then we can superimpose a sine function by multiplying the image with the function above. For example:

```
>> n = im2double(imread('newborn.png'));
>> [x,y] = meshgrid(1:256,1:256);
>> nx = n.*(0.5+0.4*sin((1.3*x+0.7*y-50)/16));
```
MATLAB/Octave

or alternatively:

```
In: n = sk.util.img_as_float(io.imread('newborn.png'))
In: ar = range(256)
In: x,y = meshgrid(ar,ar)
In: nx = n*(0.5+0.4*sin((1.3*x+0.7*y-50)/16))
```
Python

The parameters used to construct the new image were chosen by trial and error to produce bands of suitable width, and on such places as to obscure detail in our image. If the original image is in Figure 7.28(a), then the result of this is shown in Figure 7.28(b).

If we apply homomorphic filtering with

```
>> xh = homfilt(x,10,2,0.5,2);
>> imshow(xh/16)
```
MATLAB/Octave

or with

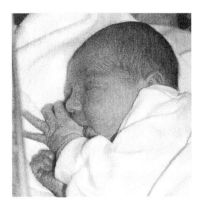

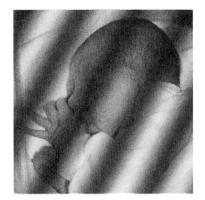

(a) The "newborn" image

(b) Altering the illumination with a sine function

FIGURE 7.28: Varying illumination across an image

```
In :   xh = homfilt(x,10,2,0.5,2)
In :   io.imshow(xh,vmax=16)
```
Python

the result is shown in Figure 7.29.

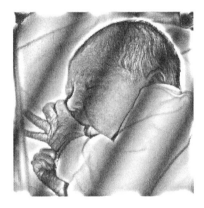

FIGURE 7.29: The result of homomorphic filtering

The result shows that detail originally unobservable in the original image—especially in regions of poor illumination—is now clear. Even though the dark bands have not been completely removed, they do not obscure underlying detail as they originally did.

Figure 7.30(a) shows a picture of a ruined archway; owing to the bright light through the archway, much of the details are too dark to be seen clearly. Given this image x, we can perform homomorphic filtering and display the result with:

```
>> xh = homfilt(x,128,2,0.5,2);
>> imshow(xh/14)
```
MATLAB/Octave

In Python, the same homfilt call can be used, with the display given by

```python
io.imshow(ah,vmax=12)
```

and this is shown in Figure 7.30(b).

(a) The arch   (b) After homomorphic filtering

FIGURE 7.30: Applying homomorphic filtering to an image

Note that the arch details are now much clearer, and it is even possible to make out the figure of a person standing at its base.

7.10   Programs

The `fftshow` program for viewing Fourier spectra, in MATLAB or Octave:

```matlab
function fftshow(f,type)

% Usage:  FFTSHOW(F,TYPE)
%
% Displays the fft matrix F using imshow, where TYPE must be one of
% 'abs' or 'log'.  If TYPE='abs', then then abs(f) is displayed; if
% TYPE='log' then log(1+abs(f)) is displayed.  If TYPE is omitted, then
% 'log' is chosen as a default.
%
%  Example:
%    c=imread('cameraman.tif');
%    cf=fftshift(fft2(c));
%    fftshow(cf,'abs')
%
```

```matlab
if nargin<2,
  type='log';
end

if (type=='log')
  fl = log(1+abs(f));
  fm = max(fl(:));
  imshow(im2uint8(fl/fm))
elseif (type=='abs')
  fa=abs(f);
  fm=max(fa(:));
  imshow(fa/fm)
else
  error('TYPE_must_be_abs_or_log.');
end;
```

MATLAB/Octave

And here is `homfilt` first for MATLAB or Octave:

```matlab
function res=homfilt(im,cutoff,order,lowgain,highgain)

% HOMFILT(IMAGE,FILTER) applies homomorphic filtering to the image IMAGE
% with the given parameters

u = im2uint8(im);

height = size(im,1);
width = size(im,2);
[x,y] = meshgrid(-floor(width/2):floor((width-1)/2),...
                 -floor(height/2):floor((height-1)/2));
lbutter = 1./(1+(sqrt(2)-1)*((x.^2+y.^2)/cutoff^2).^order);

u(find(u==0)) = 1;
lg = log(double(u));
ft = fftshift(fft2(lg));
hboost = lowgain+(highgain-lowgain)*(1-lbutter);
b = hboost.*ft;
ib = abs(ifft2(b));
res = exp(ib);
end
```

MATLAB/Octave

Moving to Python, first is `fftshow`:

```python
#
# Read this into python with: execfile('fftshow.py') OR: import fftshow
#
#

def fftshow(im,type='log'):
    if type == 'log':
        fl = log(1+abs(im))
        fm = fl.max()
```

Python

```
        io.imshow(fl/fm)
    elif type == 'abs':
        fa = abs(im)
        fm = fa.max()
        io.imshow(fa/fm)
    else:
        print("Error:_type_must_be_abs_or_log")
```

Python

and `homfilt`:

```
def homfilt(im,cutoff,order,lowgain,highgain):
    from scipy.fftpack import fft2,ifft2,fftshift
    u = im.astype('uint8')
    u[np.where(u==0)] = 1
    lg = np.log(u.astype(float))
    ft = fftshift(fft2(lg))
    rows,cols = im.shape
    rr = arange(-(rows//2),(rows+1)//2,1.0)
    cr = arange(-(cols//2),(cols+1)//2,1.0)
    y,x = np.meshgrid(cr,rr)
    bl = 1.0/(1.0+0.414*((x*x+y*y)/cutoff**2)**order)
    f = lowgain+(highgain-lowgain)*(1.0-bl)
    b = f*ft
    ib = abs(ifft2(b))
    return np.exp(ib)
```

Python

Exercises

1. By hand, compute the DFT of each of the following sequences:

   (a) $[2, \quad 3, \quad 4, \quad 5]$ (b) $[2, \quad -3, \quad 4, \quad -5]$ (c) $[-9, \quad -8, \quad -7, \quad -6]$
   (d) $[-9, \quad 8, \quad -7, \quad 6]$

   Compare your answers with those given by your system's `fft` function.

2. For each of the transforms you computed in the previous question, compute the inverse transform by hand.

3. By hand, verify the convolution theorem for each of the following pairs of sequences:

   (a) $[2, \quad 4, \quad 6, \quad 8]$ and $[-1, \quad 2 \quad -3, \quad 4]$
   (b) $[4, \quad 5, \quad 6, \quad 7]$ and $[3, \quad 1 \quad 5, \quad -1]$

4. Using your computer system, verify the convolution theorem for the following pairs of sequences:

   (a) $[2, \quad -3, \quad 5, \quad 6, \quad -2, \quad -1, \quad 3, \quad 7]$ and
   $\qquad [-1, \quad 5, \quad 6, \quad 4, \quad -3, \quad -5, \quad 1, \quad 2]$
   (b) $[7, \quad 6, \quad 5, \quad 4, \quad -4, \quad -5, \quad -6, \quad -7]$ and

$$[2, \quad 2, \quad -5, \quad -5, \quad 6, \quad 6, \quad -7, \quad -7]$$

5. Consider the following matrix:

$$\begin{bmatrix} 4 & 5 & -9 & -5 \\ 3 & -7 & 1 & 2 \\ 6 & -1 & -6 & 1 \\ 3 & -1 & 7 & -5 \end{bmatrix}$$

Using your system, calculate the DFT of each row. You can do this with the commands:

```
>> a=[4 5 -9 -5;3 -7 1 2;6 -1 -6 1;3 -1 7 -5];
>> a1=fft(a')'
```

MATLAB/Octave

or

```
In :  a = np.array([[4,5,-9,-5],[3,-7,1,2],[6,-1,-6,1],[3,-1,7,-5]])
In :  fft(a)
```

Python

(The `fft` function, applied to a matrix, produces the individual DFTs of all the columns in MATLAB or Octave, and the DFTs of all the rows in Python. So for MATLAB or Octave we transpose first, so that the rows become columns, then transpose back afterward.)

Now use similar commands to calculate the DFT of each column of `a1`.

Compare the result with the output of the command `fft2(a)`.

6. Perform similar calculations as in the previous question with the matrices produced by the commands `magic(4)` and `hilb(6)`.

7. How do you think filtering with an averaging filter will effect the output of a Fourier Transform?

Compare the DFTs of the cameraman image, and of the image after filtering with a 5×5 averaging filter.

Can you account for the result?

What happens if the averaging filter increases in size?

8. What is the result of two DFTs performed in succession? Apply a DFT to an image, and then another DFT to the result. Can you account for what you see?

9. Open up the image `engineer.png`.

Experiment with applying the Fourier Transform to this image and the following filters:

(a) Ideal filters (both low and high pass)

(b) Butterworth filters

(c) Gaussian filters

What is the smallest radius of a low pass ideal filter for which the face is still recognizable?

10. If you have access to a digital camera, or a scanner, produce a digital image of the face of somebody you know, and perform the same calculations as in the previous question.

Chapter 8

Image Restoration

8.1 Introduction

Image *restoration* concerns the removal or reduction of degradations that have occurred during the acquisition of the image. Such degradations may include *noise*, which are errors in the pixel values, or optical effects such as out of focus blurring, or blurring due to camera motion. We shall see that some restoration techniques can be performed very successfully using neighborhood operations, while others require the use of frequency domain processes. Image restoration remains one of the most important areas of image processing, but in this chapter the emphasis will be on the techniques for dealing with restoration, rather than with the degradations themselves, or the properties of electronic equipment that give rise to image degradation.

A Model of Image Degradation

In the spatial domain, we might have an image $f(x, y)$, and a spatial filter $h(x, y)$ for which convolution with the image results in some form of degradation. For example, if $h(x, y)$ consists of a single line of ones, the result of the convolution will be a motion blur in the direction of the line. Thus, we may write

$$g(x, y) = f(x, y) * h(x, y)$$

for the degraded image, where the symbol $*$ represents spatial filtering. However, this is not all. We must consider noise, which can be modeled as an additive function to the convolution. Thus, if $n(x, y)$ represents random errors which may occur, we have as our degraded image:

$$g(x, y) = f(x, y) * h(x, y) + n(x, y).$$

We can perform the same operations in the frequency domain, where convolution is replaced by multiplication, and addition remains as addition because of the linearity of the Fourier transform. Thus,

$$G(i, j) = F(i, j)H(i, j) + N(i, j)$$

represents a general image degradation, where of course F, H, and N are the Fourier transforms of f, h, and n, respectively.

If we knew the values of H and N we could recover F by writing the above equation as

$$F(i, j) = (G(i, j) - N(i, j))/H(i, j).$$

However, as we shall see, this approach may not be practical. Even though we may have some statistical information about the noise, we will not know the value of $n(x, y)$ or $N(i, j)$ for all, or even any, values. As well, dividing by $H(i, j)$ will cause difficulties if there are values that are close to, or equal to, zero.

8.2   Noise

We may define *noise* to be any degradation in the image signal, caused by external disturbance. If an image is being sent electronically from one place to another, via satellite or wireless transmission, or through networked cable, we may expect errors to occur in the image signal. These errors will appear on the image output in different ways depending on the type of disturbance in the signal. Usually we know what type of errors to expect, and hence the type of noise on the image; hence we can choose the most appropriate method for reducing the effects. Cleaning an image corrupted by noise is thus an important area of *image restoration*.

In this chapter we will investigate some of the standard noise forms and the different methods of eliminating or reducing their effects on the image.

We will look at four different noise types, and how they appear on an image.

Salt and Pepper Noise

Also called *impulse noise*, *shot noise*, or *binary noise*. This degradation can be caused by sharp, sudden disturbances in the image signal; its appearance is randomly scattered white or black (or both) pixels over the image.

To demonstrate its appearance, we will work with the image `gull.png` which we will make grayscale before processing:

```
>> g = rgb2gray(imread('gull.png'));
```
MATLAB/Octave

In Python there is also an `rgb2gray` function, in the `color` module of `skimage`:

```
In :  g = co.rgb2gray(io.imread('gull.png'))
```
Python

To add salt and pepper noise in MATLAB or Octave, we use the MATLAB function `imnoise`, which takes a number of different parameters. To add salt and pepper noise:

```
>> gsp = imnoise(g,'salt_and_pepper')
```
MATLAB/Octave

The amount of noise added defaults to 10%; to add more or less noise we include an optional parameter, being a value between 0 and 1 indicating the fraction of pixels to be corrupted. Thus, for example

```
>> imnoise(g,'salt_and_pepper',0.2);
```
MATLAB/Octave

would produce an image with 20% of its pixels corrupted by salt and pepper noise.

In Python, the method `noise.random_noise` from the `util` module of `skimage` provides this functionality:

```
In :  import skimage.util.noise as noise
In :  gn = noise.random_noise(g,mode='s&p')
In :  gn2 = noise.random_noise(g,mode='s&p',amount=0.2)
```
Python

The gull image is shown in Figure 8.1(a) and the image with noise is shown in Figure 8.1(b).

(a) Original image   (b) With added salt and pepper noise

FIGURE 8.1: Noise on an image

Gaussian Noise

Gaussian noise is an idealized form of *white noise*, which is caused by random fluctuations in the signal. We can observe white noise by watching a television that is slightly mistuned to a particular channel. Gaussian noise is white noise which is normally distributed. If the image is represented as I, and the Gaussian noise by N, then we can model a noisy image by simply adding the two:

$I + N$.

Here we may assume that I is a matrix whose elements are the pixel values of our image, and N is a matrix whose elements are normally distributed. It can be shown that this is an appropriate model for noise. The effect can again be demonstrated by the noise adding functions:

```
>> gg = inoise(g,'gaussian');
```
MATLAB/Octave

or

```
In :  gg = noise.random_noise(g,'gaussian')
```
Python

As with salt and pepper noise, the "**gaussian**" parameter also can take optional values, giving the mean and variance of the noise. The default values are 0 and 0.01, and the result is shown in Figure 8.2(a).

Speckle Noise

Whereas Gaussian noise can be modeled by random values *added* to an image; *speckle noise* (or more simply just *speckle*) can be modeled by random values *multiplied* by pixel values, hence it is also called *multiplicative noise*. Speckle noise is a major problem in some radar applications. As before, the noise adding functions can do speckle:

```
>> gs = imnoise(g,'speckle');
```
MATLAB/Octave

and

```
In :  gs = noise.random_noise(g,'speckle')
```
Python

and the result is shown in Figure 8.2(b). Speckle noise is implemented as

$$I(1+N)$$

where I is the image matrix, and N consists of normally distributed values with a default mean of zero. An optional parameter gives the variance of N; its default value in MATLAB or Octave is 0.04 and in Python is 0.01.

(a) Gaussian noise   (b) Speckle noise

FIGURE 8.2: The gull image corrupted by Gaussian and speckle noise

Although Gaussian noise and speckle noise appear superficially similar, they are formed by two totally different methods, and, as we shall see, require different approaches for their removal.

Periodic Noise

If the image signal is subject to a periodic, rather than a random disturbance, we might obtain an image corrupted by *periodic noise*. The effect is of bars over the image. Neither function `imnoise` or `random_noise` has a periodic option, but it is quite easy to create such noise, by adding a periodic matrix (using a trigonometric function) to the image. In MATLAB or Octave:

```
>> [rs,cs] = size(g);
>> [x,y] = meshgrid(1:rs,1:cs);
>> p = sin(x/3+y/5)+1;
>> gp = (2*im2double(g)+p/2)/3;
```
MATLAB/Octave

and in Python:

```
In :   r,c = g.shape
In :   x,y = np.mgrid(0:r,0:c).astype('float32')
In :   p = np.sin(x/3+y/3)+1.0
In :   gp = (2*skimage.util.img_as_float(g)+p/2)/3
```
`Python`

and the resulting image is shown in Figure 8.3.

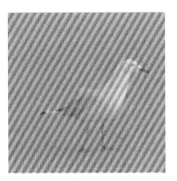

FIGURE 8.3: The gull image corrupted by periodic noise

Salt and pepper noise, Gaussian noise, and speckle noise can all be cleaned by using spatial filtering techniques. Periodic noise, however, requires the use of frequency domain filtering. This is because whereas the other forms of noise can be modeled as *local* degradations, periodic noise is a *global* effect.

8.3   Cleaning Salt and Pepper Noise

Low Pass Filtering

Given that pixels corrupted by salt and pepper noise are high frequency components of an image, we should expect a low pass filter should reduce them. So we might try filtering with an average filter:

```
>> a3 = fspecial('average');
>> g3 = imfilter(gsp,a3);
```
`MATLAB/Octave`

or

```
In :   import numpy.ndimage as ndi
In :   g3 = ndi.uniform_filter(gsp,3)
```
`Python`

and the result is shown in Figure 8.4(a). Notice, however, that the noise is not so much removed as "smeared" over the image; the result is not noticeably "better" than the noisy image. The effect is even more pronounced if we use a larger averaging filter:

```
>> a7 = fspecial('average',7);
>> g7 = imfilter(gsp,a7);
```

MATLAB/Octave

or

```
In : c7 = ndi.uniform_filter(csp,7)
```

Python

and the result is shown in Figure 8.4(b).

(a) 3×3 averaging   (b) 7×7 averaging

FIGURE 8.4: Attempting to clean salt and pepper noise with average filtering

Median Filtering

Median filtering seems almost tailor-made for removal of salt and pepper noise. Recall that the *median* of a numeric list is the middle value when the elements of the set are sorted. If there are an even number of values, the median is defined to be the mean of the middle two. A median filter is an example of a non-linear spatial filter; using a 3×3 mask, the output value is the median of the values in the mask. For example:

50	65	52
63	255	58
61	60	57

$\longrightarrow$ 50 52 57 58 $\boxed{60}$ 61 63 65 255 $\longrightarrow$ 60

The operation of obtaining the median means that very large or very small values—noisy values—will end up at the top or bottom of the sorted list. Thus, the median will in general replace a noisy value with one closer to its surroundings.

In MATLAB, median filtering is implemented by the **medfilt2** function:

```
>> gm3 = medfilt2(gsp);
```

MATLAB/Octave

and in Python by the **median_filter** method in the **ndimage** module of **numpy**:

```
In :  gm3 = ndi.median_filter(gsp,3)
```

Python

and the result is shown in Figure 8.5. The result is a vast improvement on using averaging

FIGURE 8.5: Cleaning salt and pepper noise with a median filter

filters. As with most functions, an optional parameter can be provided: in this case, a two element vector giving the size of the mask to be used.

If we corrupt more pixels with noise:

```
>> gsp2 = imnoise(g,'salt_&_pepper',0.2);
```

MATLAB/Octave

then `medfilt2` still does a remarkably good job, as shown in Figure 8.6. To remove noise

(a) 20% salt and pepper noise   (b) After median filtering

FIGURE 8.6: Using a 3 × 3 median filter on more noise

completely, we can either try a second application of the 3 × 3 median filter, the result of which is shown in Figure 8.7(a) or try a 5 × 5 median filter on the original noisy image:

```
>> gm5 = medfilt2(gsp2,[5,5]);
```

MATLAB/Octave

the result of which is shown in Figure 8.7(b).

(a) Using `medfilt2` twice   (b) Using a 5 × 5 median filter

FIGURE 8.7: Cleaning 20% salt and pepper noise with median filtering

Rank-Order Filtering

Median filtering is a special case of a more general process called *rank-order filtering*. Rather than take the median of a list, the n-th value of the ordered list is chosen, for some predetermined value of n. Thus, median filtering using a 3×3 mask is equivalent to rank-order filtering with $n = 5$. Similarly, median filtering using a 5×5 mask is equivalent to rank-order filtering with $n = 13$. MATLAB and Octave implement rank-order filtering with the `ordfilt2` function; in fact the procedure for `medfilt2` is really just a wrapper for a procedure which calls `ordfilt2`. There is only one reason for using rank-order filtering instead of median filtering, and that is that it allows the use of median filters over non-rectangular masks. For example, if we decided to use as a mask a 3×3 cross shape:

then the median would be the *third* of these values when sorted. The MATLAB/Octave command to do this is

```
>> co = ordfilt2(gsp,3,[0 1 0;1 1 1;0 1 0]);
```
MATLAB/Octave

In general, the second argument of `ordfilt2` gives the value of the ordered set to take, and the third element gives the *domain*; the non-zero values of which specify the mask. If we wish to use a cross with size and width 5 (so containing nine elements), we can use:

```
>> ordfilt2(gsp,5,[0 0 1 0 0;0 0 1 0 0;1 1 1 1 1;0 0 1 0 0;0 0 1 0 0])
```
MATLAB/Octave

In Python non-rectangular masks are handled with the `median_filter` method itself, by entering the mask (known in this context as the *footprint*) to be used:

```
In :   cross = array([[0,1,0],[1,1,1],[0,1,0]])
In :   co = ndi.median_filter(gsp,footprint=cross)
```
Python

When using `median_filter` there is no need to indicate which number to choose as output, as the output will be the median value of the elements underneath the 1's of the footprint.

An Outlier Method

Applying the median filter can in general be a slow operation: each pixel requires the sorting of at least nine values.[1] To overcome this difficulty, Pratt [35] has proposed the use of cleaning salt and pepper noise by treating noisy pixels as *outliers*; that is, pixels whose gray values are significantly different from those of their neighbors. This leads to the following approach for noise cleaning:

1. Choose a threshold value D.

2. For a given pixel, compare its value p with the mean m of the values of its eight neighbors.

3. If $|p - m| > D$, then classify the pixel as noisy, otherwise not.

4. If the pixel is noisy, replace its value with m; otherwise leave its value unchanged.

There is no MATLAB function for doing this, but it is very easy to write one. First, we can calculate the average of a pixel's eight neighbors by convolving with the linear filter

$$\frac{1}{8} \begin{bmatrix} 1 & 1 & 1 \\ 1 & 0 & 1 \\ 1 & 1 & 1 \end{bmatrix} = \begin{bmatrix} 0.125 & 0.125 & 0.125 \\ 0.125 & 0 & 0.125 \\ 0.125 & 0.125 & 0.125 \end{bmatrix}$$

We can then produce a matrix r consisting of 1's at only those places where the difference of the original and the filter are greater than D; that is, where pixels are noisy. Then $1 - r$ will consist of ones at only those places where pixels are not noisy. Multiplying r by the filter replaces noisy values with averages; multiplying $1 - r$ with original values gives the rest of the output.

In MATLAB or Octave, it is simpler to use arrays of type `double`. With $D = 0.5$, the steps to compute this method are:

```
>> gsp = im2double(gsp);
>> av = [1 1 1;1 0 1;1 1 1]/8;
>> gspa = imfilter(gsp,av);
>> D = 0.5
>> r = abs(gsp-gspa)>D;
>> imshow(r.*gspa+(1-r).*gsp)
```

MATLAB/Octave

and in Python, given that the output of `random_noise` is a floating point array, the steps are:

```
In :   av = array([[1,1,1],[1,0,1],[1,1,1]])/8.0
In :   gspa = ndi.convolve(gsp,av)
In :   D = 0.5
In :   r = (abs(gsp-gspa)>D)*1.0
In :   io.imshow(r*gspa+(1-r)*gsp)
```

Python

---

[1]In fact, this is not the case with any of our systems, all of which use a highly optimized method. Nonetheless, we introduce a different method to show that there are other ways of cleaning salt and pepper noise.

An immediate problem with the outlier method is that is it not completely automatic—the threshold D must be chosen. An appropriate way to use the outlier method is to apply it with several different thresholds, and choose the value that provides the best results.

Even with experimenting with different values of D, this method does not provide as good a result as using a median filter: the affect of the noise will be lessened, but there are still noise "artifacts" over the image. If we choose $D = 0.35$, we obtain the image in Figure 8.8(b), which still has some noise artifacts, although in different places. We can see that a lower value of D tends to remove noise from dark areas, and a higher value of D tends to remove noise from light areas. For this particular image, $D = 0.2$ seems to produce an acceptable result, although not quite as good as median filtering.

Clearly using an appropriate value of D is essential for cleaning salt and pepper noise by this method. If D is too small, then too many "non-noisy" pixels will be classified as noisy, and their values changed to the average of their neighbors. This will result in a blurring effect, similar to that obtained by using an averaging filter. If D is chosen to be too large, then not enough noisy pixels will be classified as noisy, and there will be little change in the output.

The outlier method is not particularly suitable for cleaning large amounts of noise; for such situations, the median filter is to be preferred. The outlier method may thus be considered as a "quick and dirty" method for cleaning salt and pepper noise when the median filter proves too slow.

A further method for cleaning salt and pepper noise will be discussed in Chapter 10.

8.4   Cleaning Gaussian Noise

Image Averaging

It may sometimes happen that instead of just *one* image corrupted with Gaussian noise, we have many different copies of it. An example is satellite imaging: if a satellite passes over the same spot many times, taking pictures, there will be many different images of the

(a) $D = 0.2$ (b) $D = 0.6$

FIGURE 8.8: Applying the outlier method to 10% salt and pepper noise

same place. Another example is in microscopy, where many different images of the same object may be taken. In such a case a very simple approach to cleaning Gaussian noise is to simply take the average—the mean—of all the images.

To see why this works, suppose we have 100 copies of an image, each with noise; then the i-th noisy image will be:

$$M + N_i$$

where M is the matrix of original values, and N_i is a matrix of normally distributed random values with mean 0. We can find the mean M' of these images by the usual add and divide method:

$$M' = \frac{1}{100} \sum_{i=1}^{100} (M + N_i)$$

$$= \frac{1}{100} \sum_{i=1}^{100} M + \frac{1}{100} \sum_{i=1}^{100} N_i$$

$$= M + \frac{1}{100} \sum_{i=1}^{100} N_i$$

Since N_i is normally distributed with mean 0, it can be readily shown that the mean of all the N_i's will be close to zero—the greater the number of N_i's, the closer to zero. Thus,

$$M' \approx M$$

and the approximation is closer for larger number of images $M + N_i$.

We can demonstrate this with the cat image. We first need to create different versions with Gaussian noise, and then take the average of them. We shall create 10 versions. One way is to create an empty three-dimensional array of depth 10, and fill each "level" with a noisy image, which we assume to be of type **double**:

```
>> t = zeros([size(g),10]);
>> for i=1:10 t(:,:,i)=imnoise(g,'gaussian'); end
```

MATLAB/Octave

Note here that the "**gaussian**" option of **imnoise** calls the random number generator **randn**, which creates normally distributed random numbers. Each time **randn** is called, it creates a *different* sequence of numbers. So we may be sure that all levels in our three-dimensional array do indeed contain different images. Now we can take the average:

```
>> ta = mean(t,3);
```

MATLAB/Octave

The optional parameter 3 here indicates that we are taking the mean along the *third* dimension of our array.

In Python the commands are:

```
In :  x,y = c.shape
In :  t = zeros((x,y,10))
In :  for i in range(10):
...:      t[:,:,i] = noise.random_noise(c,'gaussian')
...:
In :  ta = np.mean(t,3)
```

Python

The result is shown in Figure 8.9(a). This is not quite clear, but is a vast improvement on the noisy image of Figure 8.2(a). An even better result is obtained by taking the average of 100 images; this can be done by replacing 10 with 100 in the commands above, and the result is shown in Figure 8.9(b). Note that this method only works if the Gaussian noise has mean 0.

| (a) 10 images | (b) 100 images |

FIGURE 8.9: Image averaging to remove Gaussian noise

Average Filtering

If the Gaussian noise has mean 0, then we would expect that an average filter would average the noise to 0. The larger the size of the filter mask, the closer to zero. Unfortunately, averaging tends to blur an image, as we have seen in Chapter 5. However, if we are prepared to trade off blurring for noise reduction, then we can reduce noise significantly by this method.

Figure 8.10 shows the results of averaging with different sized masks. The results are

| (a) 4×3 averaging | (b) 5×5 averaging |

FIGURE 8.10: Using averaging filtering to remove Gaussian noise

not really particularly pleasing; although there has been some noise reduction, the "smeary" nature of the resulting images is unattractive.

Adaptive Filtering

Adaptive filters are a class of filters that change their characteristics according to the values of the grayscales under the mask; they may act more like median filters, or more like average filters, depending on their position within the image. Such a filter can be used to clean Gaussian noise by using local statistical properties of the values under the mask.

One such filter is the *minimum mean-square error filter*; this is a non-linear spatial filter; and as with all spatial filters, it is implemented by applying a function to the gray values under the mask.

Since we are dealing with additive noise, our noisy image M' can be written as

$$M' = M + N$$

where M is the original correct image, and N is the noise; which we assume to be normally distributed with mean 0. However, within our mask, the mean may not be zero; suppose the mean is m_f, and the variance in the mask is σ_f^2. Suppose also that the variance of the noise over the entire image is known to be σ_g^2. Then the output value can be calculated as

$$m_f + \frac{\sigma_f^2}{\sigma_f^2 + \sigma_g^2}(g - m_f)$$

where g is the current value of the pixel in the noisy image. Note that if the local variance σ_f^2 is high, then the fraction will be close to 1, and the output close to the original image value g. This is appropriate, as high variance implies high detail such as edges, which should be preserved. Conversely, if the local variance is low, such as in a background area of the image, the fraction is close to zero, and the value returned is close to the mean value m_f. See Lim [29] for details.

Another version of this filter [52] has output defined by

$$g - \frac{\sigma_g^2}{\sigma_f^2}(g - m_f)$$

and again the filter returns a value close to either g or m_f depending on whether the local variance is high or low.

In practice, m_f can be calculated by simply taking the mean of all gray values under the mask, and σ_f^2 by calculating the variance of all gray values under the mask. The value σ_g^2 may not necessarily be known, so a slight variant of the first filter may be used:

$$m_f + \frac{\max\{0, \sigma_f^2 - n\}}{\max\{\sigma_f^2, n\}}(g - m_f)$$

where n is the computed noise variance, and is calculated by taking the mean of all values of σ_f^2 over the entire image. This particular filter is implemented in MATLAB with the function `wiener2`. The name reflects the fact that this filter attempts to minimize the square of the difference between the input and output images; such filters are in general known as *Wiener filters*. However, Wiener filters are more usually applied in the frequency domain; see Section 8.7.

Suppose we take the noisy image `cg` shown in Figure 8.2(a), and attempt to clean this image with adaptive filtering. We will use the `wiener2` function, which can take an optional parameter indicating the size of the mask to be used. The default size is 3×3. We shall create four images:

```
>> gw1 = wiener2(gg);
>> gw2 = wiener2(gg,[5,5]);
>> gw3 = wiener2(gg,[7,7]);
>> gw4 = wiener2(gg,[9,9]);
```

MATLAB/Octave

In Python the Wiener filter is implemented by the `wiener` method of the `signal` module of `scipy`:

```
In :  from scipy.signal import wiener
In :  cw1 = wiener(cg,[3,3])
```

Python

and the other commands similarly. The results are shown in Figure 8.11. Being a low

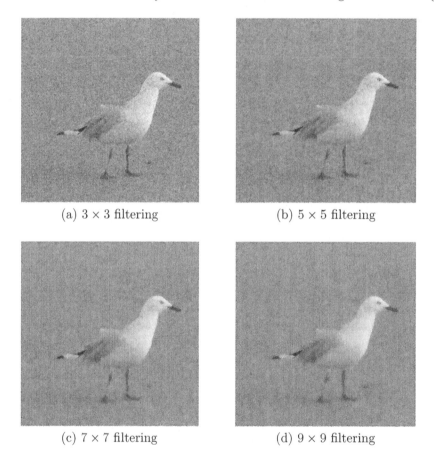

(a) 3 × 3 filtering   (b) 5 × 5 filtering

(c) 7 × 7 filtering   (d) 9 × 9 filtering

FIGURE 8.11: Examples of adaptive filtering to remove Gaussian noise

pass filter, adaptive filtering does tend to blur edges and high frequency components of the image. But it does a far better job than using a low pass blurring filter.

We can achieve very good results for noise where the variance is not as high as that in our current image.

```
>> gg2 = imnoise(g,'gaussian',0,0.005);
>> imshow(gg2)
>> gg2w = wiener2(im2double(gg2),[7,7]);
>> figure,imshow(gg2w)
```

MATLAB/Octave

The image and its appearance after adaptive filtering is shown in Figure 8.12. The result

(a) Image with variance 0.005 (b) Result after adaptive filtering

FIGURE 8.12: Using adaptive filtering to remove Gaussian noise with low variance

is a great improvement over the original noisy image. Notice in each case that there may be some blurring of the background, but the edges are preserved well, as predicted by our analysis of the adaptive filter formulas above.

8.5 Removal of Periodic Noise

Periodic noise may occur if the imaging equipment (the acquisition or networking hardware) is subject to electronic disturbance of a repeating nature, such as may be caused by an electric motor. We have seen in Section 8.2 that periodic noise can be simulated by overlaying an image with a trigonometric function. We used

```
>> p = sin(x/3+y/5)+1;
>> gp = (2*im2double(c)+p/2)/3;
```

MATLAB/Octave

where g is our gull image. The first line simply creates a sine function, and adjusts its output to be in the range 0–2. The last line first adjusts the gull image to be in the range 0–2; adds the sine function to it, and divides by 3 to produce a matrix of type **double** with all elements in the range 0.0–1.0. This can be viewed directly with **imshow**, and it is shown in Figure 8.13(a). We can produce its shifted DFT and this is shown in Figure 8.3(b). The extra two "spikes" away from the center correspond to the noise just added. In general, the tighter the period of the noise, the further from the center the two spikes will be. This is because a small period corresponds to a high frequency (large change over a small distance), and is therefore further away from the center of the shifted transform.

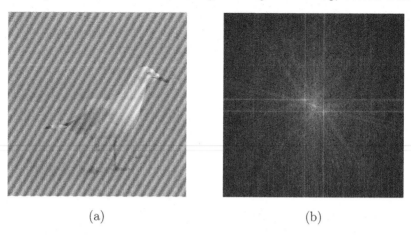

(a)   (b)

FIGURE 8.13: The gull image (a) with periodic noise and (b) its transform

We will now remove these extra spikes, and invert the result. The first step is to find their position. Since they will have the largest maximum value after the DC coefficient, which is at position $(201, 201)$, we simply have to find the maxima of all absolute values except the DC coefficient. This will be easier to manage if we turn the absolute values of the FFT into an array of type `uint8`. Note that we make use of the `find` function which returns all indices in an array corresponding to elements satisfying a certain condition; in this case, they are equal to the maximum:

```
>> gf = fftshift(fft2(gp));
>> imshow(mat2gray(log(1+abs(gf))))
>> gf2 = im2uint8(mat2gray(abs(gf)));
>> gf2(201,201) = 0;
>> [i,j] = find(gf2==max(gf2(:)))
i =

    188
    214


j =

    180
    222
```

MATLAB/Octave

Check that each point is the same distance from the center by computing the square of the distance of each point from $(201, 201)$:

```
>> (i-201).^2+(j-201).^2

    610
    610
```

MATLAB/Octave

With Python, the above operations could be performed as follows, where to find array elements satisfying a condition we use the `where` function:

```
In :   from numpy.fft import ifft2,fft2,fftshift
In :   gf = fftshift(fft2(gp))
In :   temp = ex.rescale_intensity(abs(gf),out_range=(0,1))
In :   gf2 = ut.img_as_ubyte(temp)
In :   gf2[200,200] = 0
In :   i,j = np.where(gf2 == gf2.max())
In :   i,j
Out:   (array([179, 221]), array([187, 213]))
In :   (i-200)**2+(j-200)**2
Out:   array([610, 610])
```
Python

There are two methods we can use to eliminate the spikes; we shall look at both of them.

Band reject filtering. A band reject filter eliminates a particular band: that is, an annulus (a ring) surrounding the center. If the annulus is very thin, then the filter is sometimes called a *notch filter*. In our example, we need to create a filter that rejects the band at about $\sqrt{610}$ from the center. In MATLAB/Octave:

```
>> z = sqrt((x-201).^2+(y-201).^2);
>> d = sqrt(610);
>> k = 1;
>> br = (z < floor(d-k) | z > ceil(d+k));
```
MATLAB/Octave

or in Python:

```
In :   z = np.sqrt((x-200)**2+(y-200)**2)
In :   k = 1
In :   d = sqrt(610.0)
In :   br = (z< np.floor(d-k)) | (z>np.ceil(d+k))
```
Python

This particular ring will have a thickness large enough to cover the spikes. Then as before, we multiply this by the transform:

```
>> cfr = cf.*br;
```
MATLAB/Octave

or with

```
In :   cfr = cf*br
```
Python

and finally apply the inverse transform and display its absolute value. This final result is shown in Figure 8.14(a). The result is that the spikes have been blocked out by this filter. Taking the inverse transform produces the image shown in Figure 8.14(b). Note that not all the noise has gone, but a significant amount has, especially in the center of the image. More noise can be removed by using a larger band: Figure 8.15 shows the results using $k = 2$ and $k = 5$.

Criss-cross filtering. Here the rows and columns of the transform that contain the spikes are set to zero, which can easily be done using the i and j values from above:

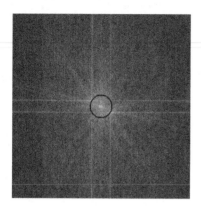

(a) A band-reject filter   (b) After inversion

FIGURE 8.14: Removing periodic noise with a band-reject filter

(a) Band-reject filter with $k = 2$ (b) Band-reject filter with $k = 5$

FIGURE 8.15: Removing periodic noise with wider band-reject filters

```
>> gf2 = gf;
>> gf2(i,:) = 0;
>> gf2(:,j) = 0;
```

MATLAB/Octave

or

```
In :  cf2 = np.copy(cf)
In :  cf2[i,:] = 0
In :  cf2[:,j] = 0
```

Python

and the result after applying the inverse is shown in Figure 8.16(a). The image after

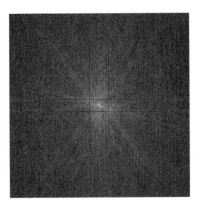

(a) A criss-cross filter

(b) After inversion

FIGURE 8.16: Removing periodic noise with a criss-cross filter

inversion is shown in Figure 8.16(b). As before, much of the noise in the center has been removed. Making more rows and columns of the transform zero would result in a larger reduction of noise.

8.6 Inverse

We have seen that we can perform filtering in the Fourier domain by multiplying the DFT of an image by the DFT of a filter: this is a direct use of the convolution theorem. We thus have

$$Y(i,j) = X(i,j)F(i,j)$$

where X is the DFT of the image; F is the DFT of the filter, and Y is the DFT of the result. If we are given Y and F, then we should be able to recover the (DFT of the) original image X simply by dividing by F:

$$X(i,j) = \frac{Y(i,j)}{F(i,j)}. \tag{8.1}$$

Suppose, for example, we take the buffalo image `buffalo.png`, and blur it using a low pass Butterworth filter. In MATLAB/Octave:

```
>> b = imread('buffalo.png');
>> bf = fftshift(fft2(b));
>> [r,c] = size(b);
>> [x,y] = meshgrid(-c/2:c/2-1,-r/2:r/2-1);
>> bworth = 1./(1+(sqrt(2)-1)*((x.^2+y.^2)/15^2).^2);
>> bw = bf.*bworth;
>> bwa = abs(ifft2(bw));
>> blur = im2uint8(mat2gray(bwa));
>> imshow(blur)
```

MATLAB/Octave

In Python:

```
In :  b = io.imread('buffalo.png')
In :  bf = fftshift(fft2(b))
In :  r,c = b.shape
In :  x,y = np.mgrid(-c/2:c/2,-r/2:r/2]
In :  bworth = 1/(1+(np.sqrt(2)-1)*((x**2+y**2)/15**2)**2)
In :  bw = bf*bworth
In :  bwa = abs(ifft2(bw))
In :  blur = ut.img_as_ubyte(ex.rescale_intensity(bwa,out_range=(0.0,1.0)))
```

Python

The result is shown on the left in Figure 8.17. We can attempt to recover the original image by dividing by the filter:

```
>> blf = fftshift(fft2(blur));
>> blfw = blf./bworth;
>> bla = abs(ifft2(blfw));
>> imshow(mat2gray(bla))
```

MATLAB/Octave

or with

```
In :  blf = fftshift(fft2(blur))
In :  blfw = blf/bworth
In :  bla = abs(ifft2(blfw))
In :  io.imshow(bla)
```

Python

and the result is shown on the right in Figure 8.17. This is no improvement! The trouble is that some elements of the Butterworth matrix are very small, so dividing produces very large values which dominate the output. This issue can be managed in two ways:

1. Apply a low pass filter L to the division:

$$X(i,j) = \frac{Y(i,j)}{F(i,j)} L(i,j).$$

This should eliminate very low (or zero) values.

FIGURE 8.17: An attempt at inverse filtering

2. "Constrained division": choose a threshold value d, and if $|F(i,j)| < d$, the original value is returned instead of a division: Thus:

$$
X(i,j) = \begin{cases} \dfrac{Y(i,j)}{F(i,j)} & \text{if } |F(i,j)| \geq d, \\[2ex] Y(i,j) & \text{if } |F(i,j)| < d. \end{cases}
$$

The first method can be implemented by multiplying a Butterworth low pass filter to the matrix **c1** above:

```
>> blf = fftshift(fft2(blur));
>> D = 40
>> bworth2 = 1./(1+(sqrt(2)-1)*((x.^2+y.^2)/D^2).^10);
>> blfb = blf./bworth.*bworth2;
>> ba = abs(ifft2(blfb));
>> imshow(mat2gray(ba))
```

MATLAB/Octave

or

```
In :  blf = fftshift(fft2(blur))
In :  D = 40
In :  bworth2 = 1/(1+(sqrt(2)-1)*((x**2+y**2)/D**2)**10)
In :  blfb = blf/bworth*bworth2
In :  ba = abs(ifft2(blfb))
In :  io.imshow(ex.rescale_intensity(ba,out_range=(0.0,1.0)))
```

Python

Figure 8.18 shows the results obtained by using a different cutoff radius of the Butterworth filter each time: (a) uses 40 (as in the MATLAB commands just given); (b) uses 60; (c) uses 80, and (d) uses 100. It seems that using a low pass filter with a cutoff round about 60 will yield the best results. After we use larger cutoffs, the result degenerates.

We can try the second method; to implement it, we simply make all values of the filter that are too small equal to 1:

(a)

(b)

(c)

(d)

FIGURE 8.18: Inverse filtering using low pass filtering to eliminate zeros

```
>> d = 0.01;
>> bw = bworth; bw(find(bw<d))=1
>> fbw = fftshift(fft2(blur))./bw;
>> w1a = abs(ifft2(w1));
>> imshow(mat2gray(w1a))
```
MATLAB/Octave

or

```
In :  d = 0.01
In :  bw = np.copy(bworth)
In :  bw[np.where(bw<d)] = 1
In :  fbw = fftshift(fft2(blur))/bw
In :  ba = abs(ifft2(fbw))
In :  io.imshow(ex.rescale_intensity(w1a,out_range=(0.0,1.0)))
```
Python

Figure 8.19 shows the results obtained by using a different cutoff radius of the Butterworth filter each time: (a) uses $d = 0.01$ (as in the MATLAB commands just given); (b) uses $d = 0.005$; (c) uses $d = 0.002$, and (d) uses $d = 0.001$. It seems that using a threshold d in the range $0.002 \leq d \leq 0.005$ produces reasonable results.

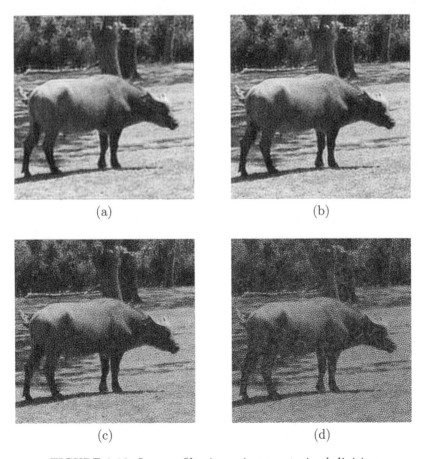

FIGURE 8.19: Inverse filtering using constrained division

Motion Deblurring

We can consider the removal of blur caused by motion to be a special case of inverse filtering. Start with the image **car.png** saved to **cr**, and which is shown in Figure 8.20(a). It can be blurred in MATLAB or Octave using the **motion** parameter of the **fspecial** function.

```
>> m = fspecial('motion',7,0);
>> cm = imfilter(cr,m);
```

MATLAB/Octave

In Python, such an effect can be obtained by creating a motion filter and then applying it:

```
>> m = np.ones((1,7))/7
>> cm = ndi.correlate(cr,m)
```

Python

and the result is shown as Figure 8.20(b). The result of the blur has effectively obliterated

(a) (b)

FIGURE 8.20: The result of motion blur

the text of the number-plate.

To deblur the image, we need to divide its transform by the transform corresponding to the blur filter. So the first step is to create a matrix corresponding to the transform of the blur:

```
>> m2 = zeros(size(cr));
>> m2(1,1:7) = m;
>> mf = fft2(m2);
```

MATLAB/Octave

or

```
In :   m2 = np.zeros_like(cr)
In :   m2[1,0:7] = m
In :   mf = fft2(m2)
```

Python

The second step is dividing by this transform.

```
>> bmi = ifft2(fft2(cm)./mf);
>> fftshow(bmi,'abs')
```

or

```
In :  bmi = ifft2(fft2(cm)/mf)
In :  io.imshow(ex.rescale_intensity(abs(bmi),out_range=(0,1)))
```

and the result is shown in Figure 8.21(a). As with inverse filtering, the result is not particularly good because the values close to zero in the matrix mf have tended to dominate the result. As above, we can constrain the division by only dividing by values that are above a certain threshold.

```
>> d = 0.02;
>> mf = fft2(m2); mf(find(abs(mf)<d)) = 1;
>> bmi = ifft2(fft2(cm)./mf);
>> imshow(mat2gray(abs(bmi))*2)
```

where the last multiplication by 2 just brightens the result, or

```
In :  d = 0.02
In :  mf = fft2(m2)
In :  mf[np.where(abs(mf)<d)] = 1
In :  bmi = abs(ifft2(fft2(cm)/mf))
In :  bmu = ut.img_as_ubyte(bmi/bmi.max())
In :  io.imshow(ex.rescale_intensity(bmu,in_range=(0,128)))
```

where the use of a limited in_range scales the image to be brighter. The output is shown in Figure 8.21(b). The image is not perfect—there are still vertical artifacts—but the number plate is now quite legible.

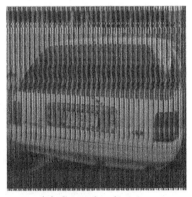

(a) Straight division   (b) Constrained division

FIGURE 8.21: Attempts at removing motion blur

8.7   Wiener Filtering

As we have seen from the previous section, inverse filtering does not necessarily produce particularly pleasing results. The situation is even worse if the original image has been corrupted by noise. Here we would have an image X filtered with a filter F and corrupted by noise N. If the noise is additive (for example, Gaussian noise), then the linearity of the Fourier transform gives us

$$Y(i,j) = X(i,j)F(i,j) + N(i,j)$$

and so

$$X(i,j) = \frac{Y(i,j) - N(i,j)}{F(i,j)}$$

as we have seen in the introduction to this chapter. So not only do we have the problem of dividing by the filter, we have the problem of dealing with noise. In such a situation, the presence of noise can have a catastrophic effect on the inverse filtering: the noise can completely dominate the output, making direct inverse filtering impossible.

To introduce Wiener filtering, we shall discuss a more general question: given a degraded image M' of some original image M and a restored version R, what measure can we use to say whether our restoration has done a good job? Clearly we would like R to be as close as possible to the "correct" image M. One way of measuring the closeness of R to M is by adding the squares of all differences:

$$\sum (m_{i,j} - r_{i,j})^2$$

where the sum is taken over all pixels of R and M (which we assume to be of the same size). This sum can be taken as a measure of the closeness of R to M. If we can minimize this value, we may be sure that our procedure has done as good a job as possible. Filters that operate on this principle of *least squares* are called *Wiener filters*. We can obtain X by

$$X(i,j) \approx \left[\frac{1}{F(i,j)} \frac{|F(i,j)|^2}{|F(i,j)|^2 + K} \right] Y(i,j) \tag{8.2}$$

where K is a constant [13]. This constant can be used to approximate the amount of noise: if the variance σ^2 of the noise is known, then $K = 2\sigma^2$ can be used. Otherwise, K can be chosen interactively (in other words, by trial and error) to yield the best result. Note that if $K = 0$, then Equation 8.2 reduces to Equation 8.1.

We can easily implement Equation 8.2 in MATLAB/Octave:

```
>> K = 0.01;
>> bf = fftshift(fft2(blur));
>> w1 = wbf.*(abs(b).^2./(abs(b).^2+K)./b);   % This is the equation
>> w1a = abs(ifft2(w1));
>> imshow(mat2gray(w1a))
```

MATLAB/Octave

or in Python:

```
In :  K = 0.01
In :  bf = fftshift(fft2(blur))
In :  w1 = bf*(abs(bworth)**2/(abs(bworth)**2+K)/bworth)
In :  w1a = abs(ifft2(w1))
In :  io.imshow(w1a)
```

Python

The result is shown in Figure 8.22(a). Images (b), (c), and (d) in this figure show the results with $K = 0.001$, $K = 0.0001$, and $K = 0.00001$ respectively. Thus, as K becomes very small, noise starts to dominate the image.

(a)　　　　　　　　　　　　　　(b)

(c)　　　　　　　　　　　　　　(d)

FIGURE 8.22: Wiener filtering

Exercises

1. The following arrays represent small grayscale images. Compute the 4×4 image that would result in each case if the middle 16 pixels were transformed using a 3×3 median filter:

8	17	4	10	15	12
10	12	15	7	3	10
15	10	50	5	3	12
4	8	11	4	1	8
16	7	4	3	0	7
16	24	19	3	20	10

1	1	2	5	3	1
3	20	5	6	4	6
4	6	4	20	2	2
4	3	3	5	1	5
6	5	20	2	20	2
6	3	1	4	1	2

7	8	11	12	13	9
8	14	0	9	7	10
11	23	10	14	1	8
14	7	11	8	9	11
13	13	18	10	7	12
9	11	14	12	13	10

2. Using the same images as in Question 1, transform them by using a 3×3 averaging filter.

3. Use the outlier method to find noisy pixels in each of the images given in Question 1. What are the reasonable values to use for the difference between the gray value of a pixel and the average of its eight 8-neighbors?

4. Pratt [35] has proposed a "pseudo-median" filter, in order to overcome some of the speed disadvantages of the median filter. For example, given a five-element sequence $\{a, b, c, d, e\}$, its pseudo-median is defined as

$$\text{psmed}(a, b, c, d, e) \;=\; \tfrac{1}{2} \max \Big[\min(a, b, c), \min(b, c, d), \min(c, d, e) \Big]$$
$$+ \; \tfrac{1}{2} \min \Big[\max(a, b, c), \max(b, c, d), \max(c, d, e) \Big]$$

So for a sequence of length 5, we take the maxima and minima of all subsequences of length three. In general, for an odd-length sequence L of length $2n + 1$, we take the maxima and minima of all subsequences of length $n + 1$.

We can apply the pseudo-median to 3×3 neighborhoods of an image, or cross-shaped neighborhoods containing 5 pixels, or any other neighborhood with an odd number of pixels.

Apply the pseudo-median to the images in Question 1, using 3×3 neighborhoods of each pixel.

5. Write a function to implement the pseudo-median, and apply it to the images above. Does it produce a good result?

6. Choose any grayscale image of your choice and add 5% salt and pepper noise to the image. Attempt to remove the noise with

   (a) Average filtering
   (b) Median filtering
   (c) The outlier method
   (d) Pseudo-median filtering

   Which method gives the best results?

7. Repeat the above question but with 10%, and then with 20% noise.

8. For 20% noise, compare the results with a 5×5 median filter, and two applications of a 3×3 median filter.

9. Add Gaussian noise to your chosen image with the following parameters:

   (a) Mean 0, variance 0.01 (the default)

   (b) Mean 0, variance 0.02

   (c) Mean 0, variance 0.05

   (d) Mean 0, variance 0.1

   In each case, attempt to remove the noise with average filtering and with Wiener filtering.

   Can you produce satisfactory results with the last two noisy images?

10. Gonzalez and Woods [13] mention the use of a *midpoint filter* for cleaning Gaussian noise. This is defined by

$$g(x, y) = \frac{1}{2}\left(\max_{(x,y) \in B} f(x, y) + \min_{(x,y) \in B} f(x, y) \right)$$

   where the maximum and minimum are taken over all pixels in a neighborhood B of (x, y). Use rank-order filtering to find maxima and minima, and experiment with this approach to cleaning Gaussian noise, using different variances. Visually, how do the results compare with spatial Wiener filtering or using a blurring filter?

11. In Chapter 5 we defined the alpha-trimmed mean filter, and the geometric mean filter. Write functions to implement these filters, and apply them to images corrupted with Gaussian noise.

   How well do they compare to average filtering, image averaging, or adaptive filtering?

12. Add the sine waves to an image of your choice by using the same commands as above, but with the final command

    ```
    s = 1+sin(x+y/1.5);
    ```

   Now attempt to remove the noise using band-reject filtering or criss-cross filtering. Which one gives the best result?

13. For each of the following sine commands:

    (a) `s = 1+sin(x/3+y/5)`

    (b) `s = 1+sin(x/5+y/1.5)`

    (c) `s = 1+sin(x/6+y/6)`

   add the sine wave to the image as shown in the previous question, and attempt to remove the resulting periodic noise using band-reject filtering or notch filtering.

   Which of the three is easiest to "clean up"?

14. Apply a 5×5 blurring filter to a grayscale image of your choice. Attempt to deblur the result using inverse filtering with constrained division. Which threshold gives the best results?

15. Repeat the previous question using a 7×7 blurring filter.

16. Work through the motion deblurring example, experimenting with different values of the threshold. What gives the best results?

Chapter 9

Image Segmentation

9.1 Introduction

Segmentation refers to the operation of partitioning an image into component parts or into separate objects. In this chapter, we shall investigate two very important topics: thresholding and edge detection.

9.2 Thresholding

Single Thresholding

A grayscale image is turned into a binary (black and white) image by first choosing a gray level T in the original image, and then turning every pixel black or white according to whether its gray value is greater than or less than T:

$$\text{A pixel becomes} \begin{cases} \text{white if its gray level is} > T, \\ \text{black if its gray level is} \leq T. \end{cases}$$

Thresholding is a vital part of image *segmentation*, where we wish to isolate objects from the background. It is also an important component of robot vision.

Thresholding can be done very simply in any of our systems. Suppose we have an 8-bit image, stored as the variable X. Then the command

    X>T or X<T

will perform the thresholding. For example, consider an image of some birds flying across a sky; it can be thresholded to show the birds alone in MATLAB or Octave:

```
>> f = imread('flying.png');
>> imshow(f); figure,imshow(f<50)
```
MATLAB/Octave

and in Python:

```
In :  f = io.imread('flying.png')
In :  io.imshow(f)
In :  fig = plt.figure(); fig.show(io.imshow(f<50))
```
Python

These commands will produce the images shown in Figure 9.1. The resulting image can

223

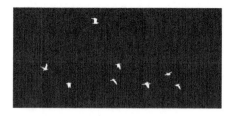

FIGURE 9.1: Thresholded image of flying birds

then be further processed to find the number, or average size of the birds.

To see how this works, recall that in each system, an operation on a single number, when applied to a matrix, is interpreted as being applied simultaneously to all elements of the matrix; this is vectorization, which we have seen earlier. In MATLAB or Octave the command X>T will thus return 1 (for true) for all those pixels for which the gray values are greater than T, and 0 (for false) for all those pixels for which the gray values are less than or equal to T. The result is a matrix of 0's and 1's, which can be viewed as a binary image. In Python, the result will be a Boolean array whose elements are True or False. Such an array can however be viewed without any further processing as a binary image. A Boolean array can be turned into a floating point array by re-casting the array to the required data type:

```
In :  (f<50).dtype
Out:  dtype('bool')

In :  f1 = (f<50).astype('float64')
In :  f1.dtype
Out:  dtype('float64')
```

Python

The flying birds image shown above has dark objects on a light background; an image with light objects over a dark background may be treated the same:

```
>> p = imread('paperclips.png')
>> imshow(p),figure,imshow(p>140)
```

MATLAB/Octave

will produce the images shown in Figure 9.2.

As well as the above method, MATLAB and Octave have the im2bw function, which thresholds an image *of any data type*, using the general syntax

```
im2bw(image,level)
```

where level is a value between 0 and 1 (inclusive), indicating the fraction of gray values to be turned white. This command will work on grayscale, colored, and indexed images of data type uint8, uint16 or double. For example, the thresholded flying and paperclip images above could be obtained using

```
>> im2bw(f,0.3);
>> im2bw(p,0.55);
```

MATLAB/Octave

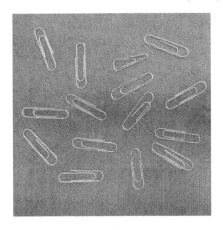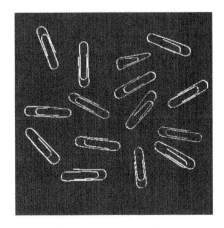

FIGURE 9.2: Thresholded image of paperclips

The `im2bw` function automatically scales the value `level` to a gray value appropriate to the image type, and then performs a thresholding by our first method.

As well as isolating objects from the background, thresholding provides a very simple way of showing hidden aspects of an image. For example, the image of some handmade paper `handmade.png` appears mostly white, as nearly all the gray values are very high. However, thresholding at a high level produces an image of far greater interest. We can use the commands

```
>> h = imread('handmade.png');
>> imshow(h),figure,imshow(h>242)
```

MATLAB/Octave

to provide the images shown in Figure 9.3.

FIGURE 9.3: The paper image and result after thresholding

Double Thresholding

Here we choose two values T_1 and T_2 and apply a thresholding operation as:

$$\text{A pixel becomes} \begin{cases} \text{white if its gray level is between } T_1 \text{ and } T_2, \\ \text{black if its gray level is otherwise.} \end{cases}$$

We can implement this by a simple variation on the above method:

```
X>T1 & X<T2
```

Since the ampersand acts as a logical "and," the result will only produce a one where both inequalities are satisfied. Consider the following sequence of commands:

```
>> x = imread('xray.png');
>> imshow(x),figure,imshow(x>50 & x<80)
```

MATLAB/Octave

The output is shown in Figure 9.4. Note how double thresholding isolates the boundaries of

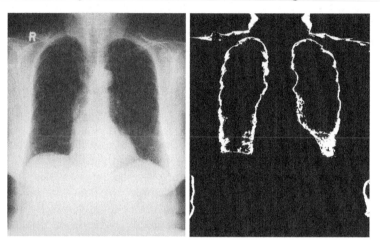

FIGURE 9.4: The image `xray.png` and the result after double thresholding

the lungs, which single thresholding would be unable to do. Similar results can be obtained using `im2bw`:

```
>> imshow(im2bw(x,0.2)&~im2bw(x,0.3))
```

MATLAB/Octave

In Python, the command is almost the same:

```
In :  io.imshow((x>40) & (x<80))
```

Python

9.3   Applications of Thresholding

We have seen that thresholding can be useful in the following situations:

1. When we want to remove unnecessary detail from an image, to concentrate on essentials. Examples of this were given in the flying birds and paperclip images: by removing all gray level information, the birds and paperclips were reduced to binary objects. But this information may be all we need to investigate sizes, shapes, or numbers of objects.

2. To bring out hidden detail. This was illustrated with paper and x-ray images. In both, the detail was obscured because of the similarity of the gray levels involved.

But thresholding can be vital for other purposes:

3. When we want to remove a varying background from an image. An example of this was the paper clips image shown previously; the paper clips are all light, but the background in fact varies considerably. Thresholding at an appropriate value completely removes the background to show just the objects.

9.4 Choosing an Appropriate Threshold Value

We have seen that one of the important uses of thresholding is to isolate objects from their background. We can then measure the sizes of the objects, or count them. Clearly the success of these operations depends very much on choosing an appropriate threshold level. If we choose a value too low, we may decrease the size of some of the objects, or reduce their number. Conversely, if we choose a value too high, we may begin to include extraneous background material.

Consider, for example, the image `pinenuts.png`, and suppose we try to threshold using the `im2bw` function and various threshold values t for $0 < t < 1$.

```
>> n = imread('pinenuts.png');
>> imshow(n);
>> n1 = im2bw(n,0.35);
>> n2 =im2bw(n,0.55);
>> figure,imshow(n1),figure,imshow(n2)
```
MATLAB/Octave

All the images are shown in Figure 9.5. One approach is to investigate the histogram of the

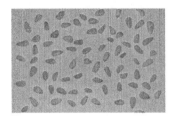  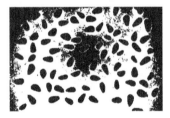

n: Original image   n1: Threshold too low   n2: Threshold too high

FIGURE 9.5: Attempts at thresholding

image, and see if there is a clear spot to break it up. Sometimes this can work well, but not always.

Figure 9.6 shows various histograms. In each case, the image consists of objects on a background. But only for some histograms is it easy to see where we can split it. In both the blood and daisies images, we could split it up about half way, or at the "valley" between the peaks, but for the paramecium and pinenut images, it is not so clear, as there would appear to be three peaks—in each case, one at the extreme right.

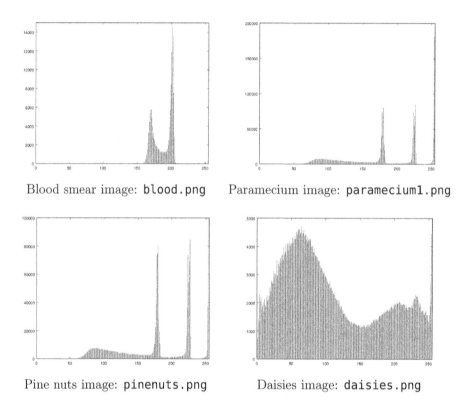

Blood smear image: `blood.png`   Paramecium image: `paramecium1.png`

Pine nuts image: `pinenuts.png`   Daisies image: `daisies.png`

FIGURE 9.6: Histograms

The trouble is that in general the individual histograms of the objects and background will overlap, and without prior knowledge of the individual histograms it may be difficult to find a splitting point. Figure 9.7 illustrates this, assuming in each case that the histograms for the objects and backgrounds are those of a normal distribution. Then we choose the

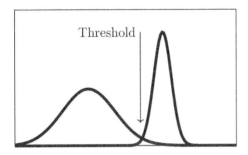 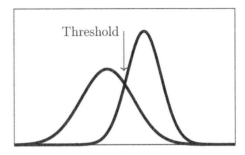

FIGURE 9.7: Splitting up a histogram for thresholding

threshold values to be the place where the two histograms cross over.

In practice though, the histograms won't be as clearly defined as those in Figure 9.7, so we need some sort of automatic method for choosing a "best" threshold. There are in fact many different methods; here are two.

Otsu's Method

This was first described by Nobuyuki Otsu in 1979. It finds the best place to threshold the image into "foreground" and "background" components so that the *inter-class variance*—also known as the *between class variance*—is maximized. Suppose the image is divided into foreground and background pixels at a threshold t, and the fractions of each are w_f and w_b respectively. Suppose also that the means are μ_f and μ_b. Then the inter-class variance is defined as

$$w_f w_b (\mu_f - \mu_b)^2.$$

If an image has n_i pixels of gray value i, then define $p_i = n_i/N$ where N is the total number of pixels. Thus, p_i is the probability of a pixel having gray value i. Given a threshold value t then

$$w_b = \sum_{k=0}^{t-1} p_k,$$
$$w_f = \sum_{k=t}^{L-1} p_k.$$

where L is the number of gray values in the image. Since by definition we must have

$$\sum_{k=0}^{L-1} p_k = 1$$

it follows that $w_k + w_f = 1$. The background weights can be computed by a cumulative sum of all the p_k values.

The means can be defined as

$$\mu_b = \frac{1}{w_b} \sum_{k=0}^{t-1} kp_k,$$

$$\mu_f = \frac{1}{w_f} \sum_{k=t}^{L-1} kp_k.$$

Note that the sums

$$\sum_{k=0}^{t-1} kp_k$$

again can be computed by cumulative sums, and

$$\sum_{k=t}^{L-1} kp_k = \sum_{k=0}^{L-1} kp_k - \sum_{k=0}^{t-1} kp_k.$$

and the first term on the left is fixed.

For a simple example, consider an 8×8 3-bit image with the following values of n_k and $p_k = n_k/N = n_k/64$

i	n_i	p_i
0	4	0.062500
1	8	0.125000
2	10	0.156250
3	9	0.140625
4	4	0.062500
5	8	0.125000
6	12	0.187500
7	9	0.140625

Then the other values can be computed easily:

k	n_k	p_k	w_b	w_f	kp_k	$\sum kp_k$	μ_b	μ_f	vars
0	4	0.06250	0.06250	0.93750	0.00000	0.00000	0.00000	4.10000	0.98496
1	8	0.12500	0.18750	0.81250	0.12500	0.12500	0.66667	4.57692	2.32935
2	10	0.15625	0.34375	0.65625	0.31250	0.43750	1.27273	5.19048	3.46246
3	9	0.14062	0.48438	0.51562	0.42188	0.85938	1.77419	5.78788	4.02348
4	4	0.06250	0.54688	0.45312	0.25000	1.10938	2.02857	6.03448	3.97657
5	8	0.12500	0.67188	0.32812	0.62500	1.73438	2.58140	6.42857	3.26296
6	12	0.18750	0.85938	0.14062	1.12500	2.85938	3.32727	7.00000	1.63013
7	9	0.14062	1.00000	0.00000	0.98438	3.84375	3.84375	–	–

So in this case, we have

$$\mu_f = \frac{1}{w_f} \left(3.84375 - \sum_{k=0}^{t-1} kp_k \right).$$

The largest value of the inter-class variance is at $k = 3$, which means that $t - 1 = 3$ or $t = 4$ is the optimum threshold.

And in fact if a histogram is drawn as shown in Figure 9.8 it can be seen that this is a reasonable place for a threshold.

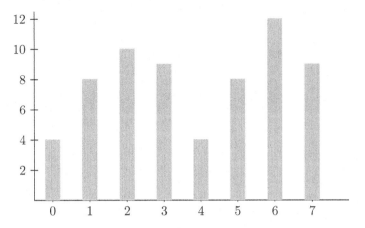

FIGURE 9.8: An example of a histogram illustrating Otsu's method

The ISODATA Method

ISODATA is an acronym for "Iterative Self-Organizing Data Analysis Technique A," where the final "A" was added, according to the original paper, "... to make ISODATA pronouncable." It is in fact a very simple method, and in practice converges very fast:

1. Pick a precision value ϵ and a starting threshold value t. Often $t = L/2$ is used.

2. Compute μ_f and μ_b for this threshold value.

3. Compute $t' = (\mu_b + \mu_f)/2$.

4. If $|t - t'| < \epsilon$ the stop and return t, else put $t = t'$ and go back to step 2.

On the cameraman image c, for example, the means can be quickly set up in MATLAB and Octave using the cumsum function, which produces the cumulative sum of a one-dimensional array, or of the columns of a two-dimensional array:

```
>> k = (0:255)';
>> n = histc(c(:),k); p = n/prod(size(c));
>> wb = cumsum(p); wf = 1-wb;
>> kpc = cumsum(k.*p);
>> mu_b = kpc./wb; mu_f = (kpc(end)-kpc)./wf;
```

MATLAB/Octave

and in Python also using its cumsum function:

```
In :   k = np.arange(256)
In :   n = ndi.histogram(c,0,255,256); p = n/(c.size+0.0)
In :   wb = np.cumsum(p); wf = 1-wb
In :   kpc = np.cumsum(k*p)
In :   mu_b = kpc/wb; mu_f = (kpc[-1]-kpc)/wf;
```

Python

Now to compute 10 iterations of the algorithm (this saves having to set up a stopping criterion with a precision value):

```
>> t = 128
>> for i = 1:10,
>     t1 = floor((mu_b(t)+mu_f(t))/2),
>     t = t1;
> end
t1 =  108
t1 =  95
t1 =  89
t1 =  88
t1 =  88
t1 =  88
t1 =  88
t1 =  88
t1 =  88
t1 =  88
```
MATLAB/Octave

```
In :  t = 128
In :  for i in range(10):
...:      t1 = int((mu_f[t]+mu_b[t])/2.0)
...:      print t1
...:      t = t1
...:
108
95
90
88
88
88
88
88
88
88
```
Python

The iteration quickly converges in only four steps.

MATLAB and Octave provide automatic thresholding with the `graythresh` function, with parameters `'otsu'` and `'intermeans'` in Octave for Otsu's method and the ISODATA method. In Python there are the methods `threshold_otsu` and `threshold_isodata` in the `filter` module of `skimage`.

Consider four images, `blood.png`, `paramecium1.png`, `pinenuts.png`, and `daisies.png` which are shown in Figure 9.9.

With the images saved as arrays `b`, `p`, `n`, and `d`, the threshold values can be computed in Octave as follows, by setting up an anonymous function to perform thresholding:

```
>> thresh = @(x) [graythresh(x,'otsu'),graythresh(x,'intermeans')];
>> ts = cellfun(@thresh, {b,p,n,d},'UniformOutput',false);
>> for i = 1:4, disp(ts{i}), end
   0.74510   0.74510
   0.75294   0.75294
   0.47451   0.47451
   0.51373   0.51373
```
Octave

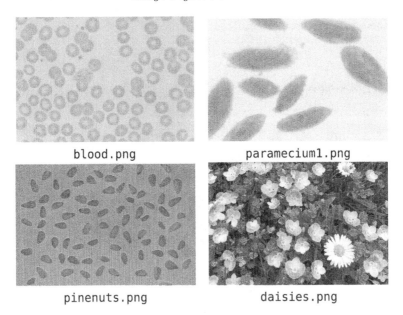

blood.png   paramecium1.png

pinenuts.png   daisies.png

FIGURE 9.9: Images to be thresholded

This is an example of putting all the images into a *cell array*, and iterating over the array using `cellfun`, using a previously described function (in this case `thresh`). A cell array is useful here as it can hold arbitrary objects.

MATLAB only supports Otsu's method:

```
>> cellfun(@(x) graythresh(x),{b,p,n,d},'UniformOutput',false)
ans =

    [0.7451]    [0.7529]    [0.4745]    [0.5137]
```
MATLAB

In Python, we can iterate over the images by putting them in a list:

```
In :  for x in [b,p,n,d]:
...:      print [fl.threshold_otsu(x), fl.threshold_isodata(x)]
...:
[190, 190]
[192, 192]
[121, 122]
[131, 131]
```
Python

Note that Python returns an 8-bit integer if that is the initial image type, whereas both MATLAB and Octave return a value of type `double`.

Once the optimum threshold value has been obtained, it can be applied to the image using `im2bw`:

```
>> imshow(im2bw(b,graythresh(b))))
```
MATLAB/Octave

or

```
In :  io.imshow(b>fl.threshold_otsu(b))
```

Python

and similarly for all the others.

The results are shown in Figure 9.10. Note that for each image the result given is quite satisfactory.

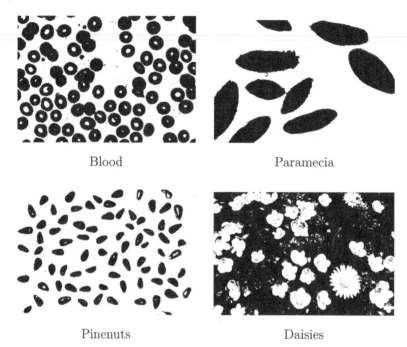

Blood   Paramecia

Pinenuts   Daisies

FIGURE 9.10: Thresholding with values obtained with Otsu's method

9.5   Adaptive Thresholding

Sometimes it is not possible to obtain a single threshold value that will isolate an object completely. This may happen if both the object and its background vary. For example, suppose we take the paramecium image and adjust it so that both the objects and the background vary in brightness across the image. This can be done in MATLAB or Octave as:

```
>> p = im2double(imread('paramecium1.png'))
>> [r,c] = size(p);
>> [y,x] = meshgrid(linspace(0,1,c),linspace(0,1,r));
>> p2 = 1-p+y/2;
>> imshow(p2)
```

MATLAB/Octave

and in Python as

```
In :   p = io.imread('paramecium1.png').astype(float)
In :   x,y = np.mgrid[0:r,0:c].astype(float)
In :   p2 = 255.0-p+y/2
In :   io.imshow(p2)
```
`Python`

Figure 9.11 shows an attempt at thresholding, using `graythresh`, with MATLAB or Octave:

```
>> t = graythresh(p2)
ans
    t =  0.43529
>> figure,imshow(im2bw(p2,t))
```
`MATLAB/Octave`

or with Python:

```
In :   t = fl.threshold_otsu(p2); t
Out:   290.7939453125
In :   io.imshow(p2>t)
```
`Python`

As you see, the result is not particularly good; not all of the objects have been isolated from their background. Even if different thresholds are used, the results are similar. To see

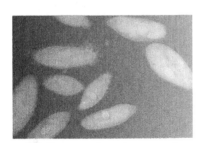

(a) Paramecium image: `p2`   (b) Thresholding attempt

FIGURE 9.11: An attempt at thresholding

why a single threshold won't work, look at a plot of pixels across the image, as shown in Figure 9.12(a).

In this figure, the line of pixels is being shown as a function; the threshold is shown on the right as a horizontal line. It can be seen that no position of the plane can cut off the objects from the background.

What can be done in a situation like this is to cut the image into small pieces, and apply thresholding to each piece individually. Since in this particular example the brightness changes from left to right, we shall cut up the image into six pieces and apply a threshold to each piece individually. The vectors `starts` and `ends` contain the column indices of the start and end of each block.

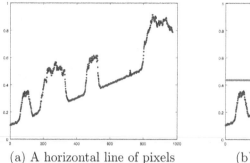

(a) A horizontal line of pixels

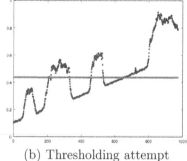

(b) Thresholding attempt

FIGURE 9.12: An attempt at thresholding—functional version

```
>> out = zeros(r,c);
>> starts = 1:c/6:c
starts =

     1   163   325   487   649   811
>> ends = 0:c/6:c
ends =

   162   324   486   648   810   972
>> for i = 1:6,
>    temp = p2(:,starts(i):ends(i))
>    out((:,starts(i):ends(i)) = im2bw(temp,graythresh(temp));
>  end
```

MATLAB/Octave

In Python, the commands are very similar:

```
In :   starts = range(0,c-1,162)
In :   ends = range(162,c+1,162)
In :   z = np.zeros((r,c))
In :   for i in range(6):
...:       temp = p2[:,starts[i]:ends[i]]
...:       z[:,starts[i]:ends[i]]=(temp>fl.threshold_otsu(temp))*1.0
...:
```

Python

Figure 9.13(a) shows how the image is sliced up, and Figure 9.13(b) shows the results after thresholding each piece. The above commands can be done much more simply by using the command **blockproc**, which applies a particular function to each block of the image. We can define our function with

```
>> thresh = @(z) im2bw(z,graythresh(z));
```

MATLAB/Octave

Notice that this uses the same as the command used above to create each piece of the previous threshold, except that now x is used to represent a general input variable.

The function can then be applied to the image **p2** with

```
>> blockproc(p2,[r,c/6],thresh);
```

MATLAB/Octave

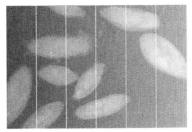

(a) Cutting up the image

(b) Thresholding each part separately

FIGURE 9.13: Adaptive thresholding

What this command means is that we apply our function **thresh** to each distinct 648×162 block of our image.

9.6 Edge Detection

Edges contain some of the most useful information in an image. We may use edges to measure the size of objects in an image; to isolate particular objects from their background; to recognize or classify objects. There are a large number of edge-finding algorithms in existence, and we shall look at some of the more straightforward of them. The general command in MATLAB or Octave for finding edges is

```
edge(image,'method',parameters...)
```

where the parameters available depend on the method used. In Python, there are many edge detection methods in the **filter** module of **skimage**. In this chapter, we shall show how to create edge images using basic filtering methods, and discuss the uses of those edge functions.

An *edge* may be loosely defined as a local discontinuity in the pixel values which exceeds a given threshold. More informally, an edge is an observable difference in pixel values. For example, consider the two 4×4 blocks of pixels shown in Figure 9.14.

51	52	53	59
54	52	53	62
50	52	53	68
55	52	53	55

50	53	155	160
51	53	160	170
52	53	167	190
51	53	162	155

FIGURE 9.14: Blocks of pixels

In the right-hand block, there is a clear difference between the gray values in the second and third columns, and for these values the differences exceed 100. This would be easily discernible in an image—the human eye can pick out gray differences of this magnitude with relative ease. Our aim is to develop methods that will enable us to pick out the edges of an image.

9.7   Derivatives and Edges

Fundamental Definitions

Consider the image in Figure 9.15, and suppose we plot the gray values as we traverse the image from left to right. Two types of edges are illustrated here: a *ramp edge*, where the

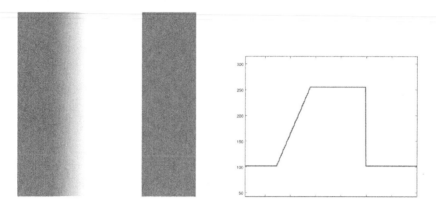

FIGURE 9.15: Edges and their profiles

gray values change slowly, and a *step edge*, or an *ideal edge*, where the gray values change suddenly.

Suppose the function that provides the profile in Figure 9.15 is $f(x)$; then its derivative $f'(x)$ can be plotted; this is shown in Figure 9.16. The derivative, as expected, returns zero

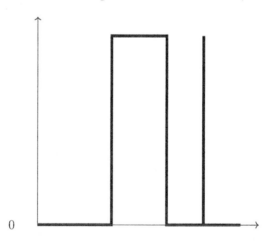

FIGURE 9.16: The derivative of the edge profile

for all constant sections of the profile, and is non-zero (in this example) only in those parts of the image in which differences occur.

Many edge finding operators are based on differentiation; to apply the continuous derivative to a discrete image, first recall the definition of the derivative:

$$\frac{df}{dx} = \lim_{h \to 0} \frac{f(x+h) - f(x)}{h}.$$

Since in an image, the smallest possible value of h is 1, being the difference between the index values of two adjacent pixels, a discrete version of the derivative expression is

$$f(x+1) - f(x).$$

Other expressions for the derivative are

$$\lim_{h \to 0} \frac{f(x) - f(x-h)}{h}, \qquad \lim_{h \to 0} \frac{f(x+h) - f(x-h)}{2h}$$

with discrete counterparts

$$f(x) - f(x-1), \qquad (f(x+1) - f(x-1))/2.$$

For an image, with two dimensions, we use partial derivatives; an important expression is the *gradient*, which is the vector defined by

$$\left[\frac{\partial f}{\partial x} \quad \frac{\partial f}{\partial y} \right]$$

which for a function $f(x, y)$ points in the direction of its greatest increase. The direction of that increase is given by

$$\tan^{-1} \left(\frac{\partial f / \partial y}{\partial f / \partial x} \right)$$

and its magnitude by

$$\sqrt{\left(\frac{\partial f}{\partial x} \right)^2 + \left(\frac{\partial f}{\partial y} \right)^2}.$$

Most edge detection methods are concerned with finding the magnitude of the gradient, and then applying a threshold to the result.

Some Edge Detection Filters

Using the expression $f(x+1) - f(x-1)$ for the derivative, leaving the scaling factor out, produces horizontal and vertical filters:

$$\begin{bmatrix} -1 & 0 & 1 \end{bmatrix} \qquad \text{and} \qquad \begin{bmatrix} -1 \\ 0 \\ 1 \end{bmatrix}$$

These filters will find vertical and horizontal edges in an image and produce a reasonably bright result. However, the edges in the result can be a bit "jerky"; this can be overcome by smoothing the result in the opposite direction; by using the filters

$$\begin{bmatrix} 1 \\ 1 \\ 1 \end{bmatrix} \qquad \text{and} \qquad \begin{bmatrix} 1 & 1 & 1 \end{bmatrix}$$

Both filters can be applied at once, using the combined filter:

$$P_x = \begin{bmatrix} -1 & 0 & 1 \\ -1 & 0 & 1 \\ -1 & 0 & 1 \end{bmatrix}$$

This filter, and its companion for finding horizontal edges:

$$P_y = \begin{bmatrix} -1 & -1 & -1 \\ 0 & 0 & 0 \\ 1 & 1 & 1 \end{bmatrix}$$

are the *Prewitt* filters for edge detection.

If p_x and p_y are the gray values produced by applying P_x and P_y to an image, then the magnitude of the gradient is obtained with

$$\sqrt{p_x^2 + p_y^2}.$$

In practice, however, it is more convenient to use either of

$$\max\{|p_x|, |p_y|\}$$

or

$$|p_x| + |p_y|.$$

This (and other) edge detection methods will be tested on the image `stairs.png`, which we suppose has been read into our system as the array `s`. It is shown in Figure 9.17. Applying each of P_x and P_y individually provides the results shown in Figure 9.18 The

FIGURE 9.17: A set of steps: A test image for edge detection

images in Figure 9.18 can be produced with the following MATLAB commands:

```
>> px = [-1 0 1;-1 0 1;-1 0 1]; py = px'
>> sx = imfilter(s,px);
>> sy = imfilter(s,py);
>> imshow(sx),figure,imshow(sy)
```

MATLAB/Octave

or these Python commands:

```
In :  sx = fl.hprewitt(s)
In :  sy = fl.vprewitt(s)
In :  io.imshow(sx)
In :  f = plt.figure(); f.show(io.imshow(sy))
```

Python

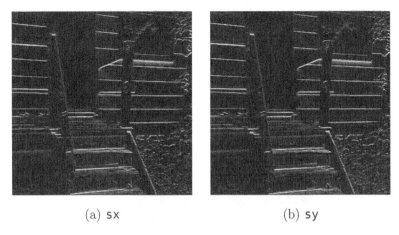

(a) sx   (b) sy

FIGURE 9.18: The result after filtering with the Prewitt filters

(There are in fact slight differences in the outputs owing to the way in which each system manages negative values in the result of a filter.) Note that the filter P_x highlights vertical edges, and P_y horizontal edges. We can create a figure containing all the edges with:

```
>> edge_p = uint8(sqrt(double(sx).^2+double(sy).^2));
>> figure,imshow(edge_p))
```
MATLAB/Octave

or

```
In :  edge_p = sqrt(sx**2+sy**2)
```
Python

and the result is shown in Figure 9.19(a). This is a grayscale image; a binary image containing edges only can be produced by thresholding. Figure 9.19(b) shows the result after thresholding with a value found by Otsu's method; this optimum threshold value is 0.3333.

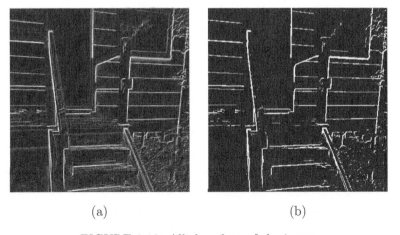

(a)   (b)

FIGURE 9.19: All the edges of the image

We can obtain edges by the Prewitt filters directly in MATLAB or Octave by using the command

```
>> edge_p = edge(s,'prewitt');
```

MATLAB/Octave

and the `edge` function takes care of all the filtering, and of choosing a suitable threshold level; see its help text for more information. The result is shown in Figure 9.20. Note that

FIGURE 9.20: The `prewitt` option of `edge`

Figures 9.19(b) and 9.20 seem different from each other. This is because the `edge` function does some extra processing over and above taking the square root of the sum of the squares of the filters.

Python does not have a function that automatically computes threshold values and cleans up the output. However, in Chapter 10, we will see how to clean up binary images.

Slightly different edge finding filters are the *Roberts cross-gradient filters*:

$$\begin{bmatrix} 1 & 0 & 0 \\ 0 & -1 & 0 \\ 0 & 0 & 0 \end{bmatrix} \text{ and } \begin{bmatrix} 0 & 1 & 0 \\ -1 & 0 & 0 \\ 0 & 0 & 0 \end{bmatrix}$$

and the *Sobel filters*:

$$\begin{bmatrix} -1 & 0 & 1 \\ -2 & 0 & 2 \\ -1 & 0 & 1 \end{bmatrix} \text{ and } \begin{bmatrix} -1 & -2 & 1 \\ 0 & 0 & 0 \\ 1 & 2 & 1 \end{bmatrix}.$$

The *Sobel filters* are similar to the Prewitt filters in that they apply a smoothing filter in the opposite direction to the central difference filter. In the Sobel filters, the smoothing takes the form

$$\begin{bmatrix} 1 & 2 & 1 \end{bmatrix}$$

which gives slightly more prominence to the central pixel. Figure 9.21 shows the respective results of the MATLAB/Octave commands

```
>> edge_r = edge(s,'roberts');
>> figure,imshow(edge_r)
```

MATLAB/Octave

and

```
>> edge_s = edge(s,'sobel');
>> figure,imshow(edge_s)
```

MATLAB/Octave

The appearance of each of these can be changed by specifying a threshold level.

(a) Roberts edge detection   (b) Sobel edge detection

FIGURE 9.21: Results of the Roberts and Sobel filters

The Python outputs, with

```
In :  edge_r = fl.roberts(s)
In :  edge_s = fl.sobel(s)
```

Python

are shown in Figure 9.22. Of the three filters, the Sobel filters are probably the best; they

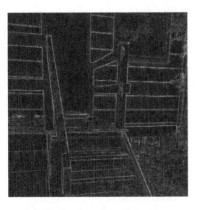

(a) Roberts edge detection   (b) Sobel edge detection

FIGURE 9.22: Results of the Roberts and Sobel filters in Python

provide good edges, and they perform reasonably well in the presence of noise.

9.8 Second Derivatives

The Laplacian

Another class of edge-detection method is obtained by considering the second derivatives. The sum of second derivatives in both directions is called the *Laplacian*; it is written as

$$\nabla^2 f = \frac{\partial^2 f}{\partial x^2} + \frac{\partial^2 f}{\partial y^2}.$$

and it can be implemented by the filter

$$\begin{bmatrix} 0 & 1 & 0 \\ 1 & -4 & 1 \\ 0 & 1 & 0 \end{bmatrix}.$$

This is known as a *discrete Laplacian*. The Laplacian has the advantage over first derivative methods in that it is an *isotropic filter* [40]; this means it is invariant under rotation. That is, if the Laplacian is applied to an image, and the image is then rotated, the same result would be obtained if the image were rotated first, and the Laplacian applied second. This would appear to make this class of filters ideal for edge detection. However, a major problem with all second derivative filters is that they are very sensitive to noise.

To see how the second derivative affects an edge, take the derivative of the pixel values as plotted in Figure 9.15; the results are shown schematically in Figure 9.23.

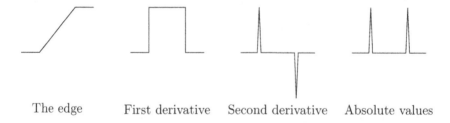

The edge   First derivative   Second derivative   Absolute values

FIGURE 9.23: Second derivatives of an edge function

The Laplacian (after taking an absolute value, or squaring) gives double edges. To see an example, suppose we enter the MATLAB/Octave commands:

```
>> lap = fspecial('laplacian',0);
>> s_lap = imfilter(s,lap);
```

MATLAB/Octave

or the Python commands

```
In :  s2 = ut.img_as_float(s)
In :  s_lap = abs(fl.laplace(s2))
```

Python

the result of which is shown in Figure 9.24.

FIGURE 9.24: Result after filtering with a discrete Laplacian

Although the result is adequate, it is very messy when compared to the results of the Prewitt and Sobel methods discussed earlier. Other Laplacian masks can be used; some are:

$$\begin{bmatrix} 1 & 1 & 1 \\ 1 & -8 & 1 \\ 1 & 1 & 1 \end{bmatrix} \text{ and } \begin{bmatrix} -2 & 1 & -2 \\ 1 & 4 & 1 \\ -2 & 1 & -2 \end{bmatrix}.$$

In MATLAB and Octave, Laplacians of all sorts can be generated using the `fspecial` function, in the form

```
fspecial('laplacian',ALPHA)
```

which produces the Laplacian

$$\frac{1}{\alpha+1} \begin{bmatrix} \alpha & 1-\alpha & \alpha \\ 1-\alpha & -4 & 1-\alpha \\ \alpha & 1-\alpha & \alpha \end{bmatrix}.$$

If the parameter `ALPHA` (which is optional) is omitted, it is assumed to be 0.2. The value 0 gives the Laplacian developed earlier.

Zero Crossings

A more appropriate use for the Laplacian is to find the *position* of edges by locating *zero crossings*. From Figure 9.23, the position of the edge is given by the place where the value of the filter takes on a zero value. In general, these are places where the result of the filter changes sign. For example, consider the the simple image given in Figure 9.25(a), and the result after filtering with a Laplacian mask in Figure 9.25(b).

We define the *zero crossings* in such a filtered image to be pixels that satisfy either of the following:

1. They have a negative gray value and are next to (by four-adjacency) a pixel whose gray value is positive

2. They have a value of zero, and are between negative and positive valued pixels

To give an indication of the way zero-crossings work, look at the edge plots and their second differences in Figure 9.26.

50	50	50	50	50	50	50	50	50	50
50	50	50	50	50	50	50	50	50	50
50	50	200	200	200	200	200	200	50	50
50	50	200	200	200	200	200	200	50	50
50	50	200	200	200	200	200	200	50	50
50	50	200	200	200	200	200	200	50	50
50	50	50	50	200	200	200	200	50	50
50	50	50	50	200	200	200	200	50	50
50	50	50	50	50	50	50	50	50	50
50	50	50	50	50	50	50	50	50	50

-100	-50	-50	-50	-50	-50	-50	-50	-50	-100
-50	0	150	150	150	150	150	150	0	-50
-50	150	-300	-150	-150	-150	-150	-300	150	-50
-50	150	-150	0	0	0	0	-150	150	-50
-50	150	-150	0	0	0	0	-150	150	-50
-50	150	-300	-150	0	0	0	-150	150	-50
-50	0	150	0	-150	0	0	-150	150	-50
-50	0	0	150	-300	-150	-150	-300	150	-50
-50	0	0	0	150	150	150	150	0	-50
-100	-50	-50	-50	-50	-50	-50	-50	-50	-100

(a) A simple image   (b) After Laplacian filtering

FIGURE 9.25: Locating zero crossings in an image

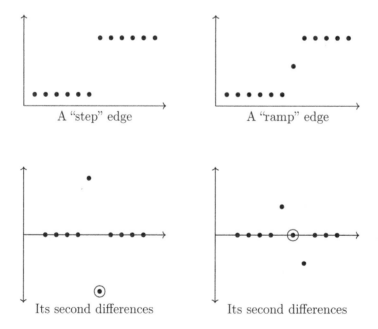

FIGURE 9.26: Edges and second differences

In each case, the zero-crossing is circled. The important point is to note that across any edge there can be only *one* zero-crossing. Thus, an image formed from zero-crossings has the potential to be very neat.

In Figure 9.25(b) the zero crossings are shaded. We now have a further method of edge detection: take the zero-crossings after a Laplacian filtering. This is implemented in MATLAB with the **zerocross** option of **edge**, which takes the zero crossings after filtering with a given filter:

```
>> lap = fspecial('laplace',0);
>> sz = edge(s,'zerocross',[],lap);
>> imshow(sz)
```
MATLAB/Octave

The result is shown in Figure 9.27(a). This is not in fact a very good result—far too many gray level changes have been interpreted as edges by this method. To eliminate them, we may first smooth the image with a Gaussian filter. This leads to the following sequence of steps for edge detection; the *Marr-Hildreth* method:

1. Smooth the image with a Gaussian filter

2. Convolve the result with a Laplacian

3. Find the zero crossings

This method was designed to provide an edge detection method to be as close as possible to biological vision. The first two steps can be combined into one, to produce a "Laplacian of Gaussian" or "LoG" filter. These filters can be created with the **fspecial** function. If no extra parameters are provided to the **zerocross** edge option, then the filter is chosen to be the LoG filter found by

```
>> fspecial('log',13,2)
```
MATLAB/Octave

This means that the following command:

```
>> edge(s,'log');
```
MATLAB/Octave

produces exactly the same result as the commands:

```
>> log = fspecial('log',13,2);
>> edge(s,'zerocross',[],log);
```
MATLAB/Octave

In fact, the **LoG** and **zerocross** options implement the same edge finding method; the difference being that the **zerocross** option allows you to specify your own filter. The result after applying an LoG filter and finding its zero crossings is given in Figure 9.27(b).

Python does not have a zero crossing detector, but it is easy to write one, and a simple one is given at the end of the chapter. Using this function, the edges can be found with

```
In :   s2 = ndi.gaussian_laplace(float64(s),3)
In :   s_edge = zerocross(s2)
```
Python

(a) Zero crossings   (b) Using an LoG filter first

FIGURE 9.27: Edge detection using zero crossings

Note that the image must be converted to type `float64` first. The filtering function applied to an image of type `uint8` will return an output of the same type, with modular "wrap around" of values outside the range 0–255. This will introduce unnecessary artifacts in the output.

9.9   The Canny Edge Detector

All the edge finding methods so far have required a straightforward application of linear filters, with the addition of finding zero crossings. All of our systems support a more complex edge detection technique, first described by John Canny in 1986 [6] and so named for him.

Canny designed his method to meet three criteria for edge detection:

1. **Low error rate of detection.** It should find *all* edges, and nothing but edges.

2. **Localization of edges.** The distance between actual edges in the image and edges found by this algorithm should be minimized.

3. **Single response.** The algorithm should not return multiple edge pixels when only a single edge exists.

Canny showed that the best filter to use for beginning his algorithm was a Gaussian (for smoothing), followed by the derivative of the Gaussian, which in one dimension is

$$\left(-\frac{x}{\sigma^2}\right)e^{-\frac{x^2}{2\sigma^2}}. \tag{9.1}$$

These have the effect of both smoothing noise and finding possible candidate pixels for edges. Since this filter is separable, it can be applied first as a column filter to the columns, next as a row filter to the rows. We can then put the two results together to form an edge image. Recall from Chapter 5 that this is more efficient computationally than applying a two-dimensional filter.

Thus at this stage we have the following sequence of steps:

1. Take our image x.

2. Create a one-dimensional Gaussian filter g.

3. Create a one-dimensional filter dg corresponding to the expression given in Equation 9.1.

4. Convolve g with dg to obtain gdg.

5. Apply gdg to x producing x1.

6. Apply gdg' to x producing x2.

We can now form an edge image with

$$\mathsf{xe} = \sqrt{\mathsf{x1}^2 + \mathsf{x2}^2}.$$

So far we have not achieved much more than we would achieve with a standard edge detection by using a filter.

The next step is that of *non-maximum suppression*. What we want to do is to threshold the edge image xe from above to keep only edge pixels and remove the others. However, thresholding alone will not produce acceptable results. The idea is that every pixel p has a direction ϕ_p (an edge "gradient") associated with it, and to be considered as an edge pixel, a p must have a greater magnitude than its neighbors in direction ϕ_p.

Just as we computed the magnitude xe above we can compute the edge gradient using the inverse tangent function:

$$\mathsf{xg} = \tan^{-1}\left(\frac{\mathsf{x2}}{\mathsf{x1}}\right).$$

In general that direction will point between pixels in its 3×3 neighborhood, as shown in Figure 9.28. There are two approaches here: we can compare the gradient of the current

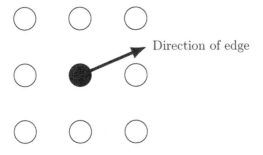

FIGURE 9.28: Non-maximum suppression in the Canny edge detector

(center) pixel with the value obtained from linear interpolation (see Chapter 6); basically we just take the weighted average of the gradients of the two pixels. So in Figure 9.28 we take the weighted average of the upper two pixels on the right.

The second approach is to quantize the gradient to one of the values $0°$, $45°$, $90°$, or $135°$, and compare the original gradient to the gradient of the pixel to which the quantized gradient points. That is, suppose the gradient at position (x, y) was $\phi(x, y)$. We quantize this to one of the four angles given to obtain $\phi'(x, y)$. Consider the two pixels in direction $\phi'(x, y)$ and $\phi'(x, y) + 180°$ from (x, y). If the edge magnitude of either of those is greater than the magnitude of the current pixel, we mark the current pixel for deletion. After we have passed over the entire image, we delete all marked pixels.

We can in fact compute the quantized gradients without using the inverse tangent: we simply compare the values in the two filter results **x1** and **x2**. Depending on the relative values of $x1(x, y)$ and $x2(x, y)$ we can place the gradient at (x, y) into one of the four gradient classes. Figure 9.29 shows how this is done. We divide the image plane into eight regions separated by lines at $45°$, as shown, with the x axis positive to the right, and the y axis positive down. Then we can assign regions, and degrees, to pixel values according to

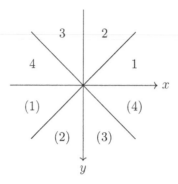

FIGURE 9.29: Using pixel locations to quantize the gradient

this table:

Region	Degree	Pixel Location
1	$0°$	$y \leq 0$ and $x > -y$
(1)	$0°$	$y \geq 0$ and $x < -y$
2	$45°$	$x > 0$ and $x \leq -y$
(2)	$45°$	$x < 0$ and $x \geq -y$
3	$90°$	$x \leq 0$ and $x > y$
(3)	$90°$	$x \geq 0$ and $x < y$
4	$135°$	$y < 0$ and $x \leq y$
(4)	$135°$	$y > 0$ and $x \geq y$

Suppose we had a neighborhood for a pixel whose gradient had been quantized to $45°$, as shown in Figure 9.30. The dashed arrow indicates the opposite direction to the current gradient. In this figure, both magnitudes at the ends of the arrows are smaller than the center magnitude, so we keep the center pixel as an edge pixel. If, however, one of those two values was *greater* than 150, we would mark the central pixel for deletion.

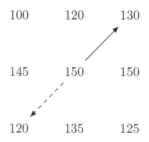

FIGURE 9.30: Quantizing in non-maximum suppression

After performing non-maximum suppression, we can threshold to obtain a binary edge image. Canny proposed, rather than using a single threshold, a technique called *hysteresis thresholding*, which uses two threshold values: a low value t_L, and a high value t_H. Any pixel with a value greater than t_H is assumed to be an edge pixel. Also, any pixel with a value p where $t_L \leq p \leq t_H$ and which is adjacent to an edge pixel is also considered to be an edge pixel.

The Canny edge detector is implemented by the `canny` option of the `edge` function in MATLAB and Octave; and in Python by the `canny` method in the `filter` module of `skimage.` by the `canny` option of the `edge` function.

MATLAB and Octave differ slightly in their parameters, but the results are very similar. With no choosing of parameters, and just

```
>> edge(s,'canny')
```
MATLAB/Octave

the result is shown in Figure 9.31(a). Python's syntax is

```
In :  fl.canny(s)
```
Python

and the result is shown in Figure 9.31(b) Two other results with different thresholds and

(a) MATLAB/Octave `canny`   (b) Python `canny`

FIGURE 9.31: Canny edge detection

Gaussian spreads are given in Figure 9.32. The higher we make the upper threshold, the less edges will be shown. We can also vary the standard deviation of the original Gaussian filter.

The Canny edge detector is the most complex of the edge detectors we have discussed; however, it is not the last word in edge detectors. A good account of edge detectors is given by Parker [32], and an interesting account of some advanced edge detection techniques can be found in Heath et al. [16].

9.10 Corner Detection

Second only to edges as means for identifying and measuring objects in images are corners, which may be loosely defined as a place where there are two edges in significantly different directions. An edge may have a very slight bend in it—a matter of only a few degrees—but that won't qualify as a corner.

$t_L = 2, t_H = 10, \sigma = 0.1$ $t_L = 0, t_H = 4, \sigma = 2$

FIGURE 9.32: Canny edge detection with different thresholds

As with edges, there are many different corner detectors; we will look at just two.

Moravec Corner Detection

This is one of the earliest and simplest detectors: fundamentally, a corner is identified as a pixel whose neighborhood is significantly different from each other local neighborhood. It can be described as a sequence of steps:

1. We suppose we are dealing with a square "window" (like a mask), of odd dimensions, placed over the current pixel p; suppose that W is the array of neighboring pixels surrounding p.

2. Move this mask one pixel in each of the eight directions from p.

3. For each shift $s = (i, j)$ compute the sum of squared differences:

$$I_s = \sum (W(x, y) - W_s(x, y))^2$$

This sum is called the *Intensity Variation*.

4. Compute the minimum M of all the I_s values.

If p is in a region with not much difference in any direction, the values I_s would be expected to be small. A corner is a place where M is maximized.

To implement this detector in MATLAB or Octave is fairly straightforward. Pad the image by zeros, and on this larger image plane move it in each direction, compute the squared distances, and sum them with a linear filter. For example:

```
>> s = im2double(imread('stairs.png'));
>> [r,c] = size(s);
>> dirs = [1 1;1 0;-1 0;0 1;0 -1;-1 -1;1 -1;-1 1];
>> sp = padarray(s,[1,1]);
>> z = zeros(r+2,c+2,8);
>> w = zeros(r+2,c+2,8);
>> for i = 1:8
>     z(2+dirs(i,1):1+dirs(i,1)+r,2+dirs(i,2):1+dirs(i,2)+c,:) = s;
>     w(:,:,i) = imfilter((z(:,:,i)-sp).^2,ones(3,3));
>   end
>> corners = min(w,3);
```

MATLAB/Octave

To show the corners, create an image that includes a darkened version of the original added to a brightened version of the corners array:

```
>> imshow(s/4+4*corners(2:r+1,2:c+1))
```

MATLAB/Octave

and the result is shown in Figure 9.33.

FIGURE 9.33: Moravec corner detection

Python has a `corner_moravec` method in `skimage.feature`; however, this implementation produces different results to the method given above. For example, suppose we take a simple image with just one corner, as shown in Figure 9.10. The MATLAB/Octave result and the Python results are shown in Figure 9.35. It can be seen that the Python results would require further processing so as not to incorrectly classify internal pixels of a region as corner points. Our very naive implementation above also incorrectly classifies points at the edge of the image as corner points, but these can be easily removed.

The Moravec algorithm is simple and fast, but its main drawback is that it cannot cope with edges not in one of the eight directions. If there is an edge at an angle, for example, of 22.5°, then one of the local intensity variations will be large, and will incorrectly identify a corner. The Moravec detector is thus *non-isotropic* as it will produce different results for different angles.

```
50  50  50  50  50   50   50   50   50   50
50  50  50  50  50   50   50   50   50   50
50  50  50  50  50   50   50   50   50   50
50  50  50  50  50   50   50   50   50   50
50  50  50  50  150  150  150  150  150  150
50  50  50  50  150  150  150  150  150  150
50  50  50  50  150  150  150  150  150  150
50  50  50  50  150  150  150  150  150  150
50  50  50  50  150  150  150  150  150  150
50  50  50  50  150  150  150  150  150  150
```

FIGURE 9.34: A simple image with one corner

```
5  0  0   0   0  0  0  0  0  5
0  0  0   0   0  0  0  0  0  0
0  0  0   0   0  0  0  0  0  0
0  0  0  10  10  0  0  0  0  20
0  0  0  10  20  0  0  0  0  20
0  0  0   0   0  0  0  0  0  0
0  0  0   0   0  0  0  0  0  0
0  0  0   0   0  0  0  0  0  0
0  0  0   0   0  0  0  0  0  0
5  0  0  20  20  0  0  0  0  45
```

```
0  0  0   0   0   0   0   0  0  0
0  0  0   0   0   0   0   0  0  0
0  0  0   0   0   0   0   0  0  0
0  0  0  10  20  30  30  30  0  0
0  0  0  20  30  30  30  30  0  0
0  0  0  30  30   0   0   0  0  0
0  0  0  30  30   0   0   0  0  0
0  0  0  30  30   0   0   0  0  0
0  0  0   0   0   0   0   0  0  0
0  0  0   0   0   0   0   0  0  0
```

Corners in MATLAB/Octave

Corners in Python

FIGURE 9.35: Moravec detection in MATLAB, Octave, and Python compared

The Harris-Stephens Detector

This corner detector is often called simply the Harris corner detector, and it was designed to alleviate some of the problems with Moravec's method. It is based in part on using the first order Taylor approximation for a function of two variables:

$$f(x+h, y+k) \approx f(x,y) + h\frac{\partial f}{\partial x}(x,y) + k\frac{\partial f}{\partial y}(x,y).$$

Suppose the derivatives in the x and y directions are computed using a linear filter.

As with the Moravec detector, the Harris detector computes the squared differences with a neighborhood and its shift. For a shift (s,t) then the sum of squares, over a mask K is:

$$\sum_{(u,v)\in K} (I(u+s, v+t) - I(u,v))^2.$$

Using the Taylor approximation above, this can be written as

$$\sum_{(u,v)\in K} (I(u,v) + sI_x(u,v) + tI_y(u,v) - I(u,v))^2$$

$$= \sum_{(u,v)\in K} (sI_x(u,v) + tI_y(u,v))^2$$

$$= \sum_{(u,v)\in K} (s^2 I_x^2 + 2st I_x I_y + t^2 I_y)$$

$$= \sum_{(u,v)\in K} \begin{bmatrix} s & t \end{bmatrix} \begin{bmatrix} I_x^2 & I_x I_y \\ I_x I_y & I_y^2 \end{bmatrix} \begin{bmatrix} s // t \end{bmatrix}$$

The Harris method looks at the matrix

$$H = \begin{bmatrix} I_x^2 & I_x I_y \\ I_x I_y & I_y^2 \end{bmatrix}$$

For a given (s,t), the expression

$$s^2 I_x^2 + 2st I_x I_y + t^2 I_y = c$$

for a constant c has the shape of an ellipse. The length of its axes a and b and their direction are given by the *eigenvalues* λ_1 and λ_2 and *eigenvectors* $\mathbf{v}_1$ and $\mathbf{v}_2$ of the matrix H as shown in Figure 9.36. These satisfy

$$H\mathbf{v}_i = \lambda_i \mathbf{v}_i.$$

In particular, it can be shown that

$$a = (\lambda_1)^{-1/2}, b = (\lambda_2)^{-1/2}.$$

Since the squares a^2 and b^2 of the axes lengths are inversely proportional to the eigenvalues, it follows that the *slowest* change of intensity is in the direction of the *largest* eigenvalue, and the fastest change in the direction of the smallest.

Given the eigenvalues then, we distinguish three cases:

1. Both are large. This corresponds to not much change in any direction, or a low frequency component of the image.

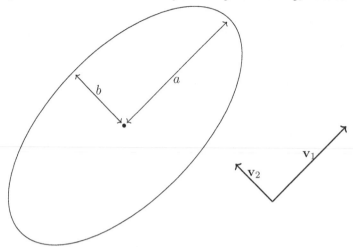

FIGURE 9.36: The ellipse associated with the Harris matrix H

2. One is large and the other is small. This corresponds to an edge in that direction.

3. Both are small. This corresponds to change in both directions, so it may be considered a corner.

For a general symmetric matrix

$$M = \begin{bmatrix} x & y \\ y & z \end{bmatrix}$$

the eigenvalues can be found to be the solutions of the equation

$$\lambda^2 - (x + z)\lambda + (xz - y^2) = 0$$

where $x + z = \text{Tr}(M)$, the *trace* of M, and $xz - y^2 = \det(M)$, the *determinant* of M. However, the solution of this equation will involve square roots, which are computationally expensive operations. Harris and Stephens suggested a simpler alternative: for a previously chosen value of k, compute

$$R = \det(M) - k(\text{Tr}(M))^2.$$

Corners will correspond to large values of R.

   The mathematics may seem complicated, but in fact the implementation is very simple. It consists of only a few steps:

1. For the given image I, compute the edge images I_x and I_y in the x and y directions. This can be done using standard linear filters.

2. Over a given mask, compute the sums

$$S = \sum I_x^2, \quad T = \sum I_x I_y, \quad U = \sum I_y^2.$$

3. Compute $R = (SU - T^2) - k(S + U)^2$.

To ensure isotropy, and to make the corner detector more robust in the presence of noise, Harris and Stephens recommended that in Step 2, instead of merely adding values over the mask, that a Gaussian filter be applied instead. Thus, the following variant would be used:

2'. Over a given mask with a Gaussian filter G, compute the convolutions

$$S = (I_x^2) * G, \quad T = (I_x I_y) * G, \quad U = (I_y^2) * G.$$

MATLAB contains a Harris-Stephens implementation in the **corner** function, and Python with the **corner_harris** method in the **skimage.feature** module.

So for Octave, here is how we might perform corner detection on the stairs image **s**, which we assume to be of data type **double**. First compute the derivatives (edges) and their products:

```
>> Ix = imfilter(s,[1 0 -1]);
>> Iy = imfilter(s,[1; 0; -1]);
>> Ixx = sx.*sx;
>> Ixy = sx.*sy;
>> Iyy = sy.*sy;
```
`Octave`

The next step is to convolve these either with a 3×3 sum filter, or its variant, a Gaussian filter. Take the first, simple option:

```
>> S = imfilter(Ixx,ones(3))
>> T = imfilter(Ixy,ones(3))
>> U = imfilter(Iyy,ones(3))
```
`Octave`

Now to compute the corner image, with a standard value $k = 0.05$.

```
>> k = 0.05
>> R = (S.*U-T.^2)-k*(S+U).^2;
```
`Octave`

This image R can be used directly by simple thresholding:

```
>> imshow(s/4+(R>k)*1)
```
`Octave`

and this is shown in Figure 9.37(a). However, non-maximum suppression, as used in the Canny edge detector, can be used to ensure that any single corner is represented by only one pixel. In this case, simply remove pixels which are not the largest in their neighborhood:

```
>> Rmax = ordfilt2(R,9,ones(3));
>> Rs = (Rmax==R).*(R>k)*1;
>> imshow(s/4+Rs)
```
`Octave`

and this is shown in Figure 9.37(b).

(a) With thresholding   (b) With non-maximum suppression

FIGURE 9.37: Harris-Stephens corner detection

9.11   The Hough and Radon Transforms

These two transforms can be used to find lines in an image. Although they seem very different on initial explanation, they provide very similar information. The Hough transform is possibly more efficient and faster; whereas the Radon transform is mathematically better behaved—in its continuous form the Radon transform is fully invertible, and in its discrete form (as in images) there are many different approximations to the inverse which can nearly completely recover the initial image.

Hough transform. First note that on the Cartesian plane a general line can be represented in the form

$$x \cos \theta + y \sin \theta = r$$

where r is the perpendicular distance to the origin, and θ is the angle of that perpendicular to the positive x axis, as shown in Figure 9.38. Note that the vector $\overrightarrow{OA}$ has direction $\langle \cos \theta, \sin \theta \rangle$ and the vector $\overrightarrow{AX} = \langle x - r \cos \theta, y - r \sin \theta \rangle$. Since these vectors are perpendicular, their dot product is zero, which means that

$$\cos \theta (x - r \cos \theta) + \sin \theta (y - r \sin \theta) = 0.$$

This can be rewritten as

$$x \cos \theta - r \cos^2 \theta + y \sin \theta - r \sin^2 \theta = 0$$

or

$$x \cos \theta + y \sin \theta = r.$$

This parameterization includes perpendicular lines, which cannot be represented in the more usual $y = ax + b$ form.

On a binary image, for every foreground point (x, y) consider the values of $x \cos \theta + y \sin \theta$ for a set of values of θ. One standard is to choose $0 \leq \theta \leq 179°$ in steps of $1°$. It will be found that many different values of (x, y) will give rise to the same $(r\theta)$ pairs. The pair (r, θ) corresponding to the greatest number of (x, y) points corresponds to the "strongest line"; that is, the line with the greatest number of points in the image.

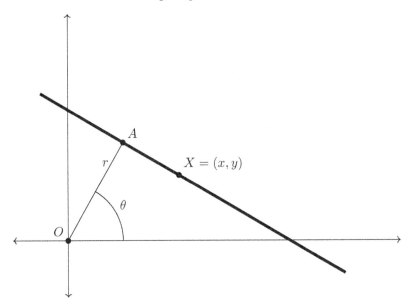

FIGURE 9.38: A line and its parameters

For example, consider the small image with just six foreground points as shown in Figure 9.39, and with the angle set $\theta = 0, 45, 90.135$.

The values of $x \cos\theta + y \sin\theta$ are:

x	y	0	45	90	135
1	2	1	2.12132	2	0.70711
3	1	3	2.82843	1	−1.41421
2	2	2	2.82843	2	0
2	3	2	3.53553	3	0.70711
3	4	3	4.94975	4	0.70711
4	4	4	5.65685	4	0

Notice that the value $r = 0.7071$ (that is, $r = 1/\sqrt{2}$) occurs most often, in the last column. This means that the strongest line has values $(r, \theta) = (1/\sqrt{2}, 135)$ which has standard Cartesian form $y = x + 1$. Value (x, y) are transformed by this procedure into (r, θ) values. We can thus set up an (r, θ) array where each value corresponds to the number of times it appears in the initial computation. For the previous table, the corresponding array is

$r =$	−1.41	0	0.71	1	2	2.12	2.83	3	3.54	4	4.95	5.66
$\theta =$ 0	0	0	0	1	2	0	0	2	0	1	0	0
45	0	0	0	0	0	1	2	0	1	0	1	1
90	0	0	0	1	2	0	0	1	0	2	0	0
135	1	2	3	0	0	0	0	0	0	0	0	0

This array is known as the *accumulator array*, and its largest values correspond to the strongest line. The *Hough transform*[1] thus transforms a Cartesian array to an (r, θ) array, from which the lines in the image can be determined.

---

[1] "Hough" is pronounced "Huff."

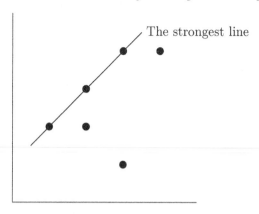

FIGURE 9.39: A simple Hough example

A quick-and-dirty, if not particularly efficient, Hough transform can be implemented by following the description above, starting with a binary image:

```
>> e = edge(c,'canny');
>> [x,y] = find(e==1);
>> th = 0:179; cos_th=cosd(th);sin_th = sind(th);
>> rtable = floor(x*cos_th+y*sin_th);
```
MATLAB/Octave

At this stage we have a table of value of r. Note that in the last command matrix products are being computed: if there are N foreground pixels in the binary image, then each product is formed from an $N \times 1$ matrix and an 1×180 matrix, resulting in an $N \times 180$ matrix. To form the accumulator array, find all the different values in the table:

```
>> u = unique(rtable);
>> N = length(u);
```
MATLAB/Octave

and then count them:

```
>> for j = 1:180,
>    for i = 1:N,
>      hough(i,j) = length(find(rtable(:,j)==u(i)));
>    end;
>  end
```
MATLAB/Octave

In Python, we need a little more care to ensure that the correct variable type is being used–list or array:

```
In :   e = fl.canny(c,2,5)
In :   x,y = np.nonzero(e)
In :   xt = np.array([x]).T
In :   yt = np.array([y]).T
In :   th = linspace(0,np.pi,180)
In :   cos_th = np.array([np.cos(th)])
In :   sin_th = np.array([np.sin(th)])
In :   rtable = np.floor(xt.dot(cos_th)+yt.dot(sin_th))
```
Python

Recall that in Python a single asterisk performs element-by-element multiplication; standard matrix products are performed with the **dot** method applied to a matrix. The **T** method is matrix transposition. Now for the accumulator array, using the **unique** function, which lists the unique elements in an array:

```
In :   u = np.unique(rtable)
In :   N = len(u)
In :   hough = np.zeros((N,180))
In :   for j in range(180):
...:       for i in range(N):
...:           hough[i,j] = np.where(rtable[:,j]==u[i])[0].shape[0]
...:
```
Python

This final array is the Hough transform of the initial binary image; it is shown (rotated 90° to fit) in Figure 9.40. The maximum value in this transform will correspond to the strongest

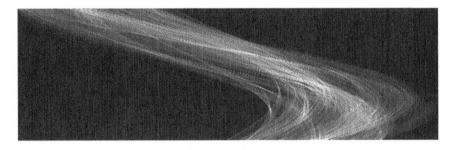

FIGURE 9.40: A Hough transform

line in the image. Obtaining this line will be discussed at the end of the next section.

The Radon transform. Whereas the Hough transform works point by point, the Radon transform works across the entire image. Given an angle θ, consider all the lines at that angle which pass across the image, as shown in Figure 9.41.

As the lines pass through the image, add up the pixel values along each line, to accumulate in a new line. The *Radon transform* is the array whose rows are the values along these new lines.

The mathematical elegance of the Radon transform is partly due to a result called the *Fourier slice theorem*, which claims that the Fourier transform of a line for an angle θ is equal to the line (a "slice") through the Fourier transform at that angle. To see this, take an image and its Fourier and Radon transforms. In MATLAB/Octave:

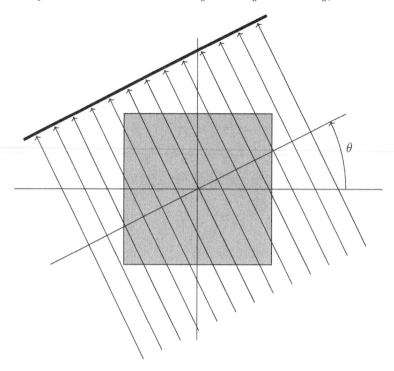

FIGURE 9.41: The Radon transform

```
>> c = imread('cameraman.png');
>> cf = fftshift(fft2(c));
>> cr = radon(c);
```
MATLAB/Octave

and in Python we can use the **radon** function from the **transform** module of **skimage**:

```
In :   from numpy.fft import fft, fft2, fftshift
In :   c = io.imread('cameraman.png')
In :   cf = fftshift(fft2(c))
In :   cr = tr.radon(float64(c),range(180))
```
Python

Pick an angle, say $\theta = 30$. Given that the first row of the Radon transform corresponds to $\theta = 0$, the thirty-first row will correspond to $\theta = 30$. Then the Fourier transform of that row can be obtained, scaled, and plotted with

```
>> c30 = fftshift(fft(cr(:,31)));
>> c30l = log(1+abs(c30));
>> plot(c30l)
```
MATLAB/Octave

or with

```
In :   c30 = fftshift(fft(cr[:,30]))
In :   c30l = np.log(1+np.abs(c30))
In :   plt.plot(c30l)
```
Python

To obtain the Fourier slice, the easiest way is to rotate the transform of the image and then read off the center row:

```
>> cl = log(1+abs(cf));
>> cl30 = imrotate(cl,30);
>> [rs,cs] = size(cl30);
>> plot(cl30(floor(rs/2),:))
```
MATLAB/Octave

or with

```
In :   cl = np.log(1+np.abs(cf))
In :   cl30 = tr.rotate(cl,30,resize="True",order=3)
In :   rs = cl30.shape[0]
In :   plt.plot(cl30[rs/2,''])
```
Python

The results are shown in Figure 9.42. The plots are remarkably similar, with differences accounted for by rounding errors and interpolation when rotating. This means, at least

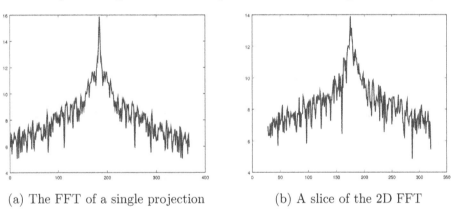

(a) The FFT of a single projection   (b) A slice of the 2D FFT

FIGURE 9.42: The Fourier slice theorem

in theory, a Radon transform can be reversed. For each projection in the transform corresponding to an angle θ, take its Fourier transform and add it into an array as a line at angle θ. The final array may be consider as the Fourier transform of the initial image, and so inverting it will provide the image.

With continuous functions the Radon transform—considered an integral transform— is indeed invertible. For images, being discrete objects, the transform is not completely invertible. However, it is possible to obtain a very good approximation of the image by using a similar method. An important consideration is preventing the DC coefficient from overpowering the rest of the result; this is usually managed by some filtering.

Finding lines using the Radon transform. To find the strongest lines, first create a binary edge image, and compute its Radon transform. This can be done in MATLAB/Octave as

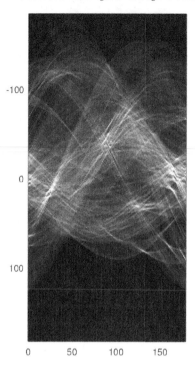

FIGURE 9.43: The Radon transform of the edges of an image

```
>> e  = edge(c,'canny');
>> angles = 0:179;
>> [rad,x] = radon(e,angles);
```

MATLAB/Octave

or in Python as

```
In :  e = fl.canny(c)
In :  rad = tr.radon(e)
```

Python

The extra output parameter x in the MATLAB/Octave implementation gives the x-axis of the rotated coordinate. The result can be displayed using `imagesc`, which displays the image as a plot:

```
>> imagesc(angles,x,rad),colormap(gray(256))
```

MATLAB/Octave

and is shown in Figure 9.43. The strongest line will be the point in the transform of greatest value:

```
>> [r,theta] = find(rad==max(rad(:)))
r =  200
theta =  24
```

MATLAB/Octave

The corresponding x value can be found using the x array:

```
>> x(r)
ans =
      16
```

If the image is centered about the origin, then the strongest line will be

$$x \cos(24) + y \sin(24) = 16$$

and this is shown superimposed on the original image in Figure 9.44, along with an enlarged view of the center.

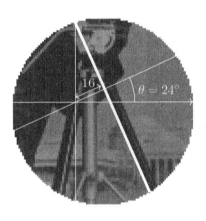

FIGURE 9.44: Finding a line in an image

Exercises

Thresholding

1. Suppose you thresholded an image at value t_1, and thresholded the result at value t_2. Describe the result if

   (a) $t_1 > t_2$

   (b) $t_1 < t_2$

2. Create a simple image with

   ```
   >> [x,y] = meshgrid(1:256,1:256);
   >> z = sqrt((x-128).^2+(y-128).^2);
   >> z2 = 1-mat2gray(z);
   ```

or with

```
In :  x,y = np.mgrid[0:256,0:256].astype(float)
In :  z = np.sqrt((x-128)**2+(y-128)**2)
In :  z2 = z.max()-z
```

`Python`

Threshold `z2` at different values, and comment on the results. What happens to the amount of white as the threshold value increases? Can you state and prove a general result?

3. Repeat the above question, but with the image `cameraman.png`.

4. Can you create a small image that produces an "X" shape when thresholded at one level, and a cross shape "+" when thresholded at another level?

   If not, why not?

5. Superimpose the image `nicework.png` onto the image `cameraman.png`. You can do this with:

```
>> n = im2uint8(imread('nicework.png'));
>> c = imread('cameraman.png');
>> m = imlincomb(0.5,c,0.5,n);
```

`MATLAB/Octave`

or with

```
In :  c = io.imread('cameraman.png')
In :  n = io.imread('nicework.png')
In :  m = c//2+n//2
```

`Python`

(Look up `imlincomb` to see what it does.) Can you threshold this new image `m` to isolate the text?

6. Try the same problem as above, but define `m` as:

```
>> m  = c.*(n==0);
```

7. Create a version of the circles image with

```
>> t = imread('circles.png');
>> [x,y] = meshgrid(1:256,1:256);
>> t2 = double(t).*(x+y)/512;
>> t3 = im2uint8(t2);
```

`MATLAB/Octave`

or

```
In :   t = io.imread('circles.png')
In :   x,y = np.mgrid[0:256,0:256].astype(float)
In :   t2 = t.astype(float)*(x+y)/2
In :   t3 = (t2/255).astype('uint8')
```

`Python`

Attempt to threshold the image t3 to obtain the circles alone, using adaptive thresholding (and if you are using MATLAB or Octave, the blkproc function). What sized blocks produce the best result?

Edge Detection

8. Enter the following matrix using either

```
>> im = 200*ones(10,10);im(3:5,3:8)=50;im(6:8,4:7)=50;
>> im = im + round(9*randn(10,10))
```

`MATLAB/Octave`

or

```
In :   im = 200*np.ones((10,10))
In :   im[2:5,2:8]=50;im[5:8,3:7]=50
In :   im = (im + np.round(8*sc.randn(10,10))).astype('uint8')
```

`Python`

This will create something like this:

201	195	203	203	199	200	204	190	198	203
201	204	209	197	210	202	205	195	202	199
205	198	46	60	53	37	50	51	194	205
208	203	54	50	51	50	55	48	193	194
200	193	50	56	42	53	55	49	196	211
200	198	203	49	51	60	51	205	207	198
205	196	202	53	52	34	46	202	199	193
199	202	194	47	51	55	48	191	190	197
194	206	198	212	195	196	204	204	199	200
201	189	203	200	191	196	207	203	193	204

and use the appropriate filter to apply each of the Roberts, Prewitt, Sobel, Laplacian, and zero-crossing edge-finding methods to the image. In the case of applying two filters (such as with Roberts, Prewitt, or Sobel) apply each filter separately, and join the results.

Apply thresholding if necessary to obtain a binary image showing only the edges.

Which method seems to produce the best results?

9. If you are using MATLAB or Octave, apply the edge function with each of its possible parameters in turns to the array above.

Which method seems to produce the best results?

10. Open up the image cameraman.png in MATLAB, and apply each of the following edge-finding techniques in turn:

(a) Roberts

(b) Prewitt

(c) Sobel

(d) Laplacian

(e) Zero-crossings of a Laplacian

(f) The Marr-Hildreth method

(g) Canny

Which seems to you to provide the best looking result?

11. Repeat the above exercise, but use the image `arch.png`.

12. Obtain a grayscale flower image with:

```
ir = imread('iris.png');
i = rgb2gray(ir);
```

MATLAB/Octave

or

```
In :  ir = io.imread('iris.png')
In :  i = co.rgb2gray(ir)
```

Python

Now repeat Question 10.

13. Pick a grayscale image, and add some noise to it using the commands introduced in Section 8.2. Create two images: `c1` corrupted with salt and pepper noise, and `c2` corrupted with Gaussian noise.

Now apply the edge finding techniques to each of the "noisy" images `c1` and `c2`.

Which technique seems to give

(a) The best results in the presence of noise?

(b) The worst results in the presence of noise?

The Hough Transform

14. Write the lines $y = x - 2$, $y = 1 - x/2$ in (r, θ) form.

15. Use the Hough transform to detect the strongest line in the binary image shown below. Use the form $x \cos \theta + y \sin \theta = r$ with θ in steps of $45°$ from $-45°$ to $90°$ and place the results in an accumulator array.

x

	−3	−2	−1	0	1	2	3
−3	0	0	0	0	0	1	0
−2	0	0	0	0	0	0	0
−1	0	1	0	1	0	1	0
0	0	0	1	0	0	0	0
1	0	0	0	0	0	0	0
2	1	0	0	0	0	1	0
3	0	0	0	0	0	0	0

(y labels the rows)

16. Repeat the previous question with the images:

y \ x	-3	-2	-1	0	1	2	3
-3	0	0	0	0	1	0	0
-2	0	0	0	0	0	0	0
-1	0	0	1	0	0	0	1
0	0	1	0	0	1	0	0
1	1	0	0	0	0	0	1
2	0	0	0	0	1	0	0
3	0	0	0	1	1	0	0

y \ x	-3	-2	-1	0	1	2	3
-3	0	0	0	1	0	0	0
-2	1	0	0	0	1	0	0
-1	0	0	0	0	0	0	0
0	0	0	1	0	0	1	0
1	0	1	0	1	0	0	0
2	1	0	0	0	0	0	1
3	0	0	1	0	0	1	0

17. Find some more lines on the cameraman image, using the Radon transform method of implementing the Hough transform.

18. Read and display the image `stairs.png`.

    (a) Where does it appear that the "strongest" lines will be?

    (b) Plot the five strongest lines.

FIGURE 1.18: A true color image

FIGURE 1.19: An indexed color image

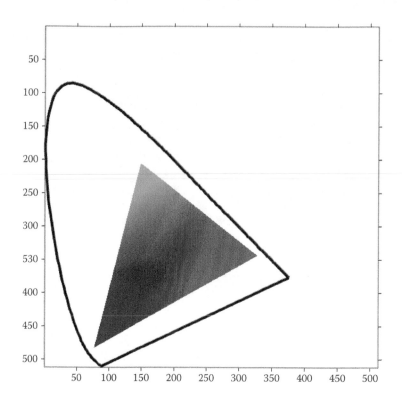

FIGURE 13.4: The RGB gamut

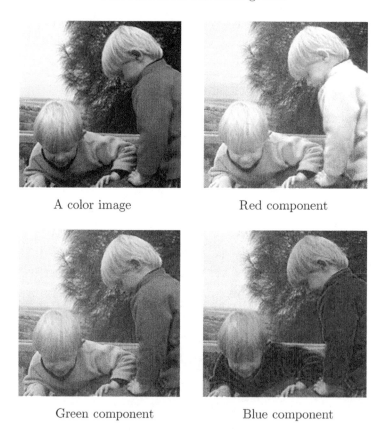

A color image

Red component

Green component

Blue component

FIGURE 13.8: An RGB color image and its components

(a)   (b)

FIGURE 13.13: Applying a color map to a grayscale image

FIGURE 13.14: An image colored with a "handmade" color map

(a) The original image   (b) Attempt at equalization

FIGURE 13.16: An attempt at color contrast enhancement

FIGURE 13.17: A better attempt at color contrast enhancement

(a) The original image

(b) Processing each RGB

(c) Processing Y only

FIGURE 13.18: Another example of color contrast enhancement

(a) The original image

(b) Processing each RGB

(c) Processing Y only

FIGURE 13.19: Unsharp masking of a color image

(a) The original image

(b) Processing each RGB

(c) Processing Y only

FIGURE 13.20: Kuwahara filtering of a color image

Salt and pepper noise · · · · · · · · · · The red component

The green component · · · · · · · · · · The blue component

FIGURE 13.21: Noise on a color image

(a) Denoising each RGB component · · · · (b) Denoising intensity only

FIGURE 13.22: Attempts at denoising a color image

FIGURE 13.25: The botanic gardens, Melbourne, Australia

(a) Applying retinex to each of RGB

(b) Applying retinex to Y only

FIGURE 13.26: Center/surround retinex on a color image

(a) twirl

(b) ripple

FIGURE 16.11: Effects on a color image

Chapter 10

Mathematical Morphology

10.1 Introduction

Mathematical morphology, or *morphology* for short, is a branch of image processing that is particularly useful for analyzing shapes in images. We shall develop basic morphological tools for investigation of binary images, and then show how to extend these tools to grayscale images. MATLAB has many tools for binary morphology in the image processing toolbox; most of which can be used for grayscale morphology as well.

10.2 Basic Ideas

The theory of mathematical morphology can be developed in many different ways. We shall adopt one standard method which uses operations on sets of points. A very solid and detailed account can be found in Haralick and Shapiro [14].

Translation

Suppose that A is a set of pixels in a binary image, and $w = (x, y)$ is a particular coordinate point. Then A_w is the set A "translated" in direction (x, y). That is

$$A_x = \{(a, b) + (x, y) : (a, b) \in A\}.$$

For example, in Figure 10.1, A is the cross-shaped set and $w = (2, 2)$. The set A has been

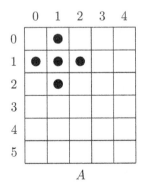
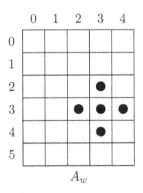

FIGURE 10.1: Translation

shifted in the x and y directions by the values given in w. Note that here we are using matrix coordinates, rather than Cartesian coordinates, so that the origin is at the top left, x goes down and y goes across.

Reflection

If A is a set of pixels, then its *reflection*, denoted $\hat{A}$, is obtained by reflecting A in the origin:

$$\hat{A} = \{(-x, -y) : (x, y) \in A\}.$$

For example, in Figure 10.2, the open and closed circles form sets which are reflections of each other.

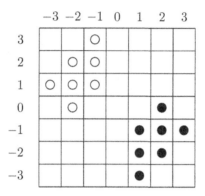

FIGURE 10.2: Reflection

10.3   Dilation and Erosion

These are the basic operations of morphology, in the sense that all other operations are built from a combination of these two.

Dilation

Suppose A and B are sets of pixels. Then the *dilation of A by B*, denoted $A \oplus B$, is defined as

$$A \oplus B = \bigcup_{x \in B} A_x.$$

What this means is that for every point $x \in B$, we translate A by those coordinates. Then we take the union of all these translations.

An equivalent definition is that

$$A \oplus B = \{(x, y) + (u, v) : (x, y) \in A, (u, v) \in B\}.$$

From this last definition, dilation is shown to be commutative; that is,

$$A \oplus B = B \oplus A.$$

An example of a dilation is given in Figure 10.3. In the translation diagrams, the gray squares show the original position of the object. Note that $A_{(0,0)}$ is of course just A itself. In this example, we have

$$B = \{(0,0), (1,1), (-1,1), (1,-1), (-1,-1)\}$$

and these are the coordinates by which we translate A. In general, $A \oplus B$ can be obtained

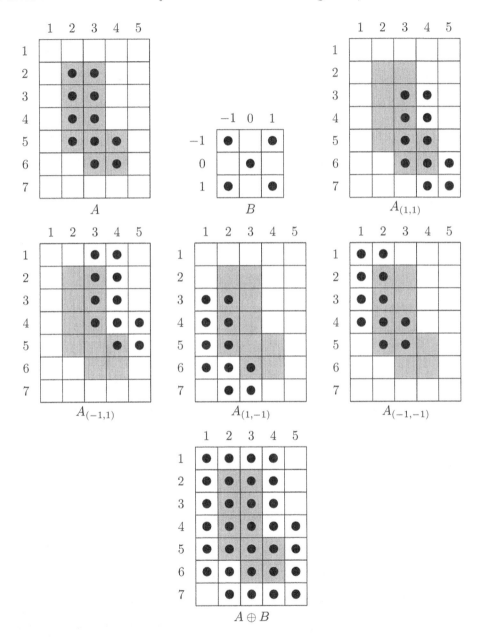

FIGURE 10.3: Dilation

by replacing every point (x, y) in A with a copy of B, and placing the $(0,0)$ point of B at (x, y). Equivalently, we can replace every point (u, v) of B with a copy of A.

Dilation is also known as *Minkowski addition*; see Haralick and Shapiro [14] for more information.

As you see in Figure 10.3, dilation has the effect of increasing the size of an object. However, it is not necessarily true that the original object A will lie within its dilation $A \oplus B$. Depending on the coordinates of B, $A \oplus B$ may end up quite a long way from A. Figure 10.4 gives an example of this: A is the same as in Figure 10.3; B has the same shape but a different position. In this figure, we have

$$B = \{(7,3), (6,2), (6,4), (8,2), (8,4)\}$$

so that

$$A \oplus B = A_{(7,3)} \cup A_{(6,2)} \cup A_{(6,4)} \cup A_{(8,2)} \cup A_{(8,4)}.$$

For dilation, we generally assume that A is the image being processed, and B is a small

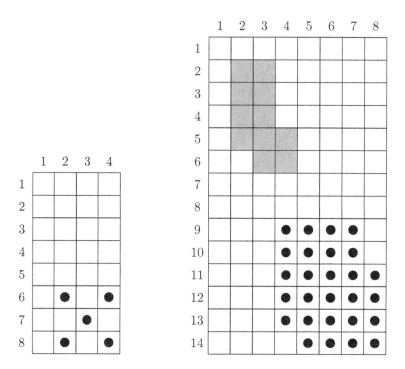

FIGURE 10.4: A dilation for which $A \nsubseteq A \oplus B$

set of pixels. In this case, B is referred to as a *structuring element* or a *kernel*.

Dilation in MATLAB or Octave can be performed with the command

```
>> imdilate(image,kernel)
```

MATLAB/Octave

and in Python with either of the commands `dilation` or `binary_dilation` from the `morphology` module of `skimage`. Both commands produce the same output for binary images, but `binary_dilation` is designed to be faster. To see an example of dilation, consider the commands:

```
>> t = imread('morph_text.png');
>> sq = ones(3,3);
>> td = imdilate(t,sq);
>> imshow(t)
>> figure,imshow(td)
```

MATLAB/Octave

or

```
In :  from skimage.morphology import binary_dilation as bwdilate
In :  t = io.imread('morph_text.png')
In :  sq = ones((3,3))
In :  td = bwdilate(t,sq)
```

Python

The result is shown in Figure 10.5. Notice how the image has been "thickened." This is really what dilation does; hence its name.

FIGURE 10.5: Dilation of a binary image

Erosion

Given sets A and B, the *erosion of A by B*, written $A \ominus B$, is defined as:

$$A \ominus B = \{w : B_w \subseteq A\}.$$

In other words, the erosion of A by B consists of all points $w = (x, y)$ for which B_w is in A. To perform an erosion, we can move B over A, and find all the places it will fit, and for each such place mark down the corresponding $(0,0)$ point of B. The set of all such points will form the erosion.

An example of erosion is given in Figure 10.6.

Note that in the example, the erosion $A \ominus B$ was a subset of A. This is not necessarily the case; it depends on the position of the origin in B. If B contains the origin (as it did in Figure 10.6), then the erosion will be a subset of the original object.

Figure 10.7 shows an example where B does not contain the origin. In this figure, the open circles in the right-hand figure form the erosion.

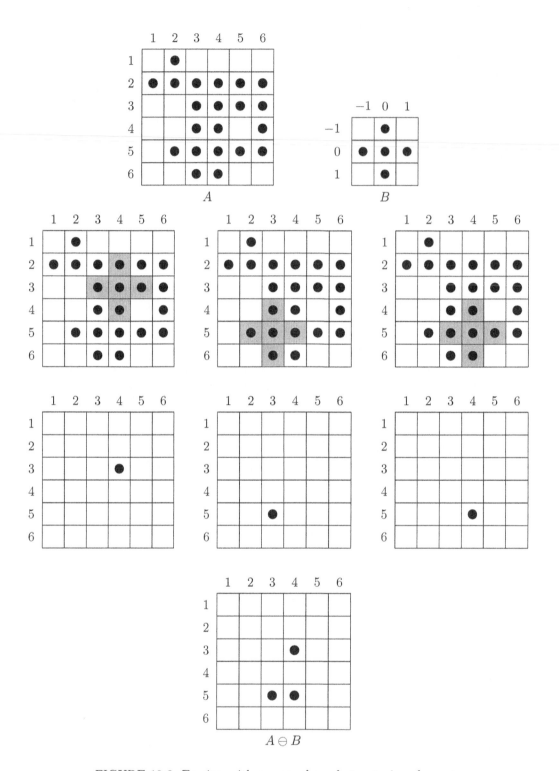

FIGURE 10.6: Erosion with a cross-shaped structuring element

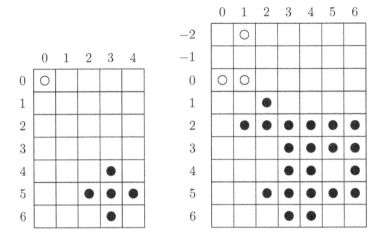

FIGURE 10.7: Erosion with a structuring element not containing the origin

Note that in Figure 10.7, the *shape* of the erosion is the same as that in Figure 10.6; however its *position* is different. Since the origin of B in Figure 10.7 is translated by $(-4, -3)$ from its position in Figure 10.6, we can assume that the erosion will be translated by the same amount. And if we compare Figures 10.6 and 10.7, we can see that the second erosion has indeed been shifted by $(-4, -3)$ from the first.

For erosion, as for dilation, we generally assume that A is the image being processed, and B is a small set of pixels: the structuring element or kernel.

Erosion is related to *Minkowski subtraction*: the Minkowski subtraction of B from A is defined as

$$A - B = \bigcap_{b \in B} A_b.$$

Erosion in MATLAB is performed with the command

```
>> imerode(image,kernel)
```

MATLAB/Octave

and in Python with either of the commands `erosion` or `binary_erosion` We shall give an example using a different binary image:

```
>> r = imread('rings.png');
>> re = imerode(c,sq);
>> imshow(r)
>> figure,imshow(re)
```

MATLAB/Octave

or

```
In :  from skimage.morphology import binary_erosion as bwerode
In :  r = io.imread('rings.png')
In :  re = bwerode(r,sq)
In :  io.imshow(r)
In :  f = plt.figure(); f.show(io.imshow(re))
```

Python

The result is shown in Figure 10.8. Notice how the image has been "thinned." This is the expected result of an erosion; hence its name. If we kept on eroding the image, we would end up with a completely black result.

FIGURE 10.8: Erosion of a binary image

Relationship between Erosion and Dilation

It can be shown that erosion and dilation are "inverses" of each other; more precisely, the complement of an erosion is equal to the dilation of the complement with the reflection of the structuring element. Thus:

$$\overline{A \ominus B} = \overline{A} \oplus \hat{B}.$$

A proof of this can be found in Haralick and Shapiro [14].

It can be similarly shown that the same relationship holds if erosion and dilation are interchanged; that

$$\overline{A \oplus B} = \overline{A} \ominus \hat{B}.$$

We can demonstrate the truth of these using MATLAB commands; all we need to know is that the complement of a binary image

b

is obtained using

```
>> ~b
```
MATLAB/Octave

and that given two images a and b, their equality can be determined in MATLAB and Octave with

```
>> isequal(a,b)
```
MATLAB/Octave

and in Python with

```
In :  np.array_equal(a,b)
```
Python

To demonstrate the equality

$$\overline{A \ominus B} = \overline{A} \oplus \hat{B}.$$

pick a binary image, say the text image, and a structuring element. Then the left-hand side of this equation is produced with

```
>> lhs=~imerode(t,sq);
```

MATLAB/Octave

and the right-hand side with

```
>> rhs=imdilate(~t,sq);
```

MATLAB/Octave

Finally, the command

```
>> isequal(lhs,rhs)
```

MATLAB/Octave

should return 1, for true. In Python, to make such commands work, the starting image must have values 0,1 only:

```
>> c = io.imread('morph_text.png').astype('bool')*1
```

Python

Then testing the equations is pretty much the same, but since the result of a morphological operation is an array of zero and ones, a logical complement cannot be used:

```
In :   lhs = 1-bwerode(c,sq)
In :   rhs = bwdilate(1-c,sq)
In :   np.array_equal(rhs,lhs)
```

Python

An Application: Boundary Detection

If A is an image, and B a small structuring element consisting of point symmetrically places about the origin, then we can define the boundary of A by any of the following methods:

(i) $A - (A \ominus B)$ "internal boundary"
(ii) $(A \oplus B) - A$ "external boundary"
(iii) $(A \oplus B) - (A \ominus B)$ "morphological gradient"

In each definition, the minus refers to set difference. For some examples, see Figure 10.9. Note that the internal boundary consists of those pixels in A that are at its edge; the external boundary consists of pixels outside A and which are just next to it, and the morphological gradient is a combination of both the internal and external boundaries.

To see some examples, choose the image **pinenuts.png**, and threshold it to obtain a binary image:

```
>> n = imread('pinenuts.png');
>> p = ~(p>130 & p<165);
```

MATLAB/Octave

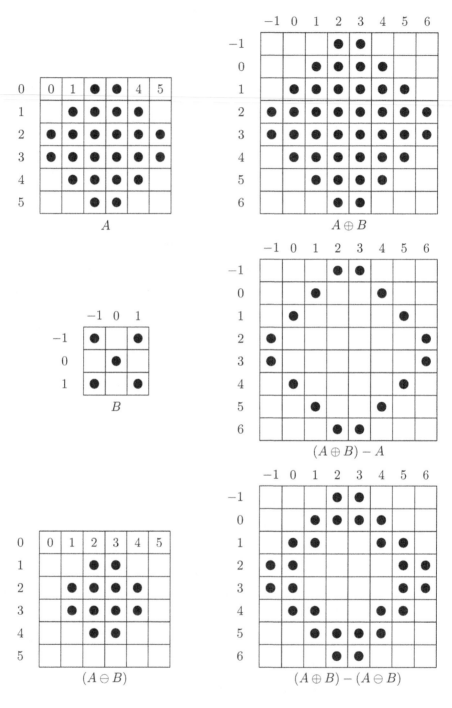

FIGURE 10.9: Boundaries

Then the internal boundary is obtained with:

```
>> pe = imerode(p,sq);
>> p_int = p & ~re;
>> imshow(p),figure,imshow(p_int)
```

MATLAB/Octave

The result is shown in Figure 10.10. The above commands can be implemented in Python

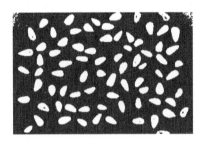 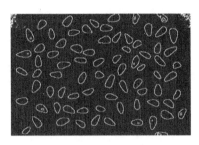

FIGURE 10.10: "Internal boundary" of a binary image

as

```
In :  n = io.imread('pinenuts.png')
In :  p = ~((n>130) & (n<165))
In :  pe = bwerode(p,sq)
In :  p_int = p-pe
In :  io.imshow(p)
In :  f = plt.figure(); f.show(io.imshow(p-pe))
```

Python

The external boundary and morphological gradients can be obtained similarly:

```
>> pd = imdilate(p,sq);
>> p_ext = pd & ~p;
>> p_grad = pd & ~pe;
>> imshow(p_ext), figure,imshow(p_grad)
```

MATLAB/Octave

or with

```
In :  pd = bwdilate(p,sq)
In :  p_exp = pd-p
In :  p_grad = pd-pe
In :  io.imshow(p_ext)
In :  f = plt.figure(); f.show(io.imshow(p_grad))
```

Python

The results are shown in Figure 10.11.

Note that the external boundaries are larger than the internal boundaries. This is because the internal boundaries show the outer edge of the image components whereas the external boundaries show the pixels just outside the components. The morphological gradient is thicker than either, and is in fact the union of both.

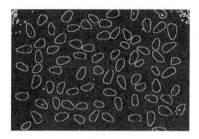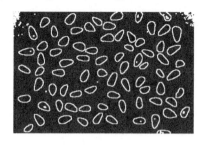

FIGURE 10.11: "External boundary" and the morphological gradient of a binary image

10.4   Opening and Closing

These operations may be considered as "second level" operations in that they build on the basic operations of dilation and erosion. They are also, as we shall see, better behaved mathematically.

Opening

Given A and a structuring element B, the *opening of A by B*, denoted $A \circ B$, is defined as:

$$A \circ B = (A \ominus B) \oplus B.$$

So, an opening consists of an erosion followed by a dilation. An equivalent definition is

$$A \circ B = \cup \{B_w : B_w \subseteq A\}.$$

That is, $A \circ B$ is the union of all translations of B that fit inside A. Note the difference with erosion: the erosion consists only of the $(0,0)$ point of B for those translations that fit inside A; the opening consists of all of B. An example of opening is given in Figure 10.12.

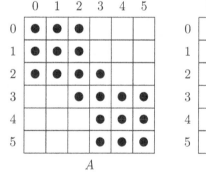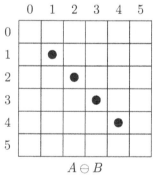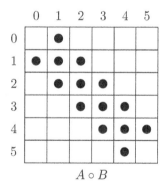

FIGURE 10.12: Opening

The opening operation satisfies the following properties:

1. $(A \circ B) \subseteq A$. Note that this is not the case with erosion; as we have seen, an erosion may not necessarily be a subset.

2. $(A \circ B) \circ B = A \circ B$. That is, an opening can never be done more than once. This property is called *idempotence*. Again, this is not the case with erosion; you can keep on applying a sequence of erosions to an image until nothing is left.

3. If $A \subseteq C$, then $(A \circ B) \subseteq (C \circ B)$.

4. Opening tends to "smooth" an image, to break narrow joins, and to remove thin protrusions.

Closing

Analogous to opening we can define *closing*, which may be considered a dilation followed by an erosion, and is denoted $A \bullet B$:

$$A \bullet B = (A \oplus B) \ominus B.$$

Another definition of closing is that $x \in A \bullet B$ if *all* translations B_w that contain x have non-empty intersections with A. An example of closing is given in Figure 10.13.

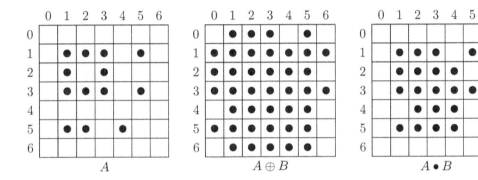

FIGURE 10.13: Closing

The closing operation satisfies the following properties:

1. $A \subseteq (A \bullet B)$.

2. $(A \bullet B) \bullet B = A \bullet B$; that is, closing, like opening, is idempotent.

3. If $A \subseteq C$, then $(A \bullet B) \subseteq (C \bullet B)$.

4. Closing also tends to smooth an image, but it fuses narrow breaks and thin gulfs, and eliminates small holes.

Opening and closing are implemented by the `imopen` and `imclose` functions, respectively. We can see the effects on a simple image using the square and cross-shaped structuring elements.

```
>> cr=[0 1 0;1 1 1;0 1 0];
>> test=zeros(10,10);test(2:6,2:4)=1;test(3:5,6:9)=1;test(8:9,4:8)=1;test
   (4,5)=1
```

MATLAB/Octave

or with

```
In :   cr = np.array([[0,1,0],[1,1,1],[0,1,0]])
In :   test = np.zeros((10,10))
In :   test[1:6,1:4] = 1; test[2:5,5:9] = 1; test[7:9,3:8] = 1; test[3,5] =
       1
```

Python

It looks like this:

```
0  0  0  0  0  0  0  0  0  0
0  1  1  1  0  0  0  0  0  0
0  1  1  1  0  1  1  1  1  0
0  1  1  1  1  1  1  1  1  0
0  1  1  1  0  1  1  1  1  0
0  1  1  1  0  0  0  0  0  0
0  0  0  0  0  0  0  0  0  0
0  0  0  1  1  1  1  1  0  0
0  0  0  1  1  1  1  1  0  0
0  0  0  0  0  0  0  0  0  0
```

Openings and closings can be done very easily:

```
>> test_open = imopen(test,cr);
>> test_close = imclose(test,cr)
```

MATLAB/Octave

or with Python, making sure to import the relevant commands first:

```
In :   from skimage.morphology import closing, opening, binary_closing as
       bwclose, binary_opening as bwopen
In :   test_open = bwopen(test,cr);
In :   test_close = bwclose(test,cr)
```

Python

The outputs are shown in Figure 10.14. Opening and closing with the square structuring

```
0  0  0  0  0  0  0  0  0  0        0  0  1  0  0  0  0  0  0  0
0  0  1  0  0  0  0  0  0  0        0  1  1  1  0  0  0  0  0  0
0  1  1  1  0  1  1  1  0  0        1  1  1  1  1  1  1  1  1  0
0  1  1  1  1  1  1  1  1  0        1  1  1  1  1  1  1  1  1  1
0  1  1  1  0  1  1  1  0  0        1  1  1  1  1  1  1  1  1  0
0  0  1  0  0  0  0  0  0  0        0  1  1  1  1  1  1  1  0  0
0  0  0  0  0  0  0  0  0  0        0  0  1  1  1  1  1  0  0  0
0  0  0  0  0  0  0  0  0  0        0  0  0  1  1  1  1  1  0  0
0  0  0  0  0  0  0  0  0  0        0  0  0  1  1  1  1  1  0  0
0  0  0  0  0  0  0  0  0  0        0  0  0  0  1  1  1  0  0  0
```

(a) Opening   (b) Closing

FIGURE 10.14: Open and closing with the cross-shaped structuring element

element are shown in Figure 10.15. With closing, the image is now fully "joined up." We can obtain a smeared effect with the text image, using a diagonal structuring element:

```
0  0  0  0  0  0  0  0  0  0        1  1  1  1  0  0  0  0  0  0
0  1  1  1  0  0  0  0  0  0        1  1  1  1  0  0  0  0  0  0
0  1  1  1  0  1  1  1  1  0        1  1  1  1  1  1  1  1  1  1
0  1  1  1  0  1  1  1  1  0        1  1  1  1  1  1  1  1  1  1
0  1  1  1  0  1  1  1  1  0        1  1  1  1  1  1  1  1  1  1
0  1  1  1  0  0  0  0  0  0        1  1  1  1  1  1  1  1  0  0
0  0  0  0  0  0  0  0  0  0        0  0  0  1  1  1  1  1  0  0
0  0  0  0  0  0  0  0  0  0        0  0  0  1  1  1  1  1  0  0
0  0  0  0  0  0  0  0  0  0        0  0  0  1  1  1  1  1  0  0
0  0  0  0  0  0  0  0  0  0        0  0  0  1  1  1  1  1  0  0
```

(a) Opening   (b) Closing

FIGURE 10.15: Opening and closing with the square structuring element

```
>> k = [0 0 0 0 1 1;0 0 1 1 0 0;1 1 0 0 0 0]
k =

    0   0   0   0   1   1
    0   0   1   1   0   0
    1   1   0   0   0   0

>> tc = imclose(t,k);
>> imshow(tc)
```
MATLAB/Octave

The result is shown in Figure 10.16.

FIGURE 10.16: An example of closing

An Application: Noise Removal

Suppose A is a binary image corrupted by impulse noise—some of the black pixels are white and some of the white pixels are back. An example is given in Figure 10.17. Then $A \ominus B$ will remove the single black pixels, but will enlarge the holes. We can fill the holes by dilating twice:

$$((A \ominus B) \oplus B) \oplus B.$$

The first dilation returns the holes to their original size; the second dilation removes them. But this will enlarge the objects in the image. To reduce them to their correct size, perform a final erosion:

$$(((A \ominus B) \oplus B) \oplus B) \ominus B.$$

The inner two operations constitute an opening; the outer two operations a closing. Thus, this noise removal method is in fact an opening followed by a closing:

$$(A \circ B) \bullet B).$$

This is called *morphological filtering*.

Suppose we take an image and apply 10% shot noise to it. This can be done in MATLAB or Octave by:

```
>> c = imread('circles.png');
>> x = rand(size(c));
>> c(find(x>0.95)) = 0;
>> c(find(x<0.05) = 1;
>> imshow(c)
```
MATLAB/Octave

and in Python using the function from the module of `numpy`:

```
In :   c = io.imread('circles.png').astype('bool')*1
In :   x = np.random.random_sample(c.shape)
In :   c[np.nonzero(x>0.95)]=0
In :   c[np.nonzero(x<=0.05)]=1
```
Python

The result is shown as Figure 10.17(a). The filtering process can be implemented with

```
>> cf1 = imclose(imopen(c,sq),sq);
>> figure, imshow(cf1)
>> cf2 = imclose(imopen(c,cr),cr);
>> figure, imshow(cf2)
```
MATLAB/Octave

and the same in Python, with `imclose` and `imopen` replaced with our aliases `bwclose` and `bwopen`. The results are shown as Figures 10.17(b) and (c). The results are rather "blocky"; although less so with the cross-shaped structuring element.

Relationship between Opening and Closing

Opening and closing share a relationship very similar to that of erosion and dilation: the complement of an opening is equal to the closing of a complement, and the complement of a closing is equal to the opening of a complement. Specifically:

$$\overline{A \bullet B} = \overline{A} \circ \hat{B}$$

and

$$\overline{A \circ B} = \overline{A} \bullet \hat{B}.$$

Again, see Haralick and Shapiro [14] for a formal proof.

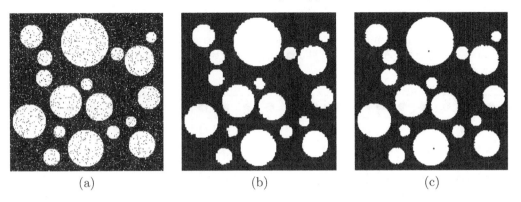

FIGURE 10.17: A noisy binary image and results after morphological filtering with different structuring elements.

10.5   The Hit-or-Miss Transform

This is a powerful method for finding shapes in images. As with all other morphological algorithms, it can be defined entirely in terms of dilation and erosion; in this case, erosion only.

Suppose we wish to locate 3×3 square shapes, such as is in the center of the image A in Figure 10.18.

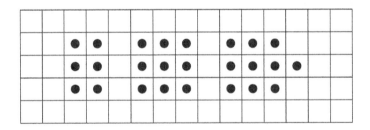

FIGURE 10.18: An image A containing a shape to be found

If we performed an erosion $A \ominus B$ with B being the square structuring element, we would obtain the result given in Figure 10.19.

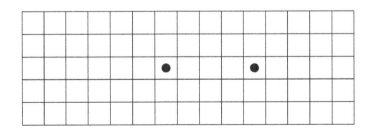

FIGURE 10.19: The erosion $A \ominus B$

The result contains two pixels, as there are exactly two places in A where B will fit. Now suppose we also erode the complement of A with a structuring element C, which fits exactly around the 3×3 square; $\overline{A}$ and C are shown in Figure 10.20. (We assume that $(0,0)$ is at the center of C.)

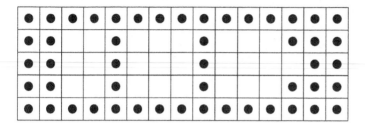 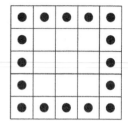

FIGURE 10.20: The complement $\overline{A}$ and the second structuring element

If we now perform the erosion $\overline{A} \ominus C$, we would obtain the result shown in Figure 10.21.

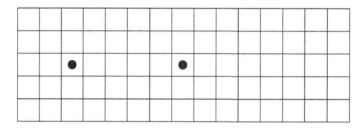

FIGURE 10.21: The erosion $\overline{A} \ominus C$

The intersection of the two erosion operations would produce just one pixel at the position of the center of the 3×3 square in A, which is just what we want. If A had contained more than one square, the final result would have been single pixels at the positions of the centers of each. This combination of erosions forms the hit-or-miss transform.

In general, if we are looking for a particular shape in an image, we design two structuring elements: B_1, which is the same shape, and B_2, which fits around the shape. We then write $B = (B_1, B_2)$ and

$$A \circledast B = (A \ominus B_1) \cap (\overline{A} \ominus B_2)$$

for the hit-or-miss transform.

As an example, we shall attempt to find the dot at the bottom of the question mark in the `morph_text.png` image. This is in fact a 4×4 square with missing corners. The two structuring elements then will be defined in MATLAB/Octave as

```
>> b1 = [0 1 1 0;1 1 1 1;1 1 1 1;0 1 1 0]
>> b2 = ones(6); b2(2:5,2:5) = ~b1
>> tb1 = imerode(t,b1);
>> tb2 = imerode(~t,b2);
>> hit_or_miss = tb1 & tb2;
>> [x,y] = find(hit_or_miss==1)
```

MATLAB/Octave

or in Python with

```
In :  b1 = np.array([[0,1,1,0],[1,1,1,1],[1,1,1,1],[0,1,1,0]])
In :  b2 = ones((6,6)); b2[1:5,1:5] = 1-b1
In :  tb1 = bwerode(t,b1);
In :  tb2 = bwerode(1-t,b2);
In :  np.where((tb1 & tb2)==1)
```
`Python`

and this returns a coordinate of $(281, 296)$ in MATLAB/Octave and $(280, 295)$ in Python, which is right in the middle of the dot. Note that eroding the text by `b1` alone is not sufficient, as there are many places in the images where `b1` can fit. This can be seen by viewing the image `tb1`, which is given in Figure 10.22.

FIGURE 10.22: Text eroded by a dot-shaped structuring element

10.6 Some Morphological Algorithms

In this section we shall investigate some simple algorithms that use some of the morphological techniques we have discussed in previous sections.

Region Filling

Suppose in an image we have a region bounded by an 8-connected boundary, as shown in Figure 10.23.

Given a pixel p within the region, we wish to fill up the entire region. To do this, we start with p, and dilate as many times as necessary with the cross-shaped structuring element B (as used in Figure 10.6), each time taking an intersection with $\overline{A}$ before continuing. We thus create a sequence of sets:

$$\{p\} = X_0, X_1, X_2, \ldots, X_k = X_{k+1}$$

for which

$$X_n = (X_{n-1} \oplus B) \cap \overline{A}.$$

Finally, $X_k \cup A$ is the filled region. Figure 10.24 shows how this is done.

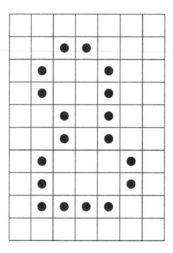

FIGURE 10.23: An 8-connected boundary of a region to be filled

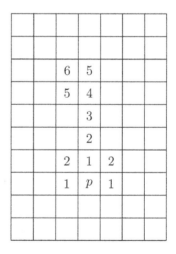

FIGURE 10.24: The process of filling the region

In the right-hand grid, we have

$$X_0 = \{p\}, X_1 = \{p, 1\}, X_2 = \{p, 1, 2\}, \ldots$$

Note that the use of the cross-shaped structuring element means that we never cross the boundary.

Connected Components

We use a very similar algorithm to fill a connected component; we use the cross-shaped structuring element for 4-connected components, and the square structuring element for 8-connected components. Starting with a pixel p, we fill up the rest of the component by creating a sequence of sets

$$X_0 = \{p\}, X_1, X_2, \ldots$$

such that

$$X_n = (X_{n-1} \oplus B) \cap A$$

until $X_k = X_{k-1}$. Figure 10.25 shows an example.

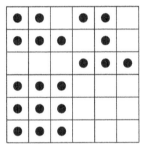 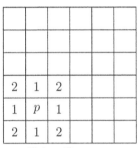 

Using the cross   Using the square

FIGURE 10.25: Filling connected components

In each case, we are starting in the center of the square in the lower left. As this square is itself a 4-connected component, the cross structuring element cannot go beyond it.

MATLAB and Octave implement this algorithm with the `bwlabel` function; Python with the `label` function in the `measure` module of `skimage`. Suppose X is the image shown in Figure 10.25. Then in MATLAB or Octave:

```
>> bwlabel(X,4)
>> bwlabel(X,8)
```
MATLAB/Octave

will show the 4-connected or 8-connected components; using the cross or square structuring elements, respectively. In Python:

```
In :  from skimage.measure import label
In :  bwlabel(X,4,background=0)+1
In :  bwlabel(X,8,background=0)+1
```
Python

The `background` parameter tells Python which pixel values to treat as background; these are assigned the value -1, all foreground components are labeled starting with the index value 0. Adding 1 puts the background back to zero, and labels starting at 1.

MATLAB and Octave differ from Python in terms of the direction of scanning, as shown in Figure 10.26.

```
[[1  1  0  2  2  0]        1  1  0  3  3  0        1  1  0  1  1  0
 [1  1  1  0  2  0]        1  1  1  0  3  0        1  1  1  0  1  0
 [0  0  0  2  2  2]        0  0  0  3  3  3        0  0  0  1  1  1
 [3  3  3  0  0  0]        2  2  2  0  0  0        1  1  1  0  0  0
 [3  3  3  0  0  0]        2  2  2  0  0  0        1  1  1  0  0  0
 [3  3  3  0  0  0]]       2  2  2  0  0  0        1  1  1  0  0  0
```

(a) Python 4-components   (b) MATLAB/Octave 4-component   (c) 8-components

FIGURE 10.26: Labeling connected components

Filling holes can be done with the `bwfill` function in MATLAB/Octave, or the function `binary_fill_holes` from the `ndimage` module of `scipy`:

```
>> n = imread('nicework.png');
>> nb = n & ~imerode(n,sq);
>> imshow(nb)
>> nf = bwfill(nb,[74,52],sq);
>> figure,imshow(nf)
```
MATLAB/Octave

Python's command by default fills all holes by dilating and intersecting with the complement. To fill just one region:

```
In :  import skimage.morphology as mo
In :  import scipy.ndimage as ndi
In :  n = io.imread('nicework.png')/255
In :  nb = n - bwerode(n,sq)
In :  nb1 = nb[0:120,0:80]
In :  nf = ndi.binary_fill_holes(nb1,sq)
In :  nb[0:120,0:80] = nf
```
Python

The results are shown in Figure 10.27. Image (a) is the original, (b) the boundary, and (c) the result of a region fill. Figure (d) shows a variation on the region filling, we just include all boundaries. This was obtained with

```
>> figure,imshow(nf|nb)
```
MATLAB/Octave

Skeletonization

Recall that the *skeleton* of an object can be defined by the "medial axis transform"; we may imagine fires burning in along all edges of the object. The places where the lines of fire meet form the skeleton. The skeleton may be produced by morphological methods.

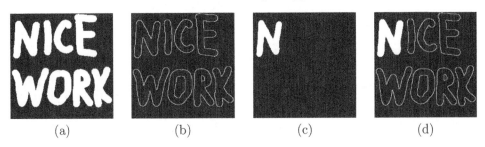

| (a) | (b) | (c) | (d) |

FIGURE 10.27: Region filling

Consider the table of operations as shown in Table 10.1.

Erosions	Openings	Set differences
A	$A \circ B$	$A - (A \circ B)$
$A \ominus B$	$(A \ominus B) \circ B$	$(A \ominus B) - ((A \ominus B) \circ B)$
$A \ominus 2B$	$(A \ominus 2B) \circ B$	$(A \ominus 2B) - ((A \ominus 2B) \circ B)$
$A \ominus 3B$	$(A \ominus 3B) \circ B$	$(A \ominus 3B) - ((A \ominus 3B) \circ B)$
$\vdots$	$\vdots$	$\vdots$
$A \ominus kB$	$(A \ominus kB) \circ B$	$(A \ominus kB) - ((A \ominus kB) \circ B)$

TABLE 10.1: Operations used to construct the skeleton

Here we use the convention that a sequence of k erosions using the same structuring element B is denoted $A \ominus kB$. We continue the table until $(A \ominus kB) \circ B$ is empty. The skeleton is then obtained by taking the unions of all the set differences. An example is given in Figure 10.28, using the cross-shaped structuring element.

Since $(A \ominus 2B) \circ B$ is empty, we stop here. The skeleton is the union of all the sets in the third column; it is shown in Figure 10.29. This method of skeletonization is called *Lantuéjoul's method*; for details, see Serra [45].

Programs are given at the end of the chapter. We shall experiment with the nice work image, either with MATLAB or Octave:

```
>> nk = imskel(n,sq);
>> imshow(nk)
>> nk2 = imskel(n,cr);
>> figure,imshow(nk2)
```

MATLAB/Octave

or with Python:

```
In :   nk = bwskel(n,sq)
In :   io.imshow(nk)
In :   nk2 = bwskel(n,cr)
In :   f = plt.figure(); f.show(io.imshow(nk2))
```

Python

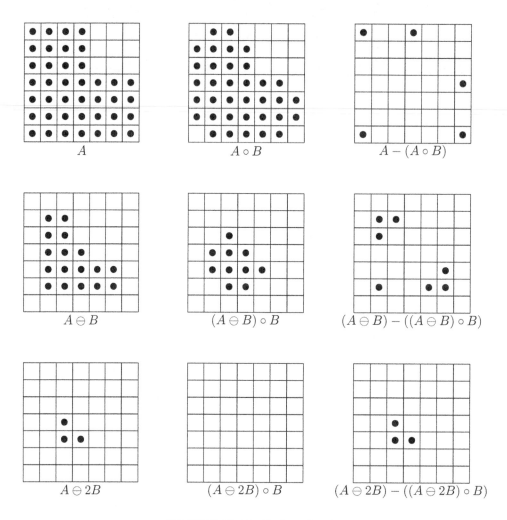

FIGURE 10.28: Skeletonization

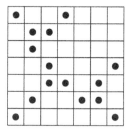

FIGURE 10.29: The final skeleton

The results are shown in Figure 10.30. Image (a) is the result using the square structuring element; image (b) is the result using the cross structuring element.

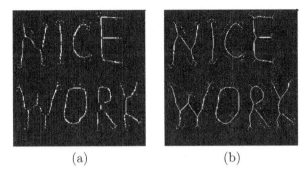

(a)   (b)

FIGURE 10.30: Skeletonization of a binary image

10.7 A Note on the **bwmorph** Function in MATLAB and Octave

The theory of morphology developed so far uses versions of erosion and dilation sometimes called *generalized* erosion and dilation; so called because the definitions allow for general structuring elements. This is the method used in the imerode and imdilate functions. However, the bwmorph function actually uses a different approach to morphology; that based on *lookup tables*. (Note however that Octave's bwmorph function implements erosion and dilation with calls to the imerode and imdilate functions.) We have seen the use of lookup tables for binary operations in Chapter 11; they can be just as easily applied to implement morphological algorithms.

The idea is simple. Consider the 3×3 neighborhood of a pixel. Since each pixel in the neighborhood can only have two values, there are $2^9 = 512$ different possible neighborhoods. We define a morphological operation to be a function that maps these neighborhoods to the values 0 and 1. Each possible neighborhood state can be associated with a numeric value from 0 (all pixels have value 0) to 511 (all pixels have value 1). The lookup table is then a binary vector of length 512; its k-th element is the value of the function for state k.

With this approach, we can define dilation as follows: a 0-valued pixel is changed to 1 if at least one of its eight neighbors has value 1. Conversely, we may define erosion as changing a 1-valued pixel to 0 if at least one of its eight neighbors has value 0.

Many other operations can be defined by this method (see the help file for bwmorph). The advantage is that any operation can be implemented extremely easily, simply by listing the lookup table. Moreover, the use of a lookup table allows us to satisfy certain requirements; the skeleton, for example, can be connected and have exactly one pixel thickness. This is not necessarily true of the algorithm presented in Section 10.6. The disadvantage of lookup tables is that we are restricted to using 3×3 neighborhoods. For more details, see "Digital Image Processing" by W. K. Pratt [35].

10.8   Grayscale Morphology

The operations of erosion and dilation can be generalized to be applied to grayscale images. But before we do, we shall reconsider binary erosion and dilation. We have defined binary erosion, $A \ominus B$, to be the union of the $(0,0)$ positions of all translations B_x for which $B_x \subseteq A$.

Suppose we take B to be a 3×3 square consisting entirely of zeros. Let A be the image as shown in Figure 10.31. Now suppose we move over the image A, and for each point p we

```
0 0 0 0 0 0 0 0
0 1 1 1 1 0 0 0
0 1 1 1 1 1 0 0              0 0 0
0 1 1 1 1 1 1 0              0 0 0
0 1 1 1 1 1 1 0              0 0 0
0 0 1 1 1 1 1 0                B
0 0 0 1 1 1 1 0
0 0 0 0 0 0 0 0
       A
```

FIGURE 10.31: An example for erosion

perform the following steps:

1. Find the 3×3 neighborhood N_p of p

2. Compute the matrix $N_p - B$

3. Find the minimum of that result.

We note that since B consists of all zeros, the second and third items could be reduced to finding the minimum of N_p. However, we shall see that for generalization it will be more convenient to have this expanded form. An immediate consequence of these steps is that if a neighborhood contains at least one zero, the output will be zero. The output is one only if the neighborhood contains all ones. For example:

```
0 0 0
0 1 1     ⟶  0
0 1 1
```

If we perform this operation, we will obtain:

```
0 0 0 0 0 0 0
0 0 0 0 0 0 0
0 0 1 1 0 0 0
0 0 1 1 1 0 0
0 0 1 1 1 0 0
0 0 0 1 1 0 0
0 0 0 0 0 0 0
0 0 0 0 0 0 0
```

which you can verify is exactly the erosion $A \ominus B$.

For dilation, we perform a sequence of steps very similar to those for erosion:

1. Find the 3×3 neighborhood N_p of p

2. Compute the matrix $N_p + B$

3. Find the maximum of that result.

We note again that since B consists of all zeros, the second and third items could be reduced to finding the maximum of N_p. If the neighborhood contains at least one $\mathbf{1}$, then the output will be $\mathbf{1}$. The output will be $\mathbf{0}$ only if the neighborhood contains all zeros.

Suppose A to be surrounded by zeros, so that neighborhoods are defined for all points in A above. Applying these steps produces:

```
1 1 1 1 1 1 0 0
1 1 1 1 1 1 1 0
1 1 1 1 1 1 1 1
1 1 1 1 1 1 1 1
1 1 1 1 1 1 1 1
1 1 1 1 1 1 1 1
0 1 1 1 1 1 1 1
0 0 1 1 1 1 1 1
```

which again can be verified to be the dilation $A \oplus B$.

If A is a grayscale image, and B is a structuring element, which will be an array of integers, we define grayscale erosion by using the steps above; for each pixel p in the image:

1. Position B so that $(0,0)$ lies over p

2. Find the neighborhood N_p of p corresponding to the shape of B

3. Find the value $\min(N_p - B)$.

We note that there is nothing in this definition which requires B to be any particular shape or size or that the elements of B be positive. And as for binary dilation, B does not have to contain the origin $(0,0)$.

We can define this more formally; let B be a set of points with associated values. For example, for our square of zeros we would have:

Point	Value
$(-1,-1)$	0
$(-1,0)$	0
$(-1,1)$	0
$(0,-1)$	0
$(0,0)$	0
$(0,1)$	0
$(1,-1)$	0
$(1,0)$	0
$(1,1)$	0

The set of points forming B is called the *domain* of B and is denoted D_B. Now we can define:

$$(A \ominus B)(x,y) = \min\{A(x+s, y+t) - B(s,t), (s,t) \in D_B\},$$
$$(A \oplus B)(x,y) = \max\{A(x+s, y+t) + B(s,t), (s,t) \in D_B\}.$$

We note that some published definitions use $s - x$ and $t - y$ instead of $x + s$ and $y + t$. This just requires that the structuring element is rotated $180°$.

	y				
	1	2	3	4	5
1	10	20	20	20	30
2	20	30	30	40	50
x 3	20	30	30	50	60
4	20	40	50	50	60
5	30	50	60	60	70

	t		
	−1	0	1
−1	1	2	3
s 0	4	5	6
1	7	8	9

A B

FIGURE 10.32: An example for grayscale erosion and dilation

An example. Suppose we take A and B as given in Figure 10.32.
In this example $(0,0) \in D_B$. Consider $(A \ominus B)(1,1)$. By the definition, we have:

$$(A \ominus B)(1,1) = \min\{A(1+s,1+t) - B(s,t), (s,t) \in D_B\}.$$

Since $D_B = \{(s,t) : -1 \le s \le 1; -1 \le t \le 1\}$, we have

$$(A \ominus B)(1,1) = \min\{A(1+s,1+t) - B(s,t) : -1 \le s \le 1; -1 \le t \le 1\}.$$

In order to ensure we do not require matrix indices that move outside A, we simply "cut off" the structuring element so that we restrict it to elements in A.

The values of $A(1+s,1+t) - B(s,t)$ can then be obtained by matrix arithmetic:

$$\begin{bmatrix} 10 & 20 \\ 20 & 30 \end{bmatrix} - \begin{bmatrix} 5 & 6 \\ 8 & 9 \end{bmatrix} = \begin{bmatrix} 5 & 14 \\ 12 & 21 \end{bmatrix}$$

and the minimum of the result is **5**. Note that to create the matrices we ensure that the $(0,0)$ point of B sits over the current point of A, in this case the point $(1,1)$. Another example:

$$(A \ominus B)(3,4) = \min\left\{ \begin{bmatrix} 30 & 40 & 50 \\ 30 & 50 & 60 \\ 50 & 50 & 60 \end{bmatrix} - \begin{bmatrix} 1 & 2 & 3 \\ 4 & 5 & 6 \\ 7 & 8 & 9 \end{bmatrix} \right\} = \min\left\{ \begin{bmatrix} 29 & 38 & 47 \\ 26 & 45 & 54 \\ 43 & 42 & 51 \end{bmatrix} \right\} = 26.$$

Finally, we have

$$A \ominus B = \begin{matrix} 5 & 6 & 14 & 15 & 16 \\ 8 & 9 & 17 & 18 & 19 \\ 12 & 13 & 25 & 26 & 39 \\ 15 & 16 & 28 & 29 & 46 \\ 18 & 19 & 39 & 48 & 49 \end{matrix}$$

Dilation is very similar to erosion, except that we add B and take the maximum of the result. As for erosion we restrict the structuring element so that it does not go outside of A. For example:

$$(A \oplus B)(1,1) = \max\left\{ \begin{bmatrix} 10 & 20 \\ 20 & 30 \end{bmatrix} + \begin{bmatrix} 5 & 6 \\ 8 & 9 \end{bmatrix} \right\} = \max\left\{ \begin{bmatrix} 15 & 26 \\ 28 & 39 \end{bmatrix} \right\} = 39,$$

$$(A \oplus B)(3,4) = \max\left\{ \begin{bmatrix} 30 & 40 & 50 \\ 30 & 50 & 60 \\ 50 & 50 & 60 \end{bmatrix} + \begin{bmatrix} 1 & 2 & 3 \\ 4 & 5 & 6 \\ 7 & 8 & 9 \end{bmatrix} \right\} = \max\left\{ \begin{bmatrix} 31 & 42 & 53 \\ 34 & 55 & 66 \\ 57 & 58 & 69 \end{bmatrix} \right\} = 69.$$

After all the calculations, we have

$$A \oplus B = \begin{matrix} 39 & 39 & 49 & 59 & 58 \\ 39 & 39 & 59 & 69 & 68 \\ 49 & 59 & 59 & 69 & 68 \\ 59 & 69 & 69 & 79 & 78 \\ 56 & 66 & 66 & 76 & 75 \end{matrix}$$

Note that an erosion may contain negative values, or a dilation value greater than 255. In order to render the result suitable for display, we have the same choices as for spatial filtering: we may apply a linear transformation or we may clip values.

In general, and this can be seen from the examples, erosion will tend to decrease and darken objects in an image, and dilation will tend to enlarge and lighten objects.

Relationship between Grayscale Erosion and Dilation

By definition of maximum and minimum we have, if X and Y are two matrices:

$$\max\{X + Y\} = -\min\{-X - Y\}.$$

Since $\max\{X + Y\}$ corresponds to $A \oplus B$ and $\min\{X - Y\}$ to $A \ominus B$, we have

$$A \oplus B = -(-A \ominus B),$$
$$A \ominus B = -(-A \oplus B),$$

or

$$-(A \oplus B) = -A \ominus B,$$
$$-(A \ominus B) = -A \oplus B.$$

We can use the `imerode` and `imdilate` functions for grayscale erosion and dilation, but we have to be more careful about the structuring element. To create a structuring element for use with grayscale morphology, we can either use a binary mask, or we have to provide both a neighborhood D_B, and its values. To do this we need to use the `strel` function.

For example, we shall use MATLAB with the previous examples. First we need to create the structuring element.

```
>> str = strel('arbitrary',ones(3,3),[1 2 3;4 5 6;7 8 9])

str =

Nonflat STREL object containing 9 neighbors.

Neighborhood:
     1     1     1
     1     1     1
     1     1     1


Height:
     1     2     3
     4     5     6
     7     8     9
```

MATLAB/Octave

Here we use the `arbitrary` parameter of `strel`; this allows us to create a structuring element containing any values we like. The first matrix: `ones(3,3)` provides the neighborhood; the second matrix provides the values. Now we can test it.

```
>> A = [10 20 20 20 30;20 30 30 40 50;20 30 30 50 60;20 40 50 50 60;30 50
   60 60 70];
>> imerode(A,str)

ans =

      5      6     14     15     16
      8      9     17     18     19
     12     13     25     26     39
     15     16     28     29     46
     18     19     39     48     49
```

MATLAB/Octave

Python provides this generalized erosion as `gray_erosion` in the `ndimage` module of `scipy`:

```
In :  A = np.array([[10,20,20,20,30],[20,30,30,40,50],[20,30,30,50,60],
   [20,40,50,50,60],[30,50,60,60,70]]).astype('uint8')
In :  struct = np.array([[1,2,3],[4,5,6],[7,8,9]])
In :  print ndi.gray_erosion(A,footprint=mo.square(3),structure=struct)
[[ 5  6 14 15 16]
 [ 8  9 17 18 19]
 [12 13 25 26 39]
 [15 16 28 29 46]
 [18 19 39 48 49]]
```

Python

For dilation MATLAB implements the convention of the structuring element being rotated by 180°. So to obtain the result we produced before, we need to rotate `str` to obtain `str2`.

```
>> str2 = strel('arbitrary',ones(3,3),[9 8 7;6 5 4;3 2 1])

str2 =

Nonflat STREL object containing 9 neighbors.

Neighborhood:
     1     1     1
     1     1     1
     1     1     1


Height:
     9     8     7
     6     5     4
     3     2     1


>> imdilate(A,str2)

ans =

    39    39    49    59    58
    39    39    59    69    68
    49    59    59    69    68
    59    69    69    79    78
    56    66    66    76    75
```

MATLAB/Octave

In Python the same result is obtained with `gray_dilation`:

```
In :  struct2 = np.array([[9,8,7],[6,5,4],[3,2,1]])
In :  print ndi.gray_dilation(A,footprint=mo.square(3),\
...:  structure=struct,mode='constant')
[[39 39 49 59 59]
 [39 39 59 69 69]
 [49 59 59 69 69]
 [59 69 69 79 79]
 [59 69 69 79 79]]
```

Python

Now we can experiment with an image. We would expect that dilation would increase light areas in an image, and erosion would decrease them.

```
>> c = imread('caribou.png');
>> str = strel('square',5);
>> cd = imdilate(c,str);
>> ce = imerode(c,str);
>> imshow(cd), figure,imshow(ce)
```

MATLAB/Octave

Using Python:

```
In :   c = io.imread('caribou.png')
In :   str = mo.square(5)
In :   cd = mo.dilation(c,str)
In :   ce = mo.erosion(c,str)
In :   io.imshow(cd)
In :   f = plt.figure(); f.show(io.imshow(ce))
```
`Python`

The results are shown in Figure 10.33. Image (a) is the dilation and image (b) the erosion.

(a)   (b)

FIGURE 10.33: Grayscale dilation and erosion

Opening and Closing

Opening and closing are defined exactly as for binary morphology: opening is an erosion followed by a dilation, and closing is a dilation followed by an erosion. The `imopen` and `imclose` functions can be applied to grayscale images, if a structuring element has been created with `strel`.

Using the caribou image and the same 5×5 square structuring element as above:

```
>> co = imopen(c,str);
>> cc = imclose(c,str);
>> imshow(co),figure,imshow(cc)
```
`MATLAB/Octave`

and the results are shown in Figure 10.34; image (a) being the opening and (b) the closing.

10.9   Applications of Grayscale Morphology

Almost all the applications we saw for binary images can be carried over directly to grayscale images.

(a)   (b)

FIGURE 10.34: Grayscale opening and closing

Edge Detection

We can use the morphological gradient

$$(A \oplus B) - (A \ominus B)$$

for grayscale edge detection. We can try this with two different structuring elements.

```
>> str1 = strel('disk' ,1);
>> str2 = strel('square',5);
>> imshow(imdilate(x,str1)-imerode(c,str1))
```
MATLAB/Octave

or

```
In :   str1 = mo.disk(1)
In :   str2 = mo.square(5)
In :   io.imshow(mo.dilation(c,str1)-mo.erosion(c,str1))
```
Python

and similarly for `str2`. The results are shown in Figure 10.35. Image (a) uses the 3×3 cross, and image (b) the 5×5 square. Another edge detection method [46, 47] applied to an image A with structuring element B is defined by

$$\min\{(A \oplus B) - A, A - (A \ominus B)\}. \tag{10.1}$$

An example using the $s \times 3$ square structuring element is given in Figure 10.36(a). Another edge filter, which works well in the presence of noise, first blurs the image using the *alpha-trimmed mean filter*. This non-linear filter first orders the elements in the neighborhood, trims off a fraction α of the lowest and highest values, and returns the mean of the rest. This filter can be implemented (with $\alpha = 0.25$) as follows:

```
function out = atmean(r)
    s = sort(r(:));
    d = floor(length(s)/4);
    out = mean(s(1+d:end-d));
end
```
MATLAB/Octave

```
def atmean(r):
    s = np.sort(r.flatten())
    d = len(s)/4
    return s[d:-d].mean()
```
Python

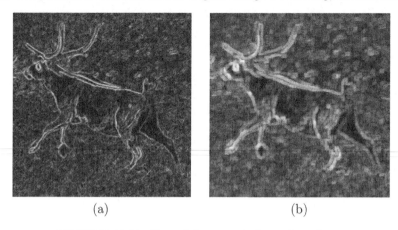

(a)   (b)

FIGURE 10.35: Use of the morphological gradient

and then applied to the image with either

```
>> ca = nlfilter(c,[3,3],@atmean)
```
MATLAB/Octave

or

```
In :   ca = ndi.generic_filter(c,atmean,size=(3,3))
```
Python

Given the result T of the filter, the edge filter is defined as

$$\min\{(T \circ B) - (T \ominus B), (T \oplus B) - (T \bullet B)\}. \tag{10.2}$$

and an example is shown in Figure 10.36(b).

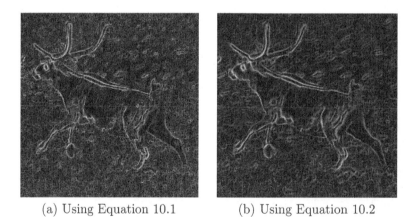

(a) Using Equation 10.1   (b) Using Equation 10.2

FIGURE 10.36: Morphological edge finding

Noise Removal

As before we can attempt to remove noise with morphological filtering: an opening followed by a closing. We can apply the following commands with `str` being either the 3×3 square or the cross:

```
>> cn = imnoise(c,'salt_&_pepper');
>> cf = imclose(imopen(cn,str),str);
>> imshow(cn),figure,imshow(cf)
```
MATLAB/Octave

or

```
In :   cn = ut.noise.random_noise(c,'s&p')
In :   cn = ut.img_as_ubyte(cn)
In :   cno = mo.opening(cn,str)
In :   cnoc = mo.closing(cno,str)
In :   io.imshow(cn)
In :   f = plt.figure(); f.show(io.imshow(cnoc))
```
Python

These are shown in Figure 10.37. The result is reasonable, if slightly blurry. Morphological

 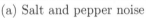  

(a) Salt and pepper noise   (b) Cleaning with 3×3 square   (c) Cleaning with cross

FIGURE 10.37: Use of morphological filtering to remove noise

filtering does not perform particularly well on Gaussian noise.

10.10 Programs

Here are programs for skeletonization by Lantuéjoul's method, first in MATLAB or Octave:

```matlab
function skel = imskel(image,kernel)
% IMSKEL - Calculates the skeleton of IMAGE using kernel KERNEL
%
skel=zeros(size(image));
e=image;
while (any(e(:))),
    o=imopen(e,kernel);
    skel=skel | (e&~o);
    e=imerode(e,kernel);
end
```

MATLAB/Octave

and in Python:

```python
def bwskel(im,ker):
    skel = np.zeros_like(im)
    from skimage.morphology import binary_erosion, binary_opening
    e = (np.copy(im)>0)*1
    while e.max()>0:
        o = binary_opening(e,ker)
        skel = skel | (e & 1-o)
        e = binary_erosion(e,ker)
    return skel
```

Python

Exercises

1. For each of the following images A_i and structuring elements B_j:

$A_1 =$ $A_2 =$ $A_3 =$

```
0 0 0 0 0 0 0 0    0 0 0 0 0 0 0 0    0 0 0 0 0 0 0 0
0 0 0 1 1 1 1 0    0 1 1 1 1 1 1 0    0 0 0 0 0 1 1 0
0 0 0 1 1 1 1 0    0 1 1 1 1 1 1 0    0 1 1 1 0 1 1 0
0 1 1 1 1 1 1 0    0 1 1 0 0 1 1 0    0 1 1 1 0 1 1 0
0 1 1 1 1 1 1 0    0 1 1 0 0 1 1 0    0 1 1 1 0 1 1 0
0 1 1 1 1 0 0 0    0 1 1 1 1 1 1 0    0 1 1 1 0 0 0 0
0 1 1 1 1 0 0 0    0 1 1 1 1 1 1 0    0 1 1 1 0 0 0 0
0 0 0 0 0 0 0 0    0 0 0 0 0 0 0 0    0 0 0 0 0 0 0 0
```

$B_1 =$ $B_2 =$ $B_3 =$

```
0 1 0    1 1 1    1 0 0
1 1 1    1 1 1    0 0 0
0 1 0    1 1 1    0 0 1
```

calculate the erosion $A_i \ominus B_j$, the dilation $A_i \oplus B_j$, the opening $A_i \circ B_j$, and the closing $A_i \bullet B_j$.

Check your answers with your computer system.

2. Suppose a square object was eroded by a circle whose radius was about one quarter the side of the square. Draw the result.

3. Repeat the previous question with dilation.

4. Using the binary images `circles.png`, `circles2.png`, `meet_text.png`, `nicework.png`, and `rings.png`, view the erosion and dilation with both the square and the cross-shaped structuring elements.

   Can you see any differences?

5. Read in the image `circles2.png`.

   (a) Erode with squares of increasing size until the image starts to split into disconnected components.

   (b) Find the coordinates of a pixel in one of the components.

   (c) Use the appropriate commands to isolate that particular component.

6. (a) With your disconnected image from the previous question, compute its boundary.

   (b) Again find a pixel inside one of the boundaries.

   (c) Use the region filling function to fill that region.

   (d) Display the image as a boundary with one of the regions filled in.

7. Using the 3×3 square structuring element, compute the skeletons of

   (a) A 7 square

   (b) A 5×9 rectangle

   (c) An L shaped figure formed from an 8×8 square with a 3×3 square taken from a corner

   (d) An H shaped figure formed from a 15×15 square with 5×5 squares taken from the centers of the top and bottom

   (e) A cross formed from an 11×11 square with 3×3 squares taken from each corner

   In each case, check your answer with your computer system.

8. Repeat the above question but use the cross-shaped structuring element.

9. For the images listed in Question 4, obtain their skeletons by both supplied functions, and by using Lantuéjoul's method. Which seems to provide the best result?

10. Use the hit-or-miss transform with appropriate structuring elements to find the dot on the "i" in the word "**Friday**" in the image `meet_text.png`.

11. Let A be the matrix:

$$
\begin{array}{cccccc}
35 & 1 & 6 & 26 & 19 & 24 \\
3 & 32 & 7 & 21 & 23 & 25 \\
31 & 9 & 2 & 22 & 27 & 20 \\
8 & 28 & 33 & 17 & 10 & 15 \\
30 & 5 & 34 & 12 & 14 & 16 \\
4 & 36 & 29 & 13 & 18 & 11 \\
\end{array}
$$

(If you are using MATLAB or Octave, this matrix can be obtained with A=magic(6). If you are using Python you will have to enter this matrix manually.)

By hand, compute the grayscale erosions and dilations with the structuring elements $B =$

$$
\begin{array}{ccc}
10 & 10 & 10 \\
10 & 10 & 10 \\
10 & 10 & 10 \\
\end{array}
\qquad
\begin{array}{ccc}
5 & 20 & 5 \\
20 & 5 & 20 \\
5 & 20 & 5 \\
\end{array}
$$

In each case, check your answers with your computer system.

12. Perform grayscale erosions, dilations, openings, and closings on the cameraman image using the above structuring elements.

13. (a) Obtain the grayscale version of the twins image as was used in Chapter 8.

(b) Apply salt and pepper noise to the image.

(c) Attempt to remove the noise using morphological techniques.

(d) Compare the results to median filtering.

14. Repeat the above question but use Gaussian noise.

15. Use morphological methods to find the edges of the image stairs.png. Do this in two ways:

(a) By thresholding first and using binary morphology.

(b) By using grayscale morphology and thresholding second.

Which seems to give the best results?

16. Compare the results of the previous question with standard edge detection techniques.

Chapter 11

Image Topology

11.1 Introduction

We are often interested in only the very basic aspects of an image: the number of occurrences of a particular object; whether there are holes, and so on. The investigation of these fundamental properties of an image is called *digital topology* or *image topology* and in this chapter we shall investigate some of the more elementary aspects of this subject.

For example, consider an image thresholded and cleaned up with a morphological opening to show a collection of blobs. Such an image can be obtained with MATLAB or Octave by:

```
>>  p = imread('pinenuts.png');
>>  p2 = ~im2bw(p,graythresh(p));
>>  p3 = imopen(p2,strel('disk',5));
```

MATLAB/Octave

or with Python by:

```
In :  import skimage.morphology as mo
In :  p = io.imread('pinenuts.png')
In :  p2 = 1.0-(p>fl.threshold_otsu(p))
In :  p3 = mo.opening(p2,mo.disk(5))
```

Python

The image n2 is shown in Figure 11.1. The number of blobs can be determined by mor-

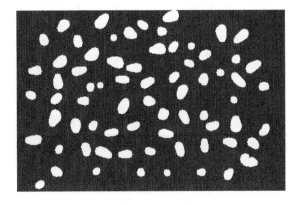

FIGURE 11.1: How many blobs?

309

phological methods. However, the study of image topology provides alternative, and very powerful, methods for such tasks as object counting. Topology provides very rigorous definitions for concepts such as adjacency and distance. As we shall see, skeletonization can be performed very efficiently using topological methods.

11.2   Neighbors and Adjacency

A first task is to define the concepts of adjacency: under what conditions a pixel may be regarded as being "next to" another pixel. For this chapter, the concern will be with binary images only, and so we will be dealing only with positions of pixels.

A pixel P has four *4-neighbors*:

and eight *8-neighbors*:

Two pixels P and Q are *4-adjacent* if they are 4-neighbors of one another, and *8-adjacent* if they are 8-neighbors of one another.

11.3   Paths and Components

Suppose that P and Q are any two (not necessarily adjacent) pixels, and suppose that P and Q can be joined by a sequence of pixels as shown in Figure 11.2.

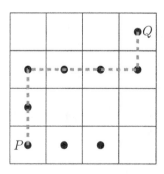   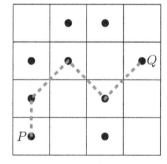

(a) 4-connectedness   (b) 8-connectedness

FIGURE 11.2: Topological connectedness

If the path contains only 4-adjacent pixels, as the path does in diagram (a), then P and Q are 4-*connected*. If the path contains 8-adjacent pixels (as shown in (b)), then P and Q are 8-*connected*.

A set of pixels all of which are 4-connected to each other is called a 4-*component*; if all the pixels are 8-connected, the set is an 8-*component*.

For example, the following image show in Figure 11.3 has two 4-components (one component containing all the pixels in the left two columns, the other component containing all the pixels in the right two columns), but only one 8-component.

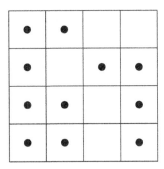

FIGURE 11.3: Components depend on connectedness

We can define *path* more formally as follows:

A 4-*path* from P to Q is a sequence of pixels

$$P = p_0, p_1, p_2, \ldots, p_n = Q$$

such that for each $i = 0, 1, \ldots, n-1$, pixel p_i is 4-adjacent to pixel p_{i+1}.

An 8-*path* is where the pixels in the sequence connecting P and Q are 8-adjacent.

11.4 Equivalence Relations

A relation $x \sim y$ between two objects x and y is an *equivalence relation* if the relation is

1. *reflexive*; $x \sim x$ for all x

2. *symmetric*; $x \sim y \iff y \sim x$ for all x and y

3. *transitive*; if $x \sim y$ and $y \sim z$ then $x \sim z$ for all x, y and z

Some examples:

1. Numeric equality: $x \sim y$ if x and y are two numbers for which $x = y$.

2. Divisors: $x \sim y$ if x and y are two numbers which have the same remainder when divided by 7.

3. Set cardinality: $S \sim T$ if S and T are two sets with the same number of elements.

4. Connectedness: $P \sim Q$ if P and Q are two connected pixels.

Here are some relations that are not equivalence relations:

1. Personal relations. Define $x \sim y$ if x and y are two people who are related to each other. This is not an equivalence relation. It is reflexive (a person is certainly related to himself or herself); symmetric; but not transitive (can you give an example?).

2. Pixel adjacency. This is not transitive.

3. Subset relation. Define $S \sim T$ if $S \subseteq T$. This is reflexive (a set is a subset of itself); transitive; but not reflexive: if $S \subseteq T$, then it is not necessarily true that $T \subseteq S$.

The importance of the equivalence relation concept is that it allows us a very neat way of dealing with issues of connectedness. We need another definition: an *equivalence class* is the set of all objects equivalent to each other.

We can now define the components of a binary image as being the equivalence classes of the connectedness equivalence relation.

11.5   Component Labeling

In this section we give an algorithm for labeling all the 4-components of a binary image, starting at the top left and working across and down. If p is the current pixel, let u be its upper 4-neighbor, and l its left-hand 4-neighbor:

	u	
l	p	

We will "scan" the image row by row moving across from left to right. We will assign labels to pixels in the image; these labels will be the numbers of the components of the image.

For descriptive purposes, a pixel in the image will be called a *foreground pixel*; a pixel not in the image will be called a *background pixel*. And now for the algorithm:

1. Check the state of p. If it is a background pixel, move on to the next scanning position. If it is a foreground pixel, check the state of u and l. If they are both background pixels, assign a new label to p. (This is the case when a new component has been encountered.)

   - If just *one* of u or l is a foreground pixel, then assign its label to p.
   - If both of u and l are foreground pixels and have the same label, assign that label to p.
   - If both of u and l are foreground pixels but have different labels, assign either of those labels to p, and make a note that those two labels are equivalent (since u and l belong to the same 4-component connected through p).

2. At the end of the scan, all foreground pixels have been labeled, but some labels may be equivalent. We now sort the labels into equivalence classes, and assign a different label to each class.

3. Do a second pass through the image, replacing the label on each foreground pixel with the label assigned to its equivalence class in the previous step.

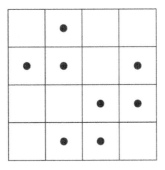

FIGURE 11.4: An example for the connectedness algorithm

We now give an example of this algorithm in practice, on the binary image shown in Figure 11.4, which has two 4-components: one the three pixels in the top left, the other the five pixels in the bottom right.

Step 1. We start moving along the top row. The first foreground pixel is the bullet in the second place, and since its upper and left neighbors are either background pixels or non-existent, we assign it the label 1, as shown in Figure 11.5(a).

In the second row, the first (foreground) pixel again has its upper or left neighbors either background or non-existent, so we assign it a new label—2, as shown in Figure 11.5(b).

The second (foreground) pixel in the second row now has both its upper and left neighbors being foreground pixels. However, they have different labels. We thus assign either of these labels to this second pixel, say label 1, and make a note that labels 1 and 2 are equivalent. This is shown in Figure 11.5(c).

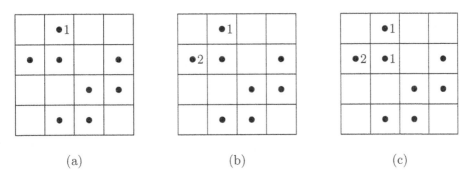

(a) (b) (c)

FIGURE 11.5: Starting the connectedness algorithm

The third foreground pixel in the second row has both its upper and left neighbors being background pixels, so we assign it a new label—3, as shown in Figure 11.6(a).

In the third row, the first foreground pixel has both its upper and left neighbors being background pixels, so we assign it a new label—4. The second (foreground) pixel in the third row now has both its upper and left neighbors being foreground pixels. However, they have different labels. We thus assign either of these labels to this second pixel, say label 3, and make a note that labels 3 and 4 are equivalent. This brings us up to 11.6(b).

In the fourth row, the first foreground pixel has both its upper and left neighbors being background pixels, so we assign it a new label—5. The second (foreground) pixel in the fourth row now has both its upper and left neighbors being foreground pixels. However, they have different labels. We thus assign either of these labels to this second pixel, say label 4, and make a note that labels 4 and 5 are equivalent, as shown in Figure 11.6(c).

This completes Step 1.

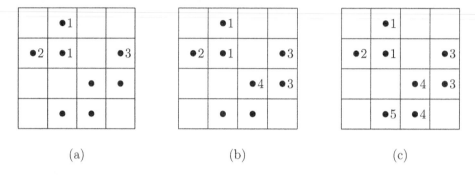

(a)   (b)   (c)

FIGURE 11.6: Continuing the connectedness algorithm

Step 2. We have the following equivalence classes of labels:

$$\{1, 2\} \quad \text{and} \quad \{3, 4, 5\}.$$

Assign label 1 to the first class, and label 2 to the second class.

Step 3. Each pixel with labels 1 or 2 from the first step will be assigned label 1, and each pixel with labels 3, 4, or 5 from the first step will be assigned label 2, as shown in Figure 11.7.

	•1		
•1	•1		•2
		•2	•2
	•2	•2	

FIGURE 11.7: Finishing the connectedness algorithm

This completes the algorithm.

The algorithm can be modified to label 8-components of an image, but in Step 1 we need to consider diagonal elements of p:

d	u	e
l	p	

The algorithm is similar to the previous algorithm; Step 1 is changed as follows:

> If p is a background pixel, move on to the next scanning position. If it is a foreground pixel, check d, u, e, and l. If they are all background pixels, assign a new label to p. If just one is a foreground pixel, assign its label to p. If two or more are foreground pixels, assign any of their labels to p, and make a note that all their labels are equivalent.

Steps 2 and 3 are as before.

This algorithm is implemented by the `bwlabel` function, and in Python by the `label` function in the `measure` module of `skimage`. We shall give an example on a small image:

```
>> im = zeros(8,8);
>> im(2:4,3:6) = 1;
>> im(5:7,2) = 1;
>> im(6:7,5:8) = 1;
>> im(8,4:5) = 1
```
MATLAB/Octave

```
In :  im = np.zeros((8,8))
In :  im[1:4,2:6] = 1
In :  im[4:7,1] = 1
In :  im[5:7,4:8] = 1
In :  im[7,3:5] = 1; print im
```
Python

In MATLAB or Octave the commands

```
>> bwlabel(im,4)
>> bwlabel(im,8)
```
MATLAB/Octave

and in Python the commands

```
In :  sk.measure.label(im,4,background=0)+1
In :  sk.measure.label(im,8,background=0)+1
```
Python

produce these results:

4-connected components:

```
0 0 0 0 0 0 0 0
0 0 2 2 2 2 0 0
0 0 2 2 2 2 0 0
0 0 2 2 2 2 0 0
0 1 0 0 0 0 0 0
0 1 0 0 3 3 3 3
0 1 0 0 3 3 3 3
0 0 0 3 3 0 0 0
```

8-connected components:

```
0 0 0 0 0 0 0 0
0 0 1 1 1 1 0 0
0 0 1 1 1 1 0 0
0 0 1 1 1 1 0 0
0 1 0 0 0 0 0 0
0 1 0 0 2 2 2 2
0 1 0 0 2 2 2 2
0 0 0 2 2 0 0 0
```

To experiment with a real image, let's try to count the number of pinenuts in the image `pinenuts.png`. The image must first be thresholded to obtain a binary image showing only the bacteria, and then `bwlabel` applied to the result. We can find the number of objects simply by finding the largest label produced.

```
>> n = imread('pinenuts.png');
>> n1 = n<125; n1 = imopen(n1,cr);
>> nl = bwlabel(n1);
>> max(nl(:))
```
MATLAB/Octave

or

```
In :   n = io.imread('pinenuts.png')
In :   n1 = n<124; n1 = mo.binary_opening(n1,cr)
In :   (sk.measure.label(n1,4,background=0)+1).max()
```

Python

which produces the result of 78, being the required number.

11.6   Lookup Tables

Lookup tables provide a very neat and efficient method of binary image processing. Consider the 3×3 neighborhood of a pixel. Since there are 9 pixels in this neighborhood, each with two possible states, the total number of different neighborhoods is $2^9 = 512$. Since the output of any binary operation will be either zero or one, a *lookup table* is a binary vector of length 512, each element representing the output from the corresponding neighborhoods. Note that this is slightly different from the lookup tables discussed in Section 4.4. In that section, lookup tables were applied to single pixel gray values; here they are applied to neighborhoods.

The trick is then to be able to order all the neighborhoods, so that we have a one-one correspondence between neighborhoods and output values. This is done by giving each pixel in the neighborhood a weighting:

1	8	64
2	16	128
4	32	256

The "value" of the neighborhood is obtained by adding up the weights of one-valued pixels. That value is then the index to the lookup table. For example, the following show neighborhoods and their values:

0	1	0
1	1	0
0	0	1

Value $= 2 + 8 + 16 + 256 = 282$

1	0	0
0	1	1
1	1	1

Value $= 1 + 4 + 16 + 32 + 128 + 256 = 437$

To apply a lookup table, we have to make it first. It would be tedious to create a lookup table element by element, so we use the `makelut` function, which defines a lookup table according to a rule. Its syntax is

```
makelut(function, n, P1, P2, ...)
```

where `function` is a string defining a MATLAB matrix function, `n` is either 2 or 3, and `P1`, `P2` and so on are optional parameters to be passed to the function. We note that `makelut` allows for lookup tables on 2×2 neighborhoods; in this case the lookup table has only $2^4 = 16$ elements.

Suppose we wish to find the 4-boundary of an image. We define a pixel to be a boundary pixel if it is a foreground pixel which is 4-adjacent to a background pixel. So the function to be used in `makelut` is a function that returns one if and only if the central pixel of the 3×3 neighborhood is a boundary pixel:

```
>> bdy = @(x) (x(5)==1) & (x(2)*x(4)*x(6)*x(8)==0)
```

MATLAB/Octave

Note that for this function we are using the single value matrix indexing scheme, so that `x(5)` is the central pixel of a 3×3 matrix `x`, and `x(2)`, `x(4)`, `x(6)`, `x(8)` are the pixels 4-adjacent to the center.

Now we can make the lookup table:

```
>> lut=makelut(f,3);
```

MATLAB/Octave

and apply it to an image:

```
>> c = imread('circles.png');
>> cw = applylut(c,lut);
>> imshow(cw)
```

MATLAB/Octave

In Python there are no lookup functions as such, but the same affect can be obtained using the `generic_filter` method from the `ndimage` module. Note that the function passed to the filter as its second argument is considered an array with only one row:

```
In :  def bdy4(x):
...:      return ((x[4]==1) and (x[1]*x[3]*x[5]*x[7]==0))*1
...:
In :  c = io.imread('circles.png').astype('bool')*1
In :  cw = ndi.generic_filter(c,bdy4,size=(3,3))
```

Python

and the result is shown in Figure 11.8(a).

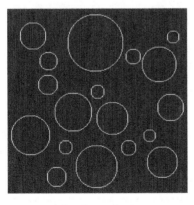
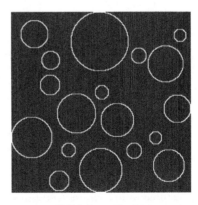

(a) 4-boundaries   (b) 8-boundaries

FIGURE 11.8: Boundaries of circles

We can easily adjust the function to find the 8-boundary pixels; these being foreground pixels which are 8-adjacent to background pixels:

```
>> bdy8 = @(x) (x(5)==1) & (prod(x(:))==0)
>> lut8 = makelut(bdy8,3);
>> cw8 = applylut(c,lut8);
>> imshow(cw)
```

MATLAB/Octave

or

```
In :  def bdy8(x):
...:      return x[4]==1 and x.prod()==0
...:
In :  cw8 = ndi.generic_filter(c,bdy8,size=(3,3))
```

Python

and the result is shown in Figure 11.8(b). Note that this is a "stronger" boundary, because more pixels are classified as boundary pixels.

As we shall see in Section 11.8, lookup tables can be used to great effect in performing skeletonization.

11.7   Distances and Metrics

It is necessary to define a function that provides a measure of *distance* between two points x and y on a grid. A distance function $d(x, y)$ is called a *metric* if it satisfies all of:

1. $d(x, y) = d(y, x)$ (symmetry)

2. $d(x, y) \geq 0$ and $d(x, y) = 0$ if and only if $x = y$ (positivity)

3. $d(x, y) + d(y, z) \leq d(x, z)$ (the triangle inequality)

A standard distance metric is provided by *Euclidean distance*, where if $x = (x_1, x_2)$ and $y = (y_1, y_1)$ then

$$d(x, y) = \sqrt{(x_1 - y_1)^2 + (x_2 - y_2)^2}.$$

This is just the length of the straight line between the points x and y. It is easy to see that the first two properties are satisfied by this metric: it is always positive, and is zero only when $x_1 = y_1$ and $x_2 = y_2$; that is, when $x = y$. The third property may be proved very easily; it simply says that given three points x, y, and z, then it is shorter to go from x to z directly than via y.

However, if we are constrained to points on a grid, then the Euclidean metric may not be applicable. Figure 11.9 shows the shortest paths for the Euclidean metric, and for 4-connected and 8-connected paths. In this figure, the Euclidean distance is $\sqrt{5^2 + 2^2} \approx 5.39$, and the 4-path (the dashed line) and 8-path (dotted line) have length, measured by the number of line segments in each, of 7 and 5, respectively.

Metrics for measuring distance by 4-paths and 8-paths are given by the following two functions:

$$d_4(x, y) = |x_1 - y_1| + |x_2 - y_2|$$
$$d_8(x, y) = \max\{|x_1 - y_1|, |x_2 - y_2|\}$$

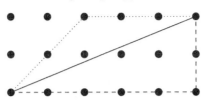

FIGURE 11.9: Comparison of three metrics

As with the Euclidean distance, the first two properties are immediate; to prove the triangle inequality takes a little effort. The metric d_4 is sometimes known as the *taxicab metric*, as it gives the length of the shortest taxicab route through a rectangular grid of streets.

The Distance Transform

In many applications, it is necessary to find the distance of every pixel from a region R. This can be done using the standard Euclidean distance defined previously. However, this means that to calculate the distance of (x, y) to R, we need to determine all possible distances from (x, y) to pixels in R, and take the smallest value as our distance. This is very computationally inefficient: if R is large, then we have to compute many square roots. A saving can be obtained by noting that since the square root function is increasing, we can write the minimum distance as

$$md(x, y) = \sqrt{\min_{(p,q)\in R} \left((x - p)^2 + (y - q)^2\right)}$$

which involves only one square root. But even this definition is slow to compute. There is a great deal of arithmetic involved, and the finding of a smallest value.

The *distance transform* is a computationally efficient method of finding such distance. We shall describe it in a sequence of steps:

Step 1. Attach to each pixel (x, y) in the image a label $d(x, y)$ giving its distance from R. Start with labeling each pixel in R with 0, and each pixel not in R with ∞.

Step 2. We now travel through the image pixel by pixel. For each pixel (x, y) replace its label with

$$\min\{d(x, y), d(x + 1, y) + 1, d(x - 1, y) + 1, d(x, y - 1) + 1, d(x, y + 1) + 1\}$$

using $\infty + 1 = \infty$.

Step 3. Repeat Step 2 until all labels have been converted to finite values.

To give some examples of Step 2, suppose we have this neighborhood:

$$\begin{matrix} \infty & \infty & \infty \\ 2 & \infty & \infty \\ \infty & 3 & \infty \end{matrix}$$

We are only interested in the center pixel (whose label we are about to change), and the four pixels above, below, and to the left and right. To these four pixels we add 1:

$$\begin{matrix} & \infty & \\ 3 & \infty & \infty \\ & 4 & \end{matrix}$$

The minimum of these five values is 3, and so that is the new label for the center pixel.

Suppose we have this neighborhood:

$$
\begin{matrix}
2 & \infty & \infty \\
\infty & \infty & \infty \\
3 & \infty & 5
\end{matrix}
$$

Again, we add 1 to each of the four pixels above, below, to the left and right, and keep the center value:

$$
\begin{matrix}
 & \infty & \\
\infty & \infty & \infty \\
 & \infty &
\end{matrix}
$$

The minimum of these values is ∞, and so at this stage the pixel's label is not changed.

Suppose we take an image whose labels after Step 1 are given below.

∞	∞	∞	∞	∞	∞
∞	∞	0	0	∞	∞
∞	∞	∞	0	∞	∞
∞	∞	∞	0	∞	∞
∞	∞	∞	0	0	∞
∞	∞	∞	∞	∞	∞

Step 1

∞	∞	1	1	∞	∞
∞	1	0	0	1	∞
∞	∞	1	0	1	∞
∞	∞	1	0	1	∞
∞	∞	1	0	0	1
∞	∞	∞	1	1	∞

Step 2 (first pass)

∞	2	1	1	2	∞
2	1	0	0	1	2
∞	2	1	0	1	2
∞	2	1	0	1	2
∞	2	1	0	0	1
∞	∞	2	1	1	2

Step 2 (second pass)

3	2	1	1	2	3
2	1	0	0	1	2
3	2	1	0	1	2
3	2	1	0	1	2
3	2	1	0	0	1
∞	3	2	1	1	2

Step 2 (third pass)

3	2	1	1	2	3
2	1	0	0	1	2
3	2	1	0	1	2
3	2	1	0	1	2
3	2	1	0	0	1
4	3	2	1	1	2

Step 2 (final pass)

At this stage we stop, as all label values are finite.

An immediate observation is that the distance values given are not in fact a very good approximation to "real" distances. To provide better accuracy, we need to generalize the above transform. One way to do this is to use the concept of a "mask"; the mask used above was

$$
\begin{matrix}
 & 1 & \\
1 & 0 & 1 \\
 & 1 &
\end{matrix}
$$

Step two in the transform then consists of adding the corresponding mask elements to labels of the neighboring pixels, and taking the minimum. To obtain good accuracy with simple arithmetic, the mask will generally consist of integer values, but the final result may require scaling.

Suppose we apply the mask

$$
\begin{matrix}
4 & 3 & 4 \\
3 & 0 & 3 \\
4 & 3 & 4
\end{matrix}
$$

to the above image. Step 1 is as above; Step 2 is:

∞	4	3	3	4	∞
∞	3	0	0	3	∞
∞	4	3	0	3	∞
∞	∞	3	0	3	∞
∞	∞	3	0	0	3
∞	∞	4	3	3	4

Step 2 (first pass)

7	4	3	3	4	7
6	3	0	0	3	6
7	4	3	0	3	6
8	6	3	0	3	6
∞	6	3	0	0	3
∞	7	4	3	3	4

Step 2 (second)

7	4	3	3	4	7
6	3	0	0	3	6
7	4	3	0	3	6
8	6	3	0	3	6
9	6	3	0	0	3
10	7	4	3	3	4

Step 2 (third)

at which point we stop, and divide all values by three:

2.3	1.3	1	1	1.3	2.3
2	1	0	0	1	2
2.3	1.3	1	0	1	2
2.7	2	1	0	1	2
3	2	1	0	0	1
3.3	2.3	1.3	1	1	1.3

These values are much closer to the Euclidean distances than those provided by the first transform.

Even better accuracy can be obtained by using the mask

	11		11	
11	7	5	7	11
	5	0	5	
11	7	5	7	11
	11		11	

and dividing the final result by 5:

11	7	5	5	7	11
10	5	0	0	5	10
11	7	5	0	5	10
14	10	5	0	5	7
16	10	5	0	0	5
16	11	7	5	5	7

Result of transform

2.2	1.4	1	1	1.4	2.2
2	1	0	0	1	2
2.2	1.4	1	0	1	2
2.8	2	1	0	1	1.4
3.2	2	1	0	0	1
3.2	2.2	1.4	1	1	1.4

After division by 5

The method we have described can in fact be very slow; for a large image we may require many passes before all distance labels become finite. A quicker method requires *two* passes only: the first pass starts at the top left of the image, and moves left to right, and top to bottom. The second pass starts at the bottom right of the image, and moves right to left, and from bottom to top. For this method we break the mask up into two halves; one half inspects only values to the left and above (this is used for the first pass), and the other half inspects only values to the right and below (this is used for the second pass).

Such pairs of masks are shown in Figures 11.10 to 11.12, and the solid lines show how the original mask is broken up into its two halves.

We apply these masks as follows: first surround the image with zeros (such as for spatial filtering), so that at the edges of the image the masks have values to work with. For the forward pass, for each pixel at position (i, j), add the values given in the forward mask to its neighbors and take the minimum, and replace the current label with this minimum.

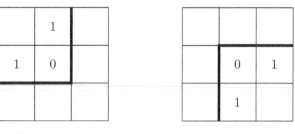

Forward mask   Backward mask

FIGURE 11.10: Simple pair of masks for the two-pass distance transform

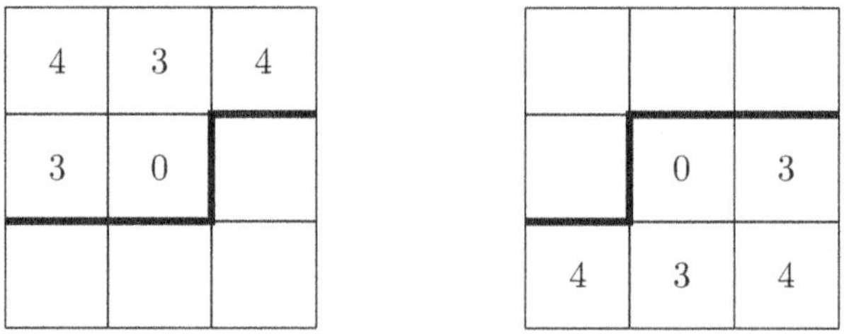

4	3	4
3	0	

Forward mask

	0	3
4	3	4

Backward mask

FIGURE 11.11: More accurate pair of masks for the two-pass distance transform

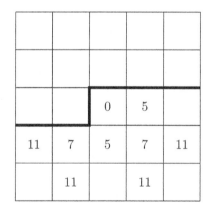

Forward mask   Backward mask

FIGURE 11.12: Masks that are more accurate still

This is similar to Step 2 in the original algorithm, except that we are using less pixels. We do the same for the backward pass, except that we use the backward mask, and we start at the bottom right.

If we apply forward masks 1 and 2 to our image, the results of the forward passes are:

∞	∞	∞	∞	∞	∞		∞	∞	∞	∞	∞	∞
∞	∞	0	0	1	2		∞	∞	0	0	3	6
∞	∞	1	0	1	2		∞	4	3	0	3	6
∞	∞	2	0	1	2		8	7	5	0	3	6
∞	∞	3	0	0	1		11	8	4	0	0	3
∞	∞	4	1	1	2		12	8	4	3	3	4
		Use of mask 1							Use of mask 2			

After applying the backward masks, we will obtain the distance transforms as above.

Implementing the Distance Transform

We can easily write a function to perform the distance transform using the second method as above. Our function will implement the transform as follows:

1. Using the size of the mask, pad the image with an appropriate number of zeros.

2. Change each zero to infinity, and each one to zero.

3. Create forward and backward masks.

4. Perform a forward pass: replace each label with the minimum of its neighborhood plus the forward mask.

5. Perform a backward pass: replace each label with the minimum of its neighborhood plus the backward mask.

Rather than provide a general function, here is how to apply the masks from Figure 11.11 to a binary image.

Suppose first that the `circles.png` image has been read in to a variable `c`. In MATLAB or Octave, the first three items above can be obtained with:

```
>> c2 = padarray(c,[1,1]);
>> c2(find(c2==0))=Inf; c2(find(c2==1))=0;
>> fmask = [4 3 4;3 0 0;0 0 0]; bmask = [0 0 0;0 0 3;4 3 4]
```

and in Python by

```
In :  c2 = np.pad(c,[1,1],mode='constant').astype('float64')
In :  c2[np.where(c2==0)]=float('inf'); c2[np.where(c2==1)]=0
In :  fmask = np.array([[4,3,3],[3,0,0],[0,0,0]])
In :  bmask = np.array([[0,0,0],[0,0,3],[4,3,3]])
```

The forward and backward passes can be performed with nested loops:

```
>> for i = 1:256,
>    for j = 1:256,
>      c2(i+1,j+1) = min(min(c2(i:i+2,j:j+2)+fmask));
>    end
> end

>> for i = 256:-1:1,
>    for j = 256:-1:1,
>      c2(i+1,j+1) = min(min(c2(i:i+2,j:j+2)+bmask));
>    end
> end
```
MATLAB/Octave

or in Python with

```
In :  for i in range(256):
...:      for j in range(256):
...:          c2[i+1,j+1] = (c2[i:i+3,j:j+3]+fmask).min()
...:

In :  for i in range(256)[::-1]:
...:      for j in range(256)[::-1]:
...:          c2[i+1,j+1] = (c2[i:i+3,j:j+3]+bmask).min()
...:
```
Python

The result can be shown as

```
>> imshow(mat2gray(2(2:257,2:257)))
```
MATLAB/Octave

or

```
In :  io.imshow(c2[1:257,1:257])
```
Python

and is shown in Figure 11.13(a). Sometimes better information can be obtained by inverting the image and finding the distance inside the objects, as shown in Figure 11.13(b).

11.8   Skeletonization

The *skeleton* of a binary object is a collection of lines and curves that encapsulate the size and shape of the object. There are in fact many different methods of defining a skeleton, and for a given object, many different possible skeletons. But to give a general idea, we show a few examples in Figure 11.14.

A problem with skeletonization is that very small changes to an object can result in large changes to the skeleton. Figure 11.15 shows what happens if we take away or add to the central image of Figure 11.14.

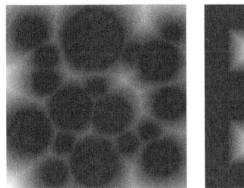

(a) "External" distances   (b) "Internal" distances

FIGURE 11.13: Examples of a distance transform

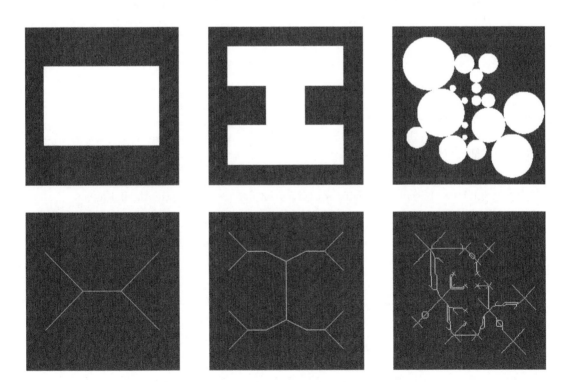

FIGURE 11.14: Examples of skeletonization

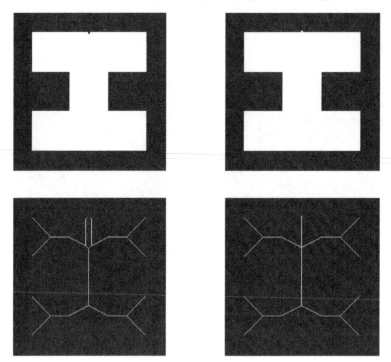

FIGURE 11.15: Skeletons after small changes to an object

One very popular way of defining a skeleton is by the *medial axis* of an object: a pixel is on the medial axis if is equidistant from at least two pixels on the boundary of the object. To implement this definition directly requires a quick and efficient method of obtaining approximate distances; one way is to use the distance transform. Other ways of approaching the medial axis are:

- Imagine the object to be burning up by a fire that advances at a constant rate from the boundary. The places where two "lines" of fire meet form the medial axis.

- Consider the set of all circles lying within the object, which touch at least two points on the boundary. The centers of all such circles form the medial axis. Figure 11.16 shows this in action.

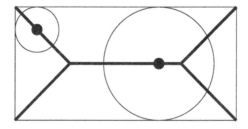

FIGURE 11.16: The medial axis of an object

A very good account of the medial axis (and of skeletonization in general) is given by Parker [32].

Topological methods provide some powerful methods of skeletonization in that we can directly define those pixels which are to be deleted to obtain the final skeleton. In general, we want to delete pixels that can be deleted without changing the *connectivity* of an object, which do not change the number of components, change the number of holes, or the relationship of objects and holes. For example, Figure 11.17 shows a non-deletable pixel; deleting the center (boxed) pixel introduces a hole into the object. In Figure 11.18, there is another

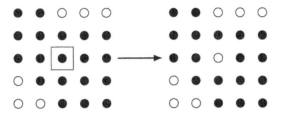

FIGURE 11.17: A non-deletable pixel: creates a hole

example of a non-deletable pixel; in this case, deletion *removes* a hole, in that the hole and the exterior become joined. In Figure 11.19, there is a further example of a non-deletable

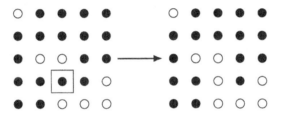

FIGURE 11.18: A non-deletable pixel: removes a hole

pixel; in this case, deletion breaks the object into two separate components. Sometimes we

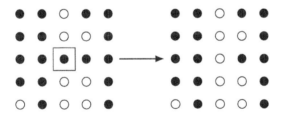

FIGURE 11.19: A non-deletable pixel: disconnects an object

need to consider whether the object is 4-connected or 8-connected. In the previous examples this was not a problem. However, look at the examples in Figure 11.20. In Figure 11.20(a), the central point can not be deleted without changing both the 4-connectivity and the 8-connectivity. In Figure 11.20(b) deleting the central pixel will change the 4-connectivity, but not the 8-connectivity. A pixel that can be deleted without changing the 4-connectivity of the object is called *4-simple*; similarly, a pixel that can be deleted without changing the 8-connectivity of the object is called *8-simple*. Thus, the central pixel in Figure 11.20(a) is neither 4-simple nor 8-simple, but the central pixel in Figure 11.20(b) is 8-simple but not 4-simple.

A pixel can be tested for deletability by checking its 3×3 neighborhood. Look again at Figure 11.20(a). Suppose the central pixel is deleted. There are two options:

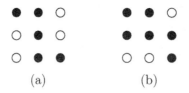

(a)   (b)

FIGURE 11.20: Are the center points deletable?

1. The top two pixels and the bottom two pixels become separated; in effect breaking up the object.

2. The top two pixels and the bottom two pixels are joined by a chain of pixels outside the shown neighborhood. In this case, all pixels will encircle a hole, and removing the central pixel will remove the hole, as in Figure 11.18.

To check whether a pixel is 4-simple or 8-simple, we introduce some numbers associated with the neighborhood of a foreground pixel p. First define N_p to be the 3×3 neighborhood of p, and N_p^* to be the 3×3 neighborhood *excluding* p. Then:

$A(p) = $ the number of 4-components in N_p^*

$C(p) = $ the number of 8-components in N_p^*

$B(p) = $ the number of foreground pixels in N_p^*

For example, in Figure 11.20(a), we have

$A(p) = 2$

$C(p) = 2$

$B(p) = 4$

and in Figure 11.20(b) we have

$A(p) = 2$

$C(p) = 1$

$B(p) = 5$

We can see the last example by deleting the central pixel and enumerating the components of the remaining foreground pixels. This is shown in Figure 11.21. Simple points can now

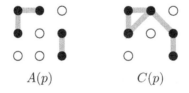

$A(p)$ $C(p)$

FIGURE 11.21: Components of N_p^*

be characterized entirely in terms of the values of $A(p)$ and $C(p)$:

A foreground pixel p is 4-simple if and only if $A(p) = 1$, and is 8-simple if and only if $C(p) = 1$.

Returning again to Figure 11.20(b), since $C(p) = 1$ the central pixel p is 8-simple and so can be deleted without affecting the 8-connectivity of the object. But since $A(p) \neq 1$, the central pixel p is *not* 4-simple and so cannot be deleted without affecting the 4-connectivity of the object. This is exemplified in Figure 11.3.

Note that $A(p)$ and $B(p)$ can be computed simply by using the various labeling functions discussed earlier. For example, consider the neighborhood shown in Figure 11.20(b):

```
>> n = [1 1 0;1 1 1;0 0 1]
>> n(5) = 0;
>> max(max(bwlabel(n,4)))
ans =
         2

>> max(max(bwlabel(n,8)))
ans =
         1
```

MATLAB/Octave

Similarly:

```
In :   from skimage.measure import label
In :   n = np.array([[1,1,0],[1,1,1],[0,0,1]])
In :   n[1,1] = 0
In :   label(n,4,background=0).max()+1
Out:   2
In :   label(n,8,background=0).max()+1
Out:   1
```

Python

How Not To Do Skeletonization

So now we know how to check if a pixel can be deleted without effecting the connectivity of the object. In general, a skeletonization algorithm works by an iteration process: at each step identifying deletable pixels and deleting them. The algorithm will continue until no further deletions are possible.

One way to remove pixels is as follows:

At each step, find all foreground pixels that are 4-simple and delete them all.

Sounds good? Let's try it on a small rectangle of size 2×4:

```
0  0  0  0  0  0
0  1  1  1  1  0
0  1  1  1  1  0
0  0  0  0  0  0
```

If we check the pixels in this object carefully, we will find that they are *all* 4-simple. Deleting them all will thus remove the object completely: a very undesirable result. Clearly we need to be a bit more careful about which pixels we can delete and when. We need to add an extra test for deletability, so that we do not delete too many pixels.

We have two options here:

1. We can provide a step-wise algorithm, and change the test for deletability at each step.

2. We can apply a different test for deletability according to where the pixel lies on the image grid.

Algorithms that work according to the first option are called *subiteration algorithms*; algorithms that work according to the second option are called *subfield algorithms*.

The Zhang-Suen Skeletonization Algorithm

This algorithm has attained the status of a modern classic; it has some faults (as we shall see), but it works fairly fast, and most of the time produces acceptable results. It is an example of a subiteration algorithm, in that we apply a slightly different deletability test for different steps of the algorithm. In each step, the neighboring pixels are indexed as

$$
\begin{array}{ccc}
p_1 & p_4 & p_7 \\
p_2 & p_5 & p_8 \\
p_3 & p_6 & p_9
\end{array}
$$

where $p_5 = p$ is the pixel being considered for deletion.

Step N

Flag a foreground pixel $p = 1$ to be deletable if

1. $2 \leq B(p) \leq 6$,

2. $X(p) = 1$,

3. If N is odd, then

$$p_4 \cdot p_8 \cdot p_6 = 0$$
$$p_8 \cdot p_6 \cdot p_2 = 0.$$

   If N is even, then

$$p_4 \cdot p_8 \cdot p_2 = 0$$
$$p_4 \cdot p_6 \cdot p_2 = 0.$$

Delete all flagged pixels.
Continue until there are no more deletable pixels in two successive iterations.

Item 3 can be written alternatively as:

---

3. If N is odd, then
$$p_6 = 0, \text{ or } p_8 = 0, \text{ or } p_2 = p_4 = 0$$
   If N is even, then
$$p_2 = 0, \text{ or } p_4 = 0, \text{ or } p_6 = p_8 = 0.$$

If we check the diagram for the neighbors of a foreground pixel p, we see that we can rephrase this item as:

For odd iterations, delete only pixels that are on the right hand side, or bottom of an object, or on a northwest corner.

For even iterations, delete only pixels that are on the left hand side, or top of an object, or on a southeast corner.

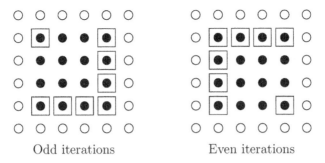

Odd iterations   Even iterations

FIGURE 11.22: Deletion in the Zhang-Suen algorithm

Figure 11.22 shows pixels that may be considered for deletion at different iterations. Item 1 in the algorithm ensures that we do not delete pixels that have only one neighbor, or have seven or more. If a pixel has only one neighbor, it would be at the end of a skeletal line, and we would not want it to be deleted. If a pixel has seven neighbors, then deleting it would start an unacceptable erosion into the object's shape. This item thus ensures that the basic shape of the object is kept by the skeleton. Item 2 is our standard connectivity condition.

A major fault with this algorithm is that there are objects that will be deleted completely. Consider a 2×2 square:

$$
\begin{array}{cccc}
0 & 0 & 0 & 0 \\
0 & 1 & 1 & 0 \\
0 & 1 & 1 & 0 \\
0 & 0 & 0 & 0
\end{array}
$$

We can check carefully that every item is satisfied by every pixel, and hence every pixel will be deleted.

An example. Consider the L shape shown in Figure 11.23, where for ease of viewing we have replaced all the zeros (background pixels) with dots. The boxed pixels show those that will be deleted by Steps 1 and 2 of the algorithm. Figure 11.24 shows Steps 3 and 4 of the skeletonization. After Step 4 no more deletions are possible, and so the skeleton consists of the unboxed foreground pixels in the right-hand diagram of Figure 11.24. Note that the skeleton does not include the corners of the original object.

Implementation in MATLAB and Octave. We can implement this algorithm easily in MATLAB or Octave; we use lookup tables: one for the odd iterations, and one for the even. We then apply these lookup tables alternately until there is no change in the image for two successive iterations. We manage this by keeping three images at any given time: the *current* image, the *previous* image, and the *last* (that is, before the previous) image. If the current and last images are equal, we stop. Otherwise, "push" the images back: the previous image becomes last, and the current image becomes the previous image. We then apply whichever lookup table is appropriate to the current image to create the new current image. That is:

$$
\begin{aligned}
last &\leftarrow previous \\
previous &\leftarrow current \\
current &\leftarrow \texttt{applylut}(current, \texttt{lut})
\end{aligned}
$$

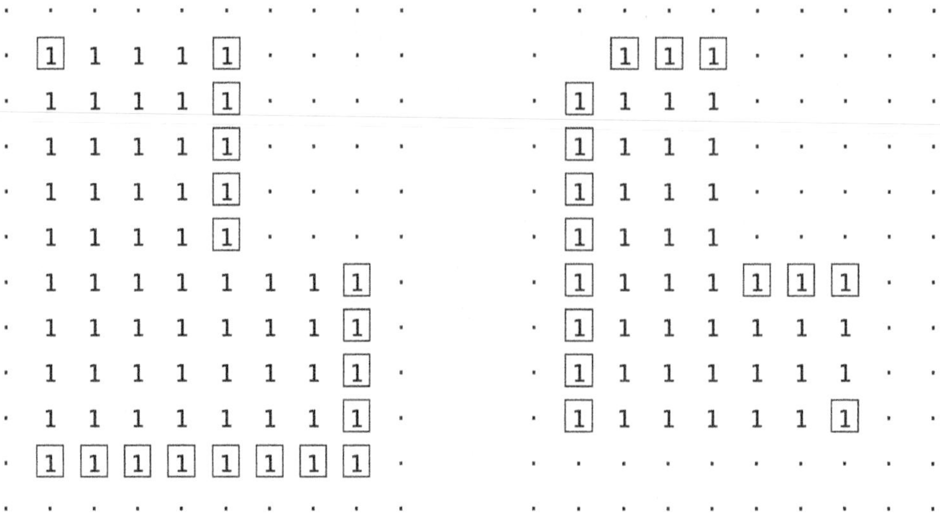

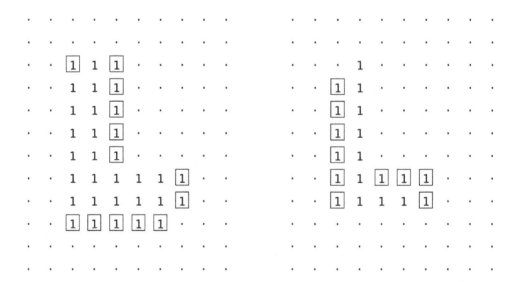

FIGURE 11.23: Steps 1 and 2 of a Zhang-Suen skeletonization

FIGURE 11.24: Steps 3 and 4 of a Zhang-Suen skeletonization

The lookup table for the odd iterations can be created as:

```
>> function out = zs_odd(x)
>    t = x;t(5)=0; B = sum(t(:));X=max(max(bwlabel(t,4)));
>      out = (B>=2) & (B<=6) & (X==1);
>      out = out&(t(6)==0 | t(8)==0 | (t(2)==0 & t(4)==0));
>  end
>> lut_odd = makelut(@zs_odd,3);
```
MATLAB/Octave

and for the even iterations as:

```
>> function out = zs_even(x)
>    t = x;t(5)=0; B = sum(t(:));X=max(max(bwlabel(t,4)));
>      out = (B>=2) & (B<=6) & (X==1);
>      out = out&(t(2)==0 | t(4)==0 | (t(6)==0 & t(8)==0));
>  end
>> lut_even = makelut(@zs_even,3);
```
MATLAB/Octave

Applying these lookup tables to the L shape from the previous figures can be easily done, but to simplify the work we shall perform two iterations at a time:

```
>> L = zeros(12,10); L(2:11,2:6)=1; L(7:11,7:9)=1;
>> previous = L;
>> current = L & ~applylut(L,lut_odd);
>> while true,
>    current = current & ~applylut(current,lut_even);
>    current = current & ~applylut(current,lut_odd);
>    if isequal(previous,current), break;end
>    previous = current;
> end
```
MATLAB/Octave

Note that the results of applying the lookup tables is an array A of zeros and ones, where the ones give the positions of the deletable foreground pixels. To delete these pixels from the current array X we need to perform a set difference $X \backslash A$, which can be implemented with X & ~A.

At this stage, the value of `current` will be the skeleton shown in Figure 11.24.

Implementation in Python. In fact, the Zhang-Suen algorithm is implemented as the `skeletonize` method in the `skimage.morphology` module. But as an example, we will show how we can do it ourselves. Start with writing a function for both the odd and even iterations, using a separate argument (here called `parity`) to distinguish them:

```
In :    def zs_fun(x,parity):
...:        n = np.reshape(x,[3,3])
...:        n[1,1] = 0
...:        B = n.sum()
...:        X = sk.measure.label(n,background=0).max()+1
...:        out = (B>=2) and (B<=6) and (X==1)
...:        if parity==0:
...:            out = out & (x[1]==0 or x[3]==0 or (x[5]==0 and x[7]==0))
...:        elif parity==1:
...:            out = out & (x[5]==0 or x[7] ==0 or (x[1]==0 and x[3]==0))
...:        return out
```
Python

The functions will be applied using the `ndimage.generic_filter` method:

```
In :    from scipy.ndimage import generic_filter as gfilt
In :    L = zeros((12,10)); L[1:11,1:6]=1; L[6:11,6:9]=1;
In :    prev = np.copy(L)
In :    curr = prev*(1-gfilt(prev,zs_fun,size=(3,3),extra_arguments=(1,)))
```
Python

Then as with MATLAB or Octave we apply the even and odd iterations until there is no more change:

```
In :    while True:
...:        curr = curr*(1-gfilt(curr,zs_fun,size=(3,3),extra_arguments=(0,)
)
...:        curr = curr*(1-gfilt(curr,zs_fun,size=(3,3),extra_arguments=(1,)
)
...:        if np.array_equal(prev,curr):
...:            break
...:        else:
...:            prev = np.copy(curr)
...:
```
Python

Two more examples; the circles image and the "nice work" image. Both the images and their skeletonizations are shown in Figure 11.25.

The Guo-Hall Skeletonization Algorithm

There are in fact a number of Guo-Hall algorithms; we will investigate one that has the advantages of being simple to describe, easy to implement, fast to run, and gives good results. What could be better?

The Guo-Hall algorithm is an example of a subfield algorithm. We imagine the image grid being labeled with 1's and 2's in a chessboard configuration:

```
1  2  1  2  ···
2  1  2  1  ···
1  2  1  2  ···
2  1  2  1  ···
⋮  ⋮  ⋮  ⋮  ⋱
```

FIGURE 11.25: Examples of the Zhang-Suen skeletonization

At Step 1, we only consider foreground pixels whose labels are 1. At Step 2, we only consider foreground pixels whose labels are 2. We continue alternating between pixels labeled 1 and pixels labeled 2 from step to step until no more deletions are possible. Here is the algorithm:

Flag a foreground pixel p as deletable if all of these conditions are met:

1. $C(p) = 1$,

2. $B(p) > 1$,

3. p is 4-adjacent to a background pixel

then delete all flagged pixels. This is done, in parallel, at alternate iterations for each of the two subfields.

Continue until no deletions are possible in two successive iterations.

Consider our "L" example from above. We first superimpose a chessboard of 1's and 2's. In Step 1 we just consider 1's only. Step 1 shown in Figure 11.26 illustrates the first step: we delete only those 1's satisfying the Guo-Hall deletability conditions. These pixels are shown in squares. Having deleted them, we now consider 2's only; the deletable 2's are shown in Step 2.

```
 .   .   .   .   .   .   .   .   .   .          .   .   .   .   .   .   .   .   .   .
 .  [1]  2  [1]  2  [1]  .   .   .   .          .   .  [2]  .  [2]  .   .   .   .   .
 .   2   1   2   1   2   .   .   .   .          .  [2]  1  [2]  1  [2]  .   .   .   .
 .  [1]  2   1   2  [1]  .   .   .   .          .   .  [2]  1  [2]  .   .   .   .   .
 .   2   1   2   1   2   .   .   .   .          .  [2]  1   2   1  [2]  .   .   .   .
 .  [1]  2   1   2  [1]  .   .   .   .          .   .  [2]  1  [2]  .   .   .   .   .
 .   2   1   2   1   2  [1]  2  [1]  .          .  [2]  1   2   1  [2]  .  [2]  .   .
 .  [1]  2   1   2   1   2   1   2   .          .   .  [2]  1   2   1  [2]  1  [2]  .
 .   2   1   2   1   2   1   2  [1]  .          .  [2]  1   2   1   2   1  [2]  .   .
 .  [1]  2   1   2   1   2   1   2   .          .   .   2   1  [2]  1  [2]  1  [2]  .
 .   2  [1]  2  [1]  2  [1]  2  [1]  .          .   2   .  [2]  .  [2]  .  [2]  .   .
 .   .   .   .   .   .   .   .   .   .          .   .   .   .   .   .   .   .   .   .
```

FIGURE 11.26: Steps 1 and 2 of a Guo-Hall skeletonization

Steps 3 and 4 as shown in Figure 11.27 continue the work; by Step 4 there are no more deletions to be done, and we stop. We notice two aspects of the Guo-Hall algorithm as compared with Zhang-Suen:

1. More pixels may be deleted at each step, so we would expect the algorithm to work faster.

2. The final result includes more corner information than the Zhang-Suen algorithm.

We can implement this in using very similar means to our implementation of Zhang-Suen.

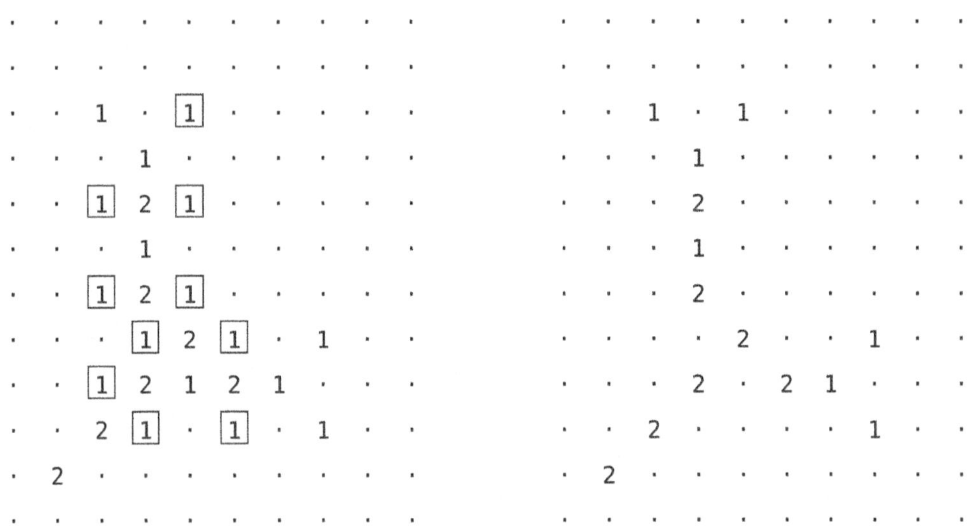

```
.   .   .   .   .   .   .       .   .   .   .   .   .   .

.   .   .   .   .   .   .       .   .   .   .   .   .   .

.   .  1   . [1]  .   .   .   .       .   .  1   .  1   .   .   .   .

.   .   .  1   .   .   .   .   .       .   .   .  1   .   .   .   .

.   . [1] 2 [1]  .   .   .   .       .   .   .  2   .   .   .   .

.   .   .  1   .   .   .   .   .       .   .   .  1   .   .   .   .

.   . [1] 2 [1]  .   .   .   .       .   .   .  2   .   .   .   .

.   .   . [1] 2 [1]  . 1   .   .       .   .   .   .  2   .   . 1   .   .

.   . [1] 2  1  2  1   .   .   .       .   .   .   .  2   . 2  1   .   .

.   .  2 [1]  . [1]  . 1   .   .       .   .  2   .   .   .   . 1   .   .

.  2   .   .   .   .   .   .   .       .  2   .   .   .   .   .   .   .

.   .   .   .   .   .   .       .   .   .   .   .   .   .
```

FIGURE 11.27: Steps 3 and 4 of a Guo-Hall skeletonization

Implementation in Python. Start by creating a version of the image tiled with ones and twos. Using the "nice work" image stored as binary array `n`:

```
In :  r,c = n.shape
In :  n2 = np.tile(np.array([[1,2],[2,1]]),[r/2,c/2])*n
```
Python

As with the Zhang-Suen algorithm, create a function to flag deletable pixels:

```
In : def gh_fun(x,parity):
...:     n = (np.reshape(x,[3,3])>0)*1
...:     n[1,1] = 0
...:     B = n.sum()
...:     X = sk.measure.label(n,8,background=0).max()+1
...:     return ((B>=2) and (X==1) and (x[1]*x[3]*x[5]*x[7]==0) and (x
    [4]==parity))
```
Python

Then we can create the current and previous images:

```
In :  prev = np.copy(n2)
In :  curr = prev*(1-gfilt(prev,gh_fun,size=(3,3),extra_arguments=(1,)))
```
Python

Now apply the algorithm until there are no further changes:

```
In :   while True:
...:        curr = curr*(1-gfilt(curr,gh_fun,size=(3,3),extra_arguments=(2,))
            )
...:        curr = curr*(1-gfilt(curr,gh_fun,size=(3,3),extra_arguments=(1,))
            )
...:        if np.array_equal(prev,curr):
...:            break
...:        else:
...:            prev = np.copy(curr)
...:
```

Python

Implementation in MATLAB and Octave. Start as before by defining the function to apply to the subfields:

```
>> function out = gh(x)
>    n = x; n(5) = 0;
>    B = sum(n(:)); X = max(max(bwlabel(n,8)));
>    S = n(2)*n(4)*n(6)*n(8);
>    out = (B>=2 & X==1 & S==0);
> end
>> gh_lut = makelut(@gh,3);
```

MATLAB/Octave

Next, create a chessboard version of the image:

```
>> [r,c] = size(n);
>> n2 = repmat([1 2;2 1],r/2,c/2).*n;
```

MATLAB/Octave

Now apply the lookup table to the ones and twos in turn, until there is no change:

```
>> prev = n2;
>> g = apply(n2>0, gh_lut);
>> curr = prev .* ~((prev==1) & (g==1));
>> while true,
>    g = apply(curr>0, gh_lut);
>    curr = curr .* ~((curr==2) & (g==1));
>    g = applylut(curr>0,gh_lut);
>    curr = curr .* ~((curr==1) & (g==1));
>    if isequal(prev,curr), break;end
>    prev = curr;
> end
>> imshow(curr>0)
```

MATLAB/Octave

The result is shown in Figure 11.28(a).

Note the differences between these skeletons and those produced by the Zhang-Suen algorithm as shown in Figure 11.25.

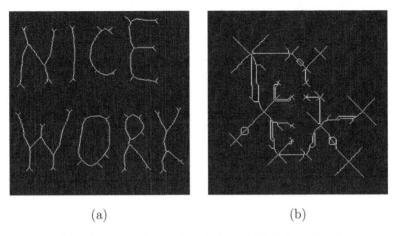

(a)   (b)

FIGURE 11.28: Examples of Guo-Hall skeletonization

Skeletonization Using the Distance Transform

The distance transform can be used to provide the skeleton of a region R. We apply the distance transform, using mask 1, to the image negative, just as we did for the circle previously. Then the skeleton consists of those pixels (i, j) for which

$$d(i, j) \geq \max\{d(i-1, j), d(i+1, j), d(i, j-1), d(i, j+1)\}.$$

For example, suppose we take a small region consisting of a single rectangle, and find the distance transform of its negative:

```
>> c = zeros(7,9); c(2:6,2:8)=1;
>> mask = [Inf 1 Inf;1 0 Inf;Inf Inf Inf];
>> cd=disttrans(1-c,mask1)

cd =

     0    0    0    0    0    0    0    0    0
     0    1    1    1    1    1    1    1    0
     0    1    2    2    2    2    2    1    0
     0    1    2    3    3    3    2    1    0
     0    1    2    2    2    2    2    1    0
     0    1    1    1    1    1    1    1    0
     0    0    0    0    0    0    0    0    0
```
MATLAB/Octave

The skeleton can be obtained by using MATLAB's `ordfilt2` function, which was introduced in Chapter 5. This can be used to find the largest value in a neighborhood, and the neighborhood can be very precisely defined:

```
>> cd2=ordfilt2(cd,5,[0 1 0;1 1 1;0 1 0])
cd2 =

        0   1   1   1   1   1   1   1   0
        1   1   2   2   2   2   2   1   1
        1   2   2   3   3   3   2   2   1
        1   2   3   3   3   3   3   2   1
        1   2   2   3   3   3   2   2   1
        1   1   2   2   2   2   2   1   1
        0   1   1   1   1   1   1   1   0

>> cd2 <= cd
ans =

        1   0   0   0   0   0   0   0   1
        0   1   0   0   0   0   0   1   0
        0   0   1   0   0   0   1   0   0
        0   0   0   1   1   1   0   0   0
        0   0   1   0   0   0   1   0   0
        0   1   0   0   0   0   0   1   0
        1   0   0   0   0   0   0   0   1
```

MATLAB/Octave

In Python, all of this can be done very similarly. First define the image and compute its distance transform.

```
In :   c = np.zeros((7,9)); c[1:6,1:8] = 1
In :   inf = float('inf')
In :   mask = np.array([[inf,1,inf],[1,0,inf],[inf,inf,inf]])
In :   cd = disttrans(1-c,mask)
```

Python

Next the largest elements:

```
In :   cd2 = ndi.rank_filter(cd,rank=4,footprint=np.array
        ([[0,1,0],[1,1,1],[0,1,0]]))
In :   print (cd2 <= cd)*1
```

Python

We can easily restrict the image so as not to obtain the extra 1's in the corners of the result. Now let's do the same thing with our circles image:

```
>> c = 1-imread('circles2.png');
>> cd = disttrans(c,mask);
>> cd2 = ordfilt2(cd,5,[0 1 0;1 1 1;0 1 0]);
>> imshow((cd2 <= cd) & ~c)
```

MATLAB/Octave

or

```Python
In :   c = 1 - io.imread('circles2.png')/255
In :   cd = disttrans(c,mask)
In :   cd2 = ndi.rank_filter(cd,rank=4,footprint=np.array
       ([[0,1,0],[1,1,1],[0,1,0]]))
In :   io.imshow((cd2 <= cd)*(1-c))
```

and the result is shown in Figure 11.29(a). Image (b) shows the distance transform applied to the central image in Figure 11.14. The use of the commands (`cd2 <= cd`) `& ~c` or

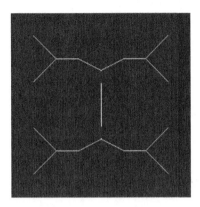

(a) Circles skeleton   (b) Skeleton from Figure 11.14

FIGURE 11.29: Skeletonization using the distance transform

(`cd2 <= cd`)`*(1-c)` blocks out the outside of the circles, so that just the skeleton is left. We will see in Chapter 10 how to thicken this skeleton, and also other ways of obtaining the skeleton.

Details about both the Guo-Hall and Zhang-Suen algorithms, and others, with discussions and examples, can be found in [33].

11.9   Programs

First, programs in MATLAB/Octave for the distance transform:

```MATLAB/Octave
function res=disttrans(image,mask)
%
% This function implements the distance transform by applying MASK to
% IMAGE, using the two step algorithm with "forward" and "backwards" masks
       .
%
backmask = rot90(rot90(mask));
[mr,mc] = size(mask);
if (mod(mr,2)==0 || mod(mc,2)==0) then
     error('The_mask_must_have_odd_dimensions.')
     end;
[r,c] = size(image);
```

```matlab
nr = (mr-1)/2;
nc = (mc-1)/2;
image2 = padarray(image,[nr,nc],'replicate');


%
% This is the first step; replacing R values with 0 and other values
% with infinity
%
image2(find(image2==0))=Inf;
image2(find(image2==1))=0;
%
% Forward pass
%
for i=nr+1:nr+r,
  for j=nc+1:nc+c,
    temp = image2(i-nr:i+nr,j-nc:j+nc)+mask;
    image2(i,j) = min(temp(:));
  end;
end;
%
% Backward pass
%
for i=r+nr:-1:nr+1,
  for j=c+nc:-1:nc+1,
    temp = image2(i-nr:i+nr,j-nc:j+nc)+backmask;
    image2(i,j) = min(temp(:));
  end;
end;
%
res=image2(nr+1:r+nr,nc+1:c+nc);
end
```

MATLAB/Octave

The Zhang-Suen skeletonization algorithm:

```matlab
function out = zs(image)
  lut_odd = makelut(@zs_odd,3);
  lut_even = makelut(@zs_even,3);
  previous = image;
  current = image & ~applylut(image,lut_odd);
  while true,
    current = current & ~applylut(current,lut_even);
    if isequal(previous,current),
      break;
    end
    current = current & ~applylut(current,lut_odd);
    if isequal(previous,current),
      break;
    end
  previous = current;
  end
  out = current;
end
```

MATLAB/Octave

```matlab
function out = zs_odd(x)
  t = x;
  t(5) = 0;
  B = sum(t(:));
  X = max(max(bwlabel(t,4)));
  out = (B>=2) & (B<=6) & (X==1);
  out = out & (t(6)==0 | t(8)==0 | (t(2)==0 & t(4)==0));
end

function out = zs_even(x)
  t = x;
  t(5)=0;
  B = sum(t(:));
  X = max(max(bwlabel(t,4)));
  out = (B>=2) & (B<=6) & (X==1);
  out = out & (t(2)==0 | t(4)==0 | (t(6)==0 & t(8)==0));
end
```

MATLAB/Octave

and the Guo-Hall skeletonization algorithm:

```matlab
function out = gh(image)
  gh_lut = makelut(@gh_fun,3);
  [r,c] = size(image);
  image2 = repmat([1 2;2 1],r/2,c/2).*image;
  prev = image2;
  g = applylut(image2>0, gh_lut);
  curr = prev .* ~((prev==1) & (g==1));
  while true,
    g = applylut(curr>0, gh_lut);
    curr = curr .* ~((curr==2) & (g==1));
    g = applylut(curr>0,gh_lut);
    curr = curr .* ~((curr==1) & (g==1));
    if isequal(prev,curr),
      break;
    end
    prev = curr;
  end
  out = curr>0;
end

function out = gh_fun(x)
  n = x;
  n(5) = 0;
  B = sum(n(:));
  X = max(max(bwlabel(n,8)));
  S = n(2)*n(4)*n(6)*n(8);
  out = (B>=2 & X==1 & S==0);
end
```

MATLAB/Octave

In Python, a distance transform program is

```python
def disttrans(image,mask):
    backmask = np.rot90(np.rot90(mask))
    mr,mc = mask.shape
    if mr%2 == 0 or mc%2 == 0:
        raise ValueError("Error:_Mask_must_have_odd_numbers_of_rows_and_
            columns")
    r,c = image.shape
    nr = (mr-1)/2
    nc = (mc-1)/2
    image2 = np.float64(np.pad(image,[nr,nc],mode='edge'))
    image2[np.where(image2==0)]=float('inf')
    image2[np.where(image2==1)]=0
    for i in range(r):
        for j in range(c):
            image2[i+nr,j+nc] = (image2[i:i+mr,j:j+mc]+mask).min()
    for i in range(r)[::-1]:
        for j in range(c)[::-1]:
            image2[i+nr,j+nc] = (image2[i:i+mr,j:j+mc]+backmask).min()
    return image2[nr:nr+r,nc:nc+c]
```
Python

A program for Zhang-Suen skeletonization:

```python
from numpy import array_equal,copy,reshape
from scipy.ndimage import generic_filter as gfilt
from skimage.measure import label as bwlabel

def zs(im):
    prev = copy(im)
    curr = prev*(1-gfilt(prev,zs_fun,size=(3,3),extra_arguments=(1,)))
    while True:
        curr = curr*(1-gfilt(curr,zs_fun,size=(3,3),extra_arguments=(0,)))
        curr = curr*(1-gfilt(curr,zs_fun,size=(3,3),extra_arguments=(1,)))
        if array_equal(prev,curr):
            break
        else:
            prev = copy(curr)
    return curr

def zs_fun(x,parity):
    n = np.reshape(x,[3,3])
    n[1,1] = 0
    B = n.sum()
    X = bwlabel(n,background=0).max()+1
    out = (B>=2) and (B<=6) and (X==1)
    if parity==0:
        out = out & (x[1]==0 or x[3]==0 or (x[5]==0 and x[7]==0))
    elif parity==1:
        out = out & (x[5]==0 or x[7] ==0 or (x[1]==0 and x[3]==0))
    return out
```
Python

And a program for Guo-Hall skeletonization:

```python
from numpy import reshape,zeros,copy,array_equal,array,tile
from scipy.ndimage import generic_filter as gfilt
from skimage.measure import label as bwlabel

def gh(im):
    r,c = im.shape
    im2 = tile(array([[1,2],[2,1]]),[r/2,c/2])*im
    prev = copy(im2)
    curr = prev*(1-gfilt(prev,gh_fun,size=(3,3),extra_arguments=(1,)))
    while True:
        curr = curr*(1-gfilt(curr,gh_fun,size=(3,3),extra_arguments=(2,)))
        curr = curr*(1-gfilt(curr,gh_fun,size=(3,3),extra_arguments=(1,)))
        if array_equal(prev,curr):
            break
        else:
            prev = copy(curr)
    return curr

def gh_fun(x,parity):
    n = (reshape(x,[3,3])>0)*1
    n[1,1] = 0
    B = n.sum()
    X = bwlabel(n,8,background=0).max()+1
    return ((B>=2) and (X==1) and (x[1]*x[3]*x[5]*x[7]==0) and (x[4]==
        parity))
```

Python

Exercises

1. What are the coordinates of the 4-neighbors of the pixel (i, j)? What are the coordinates of its 8-neighbors?

2. Find the length of the shortest 4-path from

   (a) pixel $(1, 1)$ to pixel $(5, 4)$

   (b) pixel $(3, 1)$ to pixel $(1, 6)$

   (c) pixel (i, j) to pixel (l, m)

   For this question, if a path between pixels p and q consists of $p = p_0, p_1, p_2, \ldots, p_n = q$ where all of the pixels in the path are different, then the length of that path is defined to be $n - 1$.

3. Find the shortest 8-paths between each pair of pixels in the preceding question.

4. Consider the two images

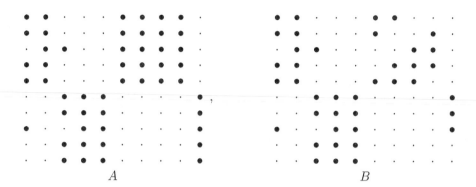

A   B

Find the 4-components and 8-components of each image.

5. The above matrices were obtained with the MATLAB/Octave commands

```
>> A=magic(10)>50
>> B=magic(10)>60
```

MATLAB/Octave

If you use Python you will have to enter the matrices manually. Check your answers of the previous question with the **bwlabel** function of MATLAB/Octave or **sk.measure.label** in Python.

6. We can define the 6-*neighbors* of a pixel p with coordinates (x, y) to be the pixels with coordinates $(x+1, y)$, $(x-1, y)$, $(x, y+1)$, $(x, y-1)$, $(x+1, y+1)$, and $(x-1, y-1)$. Draw a diagram showing the six 6-neighbors of a pixel.

7. Prove the triangle inequality for the Euclidean distance metric, and for the 4-path and 8-path metrics d_4 and d_8.

8. Define the relation 6-*connectedness* as:

p is 6-connected to q if there is a path $p = p_1, p_2, \ldots, p_n = q$ such that for each $i = 1, 2, \ldots, n-1$, p_i is 6-adjacent to p_{i+1}.

Show that this is an equivalence relation.

9. Find the lengths of the shortest 6-paths between pixels with the following coordinates:

    (a) $(0, 0)$ and $(2, 2)$,   (b) $(1, 2)$ and $(5, 4)$,   (c) $(2, 1)$ and $(6, 8)$,

    (d) $(3, 1)$ and $(7, 4)$

Can you develop an expression for a 6-path metric?

10. Show how to refine the algorithms for component labeling to label the 6-components of a binary image.

11. For the following image:

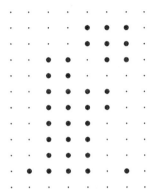

use your algorithm developed in the previous question to label the 6-components.

12. Use `bwlabel` or `sk.measure.label` to determine the number of blobs in the example given at the beginning of this chapter. Is there any difference between the results using 4 and 8 adjacency? Can you account for your answer?

13. Obtain the 6-components of the images A and B given in Question 4.

14. Let C be the image:

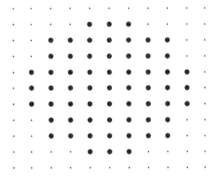

Obtain the 4-boundary and 8-boundary of this image.

15. The image in the previous question was obtained with the MATLAB commands:

```
>> [x,y] = meshgrid(-5:5,-5:5);
>> C = (x.^2+y.^2)<20
```

MATLAB/Octave

or with the Python commands

```
In :  x,y = np.mgrid[-5:6,-5:6]
In :  C = ((x**2+y**2)<20)*1
```

Python

Check your answers to Question 14 with the `bwperim` or the `sk.measure.perimeter` function.

16. Determine if each of the following foreground pixels is (a) 4-simple, (b) 8-simple.

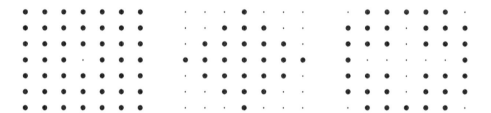

17. Calculate $C_S(p)$ for each configuration in the previous question, and show that the result is equal to $C(p)$ in each case.

18. Find configurations of pixels surrounding p for which

    (a) $C(p) = A(p) - 2$
    (b) $C(p) = A(p) - 3$
    (c) $C(p) = A(p) - 4$

19. Show that $C(p) \leq A(p)$ for every configuration of pixels.

20. Skeletonize each of the following images using (a) the Zhang-Suen algorithm, and (b) the Guo-Hall algorithm:

21. Consider the cross-shaped object formed by starting with an 11×11 square and removing a 3×3 square from each corner. Skeletonize it by each of the algorithms.

22. Check your answers to the previous two questions with your computer system.

23. Which of the algorithms seems to be the fastest, in requiring the least number of iterations, or removing the greatest number of pixels at each step?

24. Sketch the medial axes of: (a) a 2×1 rectangle, and (b) a triangle.

25. Apply both algorithms to the image `circles2.png`.

    (a) Which one works faster?
    (b) Which gives the best looking results?

26. Repeat the previous question on some other binary images.

27. For each of the following images:

0	0	0	0	0	0	0	0
0	0	0	0	1	0	0	0
0	1	1	1	1	0	0	0
0	0	1	1	1	1	0	0
0	0	0	1	1	1	1	1
0	1	1	1	1	0	0	0
0	0	0	0	1	0	0	0
0	0	0	0	0	0	0	0

0	0	0	0	0	0	0	0
0	1	1	1	1	1	1	0
0	1	0	0	0	0	1	0
0	1	0	0	0	0	1	0
0	1	0	0	0	0	1	0
0	1	0	0	0	0	1	0
0	1	1	1	1	1	1	0
0	0	0	0	0	0	0	0

0	0	0	0	0	0	0	0
0	0	0	1	1	0	0	0
0	0	0	1	1	0	0	0
0	0	1	1	1	1	0	0
0	1	1	1	1	1	1	0
0	1	0	0	0	0	1	0
0	1	0	0	0	0	1	0
0	0	0	0	0	0	0	0

apply the distance transform to approximate distances from the region containing 1's to all other pixels in the image, using the masks:

		1	
(i)	1	0	1
		1	

(ii)	1	1	1
	1	0	1
	1	1	1

(iii)	4	3	4
	3	0	3
	4	3	4

(iv)

	11		11	
11	7	5	7	11
	5	0	5	
11	7	5	7	11
	11		11	

and applying any necessary scaling at the end.

28. Apply the distance transform and use it to find the skeleton in the images `circles2.png` and `nicework.png`.

29. Compare the result of the previous question with the results given by the Zhang-Suen and Guo-Hall methods.

Which method produces the most visually appealing results?

Which method seems fastest?

Chapter 12

Shapes and Boundaries

12.1 Introduction

In this chapter we shall investigate tools for examining image shapes. Some tools have already been discussed in Chapter 10; now we look at specific methods for examining shapes of objects. Questions we might ask about shapes include:

- How do we tell if two objects have the same shape?

- How can we classify shape?

- How can we describe the shape of an object?

Formal means of describing shapes are called *shape descriptors*. Shape descriptors may include size, symmetry, and length of perimeter. A precise definition of the exact shape in some efficient manner is a *shape representation*.

In this chapter we shall be concerned with the *boundary* of objects. A boundary differs from an edge in that whereas an edge is an example of a local property of an image, a boundary is a global property.

12.2 Chain Codes and Shape Numbers

The idea of a chain code is quite straightforward; we walk around the boundary of an object, taking note of the directions we take. The resulting list of direction is the chain code.

We need to consider two types of boundaries: 4-connected and 8-connected (see Chapter 11. If the boundary is 4-connected, there are four possible directions in which to walk; if the boundary is 8-connected, there are eight possible directions. These are shown in Figure 12.1.

To see how they work, consider the object and its boundary shown in Figure 12.2.

Suppose we walk along the boundary in a clockwise direction starting at the left-most point in the bottom row, and list the directions as we go. This is shown in Figure 12.3.

We can thus read off the chain code as:

$$3 \quad 3 \quad 3 \quad 2 \quad 3 \quad 3 \quad 0 \quad 0 \quad 0 \quad 1 \quad 0 \quad 1 \quad 1 \quad 1 \quad 2 \quad 1 \quad 2 \quad 2$$

If we treat the object in Figure 12.2 as being 8-connected, then its boundary and chain code are generated as in the right-hand diagram in Figure 12.3. In this case, the resulting chain code is:

$$6 \quad 6 \quad 5 \quad 6 \quad 6 \quad 0 \quad 0 \quad 0 \quad 1 \quad 5 \quad 5 \quad 5 \quad 3 \quad 4 \quad 4$$

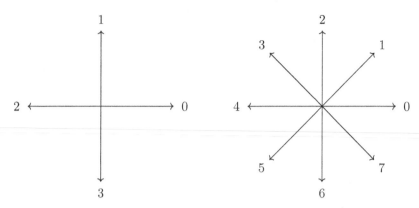

Directions for 4-connectedness   Directions for 8-connectedness

FIGURE 12.1: Directions for chain codes

0	1	1	1	0
0	1	1	1	1
0	1	1	1	1
1	1	1	1	1
1	1	1	1	1
1	1	1	1	0

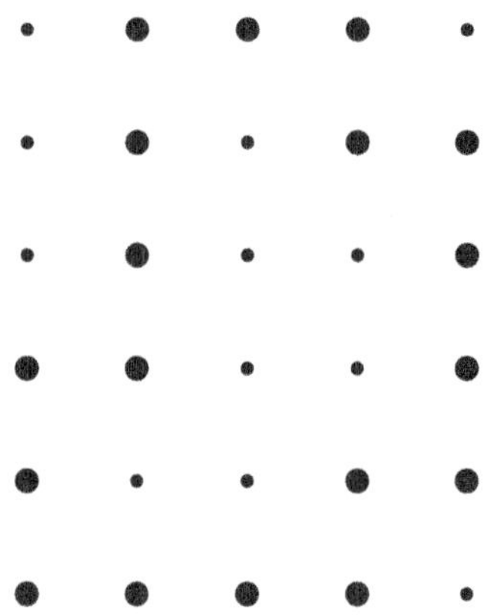

FIGURE 12.2: A 4-connected object and its boundary

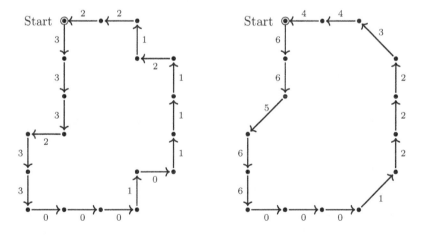

FIGURE 12.3: Obtaining the chain code from the object in Figure 12.2

To obtain the chain code in MATLAB, we have to write a small function to do the job for us. We have to first be able to trace out the boundary of our object, and once we can do that, we can write down our directions to form the chain code. A simple boundary following algorithm has been given by Sonka et al. [49]. For simplicity, we shall just give the version for 4-connected boundaries:

1. Start by finding the pixel in the object that has the left-most value in the topmost row; call this pixel P_0. Define a variable `dir` (for direction), and set it equal to 3. (Since P_0 is the top left pixel in the object, the direction to the next pixel must be 3.)

2. Traverse the 3×3 neighborhood of the current pixel in an anticlockwise direction, beginning the search at the pixel in direction

$$\mathtt{dir} + 3 \pmod 4$$

This simply sets the current direction to the first direction anticlockwise from `dir`:

dir	0	1	2	3
dir + 3 (mod 4)	3	0	1	2

The first foreground pixel will be the new boundary element. Update `dir`.

3. Stop when the current boundary element P_n is equal to the second element P_1 and the previous boundary pixel P_{n-1} is equal to the first boundary element P_0.

Suppose we have a binary image `im` consisting of a single object. We can find the top left pixel with the following MATLAB commands:

```
>> [x,y] = find(im==1);
>> x = min(x)
>> imx = im(x,:);
>> imy = min(imx)
```
MATLAB/Octave

or in Python with

```
In :  x,y = np.where(im==1)
In :  mx = mx.min()
In :  my = y[np.where(x==mx)].min()
```
Python

pixels. The second command finds the minimum of the first coordinates. Thus, `mx` is the top row of the object. The third command isolates this top row, and the final command finds the left-most column in it.

Given the indices `x` and `y` of the current pixel, and the value `dir`, how do we implement Step 2? We give a simple example, shown in Figure 12.4. Suppose that in this particular example, the current value of `dir` is 0 so that

$$\mathtt{dir} + 3 \pmod 4 = 3.$$

The dotted arrows indicate the direction of traversing (starting from the pixel in direction 3 from P_k) until we reach a new foreground pixel. This pixel is then P_{k+1}.

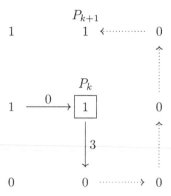

FIGURE 12.4: Traversing a neighborhood

We start by noting the row and column increments to the four-neighbors of a pixel:

$$-1 \quad 0$$

$$0 \quad -1 \qquad 0 \quad 0 \qquad 0 \quad 1$$

$$1 \quad 0$$

Since we are looking at four-connected boundaries, these are the only neighbors we need. We can put these into a matrix, with the first row corresponding to the increments in direction 0:

$$n = \begin{bmatrix} 0 & 1 \\ -1 & 0 \\ 0 & -1 \\ 1 & 0 \end{bmatrix}$$

This means that the indices in the j-th row correspond to the increments in indices in direction $j - 1$. Thus, given a direction `dir`, we can enter:

```
>> newdir = mod(dir+3,4);
>> for i=0:3,
>    j=mod(newdir+i,4)+1;
>    tt(i+1) = im(x+n(j,1),y+n(j,2)),
> end
```

MATLAB/Octave

or

```
In :   newdir = mod(dir+3,4)
In :   for i in range(4):
...:       j = mod(newdir+i,4)
...:       tt[i] = im[x+n[j,0],y+n[j,1]]
```

Python

and this will traverse the neighborhood of our image `im` at position (x, y) starting from the correct direction. Notice that for MATLAB and Octave we need to set

```
>> j = mod(newdir+i,4)+1;
```
MATLAB/Octave

The extra "**+1**" takes account of the fact that the modulus function returns values $0, 1, 2, 3$, but the rows of n are $1, 2, 3, 4$. The vector tt contains the values of the neighborhood as it is traversed.

The first non-zero value is easily found with

```
>> d = min(find(tt==1));
```
MATLAB/Octave

or with

```
In :  d = min(flatnonzero(tt==uint8(1)))
```
Python

and now we can update dir, and the position of the current pixel:

```
>> dir = mod(newdir+d-1,4);
>> x = x+n(dir+1,1);
>> y = y+n(dir+1,2);
```
MATLAB/Octave

or

```
In :  dir = mod(newdir+d,4)
In :  x = x+n[dir,0]
In :  y = y+n[dir,1]
```
Python

This most recent value of dir is placed into a vector, and this is the vector which will be the final chain code.

Functions for implementing the full chaincode are given at the end of the chapter.

These programs can be tested with the shape in Figure 12.2 (surrounding the image with zeros first):

```
>> test = zeros(8,7); test(2:7,3:5) = 1; test(5:7,2) = 1; test(3:6,6) = 1;
>> [cc,bdy] = shape(test,4);
>> uint8(cc)
ans =

   3  3  3  2  3  3  0  0  0  1  0  1  1  1  2  1  2  2
```
MATLAB/Octave

and comparing this with the code given earlier, the function has indeed returned the correct chain code.

We can easily modify the program to perform chain codes for 8-connected boundaries. The above algorithm must be slightly changed; again, see Sonka et al. [49].

1. Start by finding the pixel in the object that has the left most value in the topmost row; call this pixel P_0. Define a variable dir (for direction), and set it equal to 7. (Since P_0 is the top left pixel in the object, the direction to the next pixel must be 7.)

2. Traverse the 3×3 neighborhood of the current pixel in an anticlockwise direction, beginning the search at the pixel in direction

$$\text{dir} + 7 \quad (\text{mod } 8) \quad \text{if } \texttt{dir} \text{ is even}$$
$$\text{dir} + 6 \quad (\text{mod } 8) \quad \text{if } \texttt{dir} \text{ is odd}$$

This simply sets the current direction to the first direction anticlockwise from `dir`:

dir	0	1	2	3	4	5	6	7
dir + 7 (mod 8)	7	0	1	2	3	4	5	6
dir + 6 (mod 8)	6	7	0	1	2	3	4	5

The first foreground pixel will be the new boundary element. Update `dir`.

3. Stop when the current boundary element P_n is equal to the second element P_1 and the previous boundary pixel P_{n-1} is equal to the first boundary element P_0.

Note that the choosing of the starting direction in Step 2 can be implemented by

```
newdir = mod(dir+7-mod(dir,2),8);
```
MATLAB/Octave

or

```
newdir = (dir+7-dir%2)%8
```
Python

which produces $\text{dir} + 7$ if `dir` is even, and $\text{dir} + 6$ if `dir` is odd. As well as this, we need to take into account *all* possible directions from a pixel; the index increments are:

$$
\begin{array}{ccc}
-1 \ -1 & -1 \ \ 0 & -1 \ \ 1 \\
\\
0 \ -1 & 0 \ \ 0 & 0 \ \ 1 \\
\\
1 \ -1 & 1 \ \ 0 & 1 \ \ 1
\end{array}
$$

These will all be placed in the array

$$
n = \begin{bmatrix}
0 & 1 \\
-1 & 1 \\
-1 & 0 \\
-1 & -1 \\
0 & -1 \\
1 & -1 \\
1 & 0 \\
1 & 1
\end{bmatrix}
$$

where as before, direction 0 corresponds to the top row of n. This can be tested on the test image `test`:

```
>> uint8(shape(test,8))
ans =

  6  6  5  6  6  0  0  0  1  2  2  2  3  4  4
```

MATLAB/Octave

and the answer is indeed the code obtained earlier by following the arrows around the object.

Programs are given at the end of the chapter.

Normalization of Chain Codes

There are two problems with the definition of the chain code as given in previous sections:

1. The chain code is dependent on the starting pixel

2. The chain code is dependent on the orientation of the object

First look at Problem (1). The idea is to "normalize" the chain code as follows: imaging the code to be written around the edge of a circle. We choose as our starting place the position for which the code, when read off, will be the lowest possible integer. The result is the *normalized chain code* for our object.

For example, suppose we have an object consisting of a 3×3 square:

```
>> a=zeros(5,5);a(2:4,2:4)=1

    0    0    0    0    0
    0    1    1    1    0
    0    1    1    1    0
    0    1    1    1    0
    0    0    0    0    0

>> shape(a,4)

ans =

    3    3    0    0    1    1    2    2
```

MATLAB/Octave

Now let's put these codes around a circle as shown in Figure 12.5.

The arrow indicates where we should start reading off the code to obtain the lowest integer; in this case it is

```
0  0  1  1  2  2  3  3
```

But we can do this easily in MATLAB or Octave by using the `circshift` function to create every possible shift. For our chain code above, we can do this:

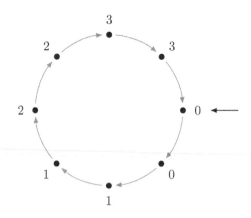

FIGURE 12.5: A chain code written cyclically

```
>> N = length(c);
>> z = zeros(N);
>> for i = 1:N, z(i,:) = circshift(c',i)';end
>> z
ans =

     2     3     3     0     0     1     1     2
     2     2     3     3     0     0     1     1
     1     2     2     3     3     0     0     1
     1     1     2     2     3     3     0     0
     0     1     1     2     2     3     3     0
     0     0     1     1     2     2     3     3
     3     0     0     1     1     2     2     3
     3     3     0     0     1     1     2     2
```

MATLAB/Octave

To find the row that contains the least integer, we use the handy **sortrows** function, which sorts the rows "lexicographically": first on the first element, then on the second element, and so on:

```
>> zs = sortrows(z)

zs =

     0     0     1     1     2     2     3     3
     0     1     1     2     2     3     3     0
     1     1     2     2     3     3     0     0
     1     2     2     3     3     0     0     1
     2     2     3     3     0     0     1     1
     2     3     3     0     0     1     1     2
     3     0     0     1     1     2     2     3
     3     3     0     0     1     1     2     2
```

MATLAB/Octave

Now the normalized chain code is just the first row:

```
>> zs(1,:)

ans =

     0     0     1     1     2     2     3     3
```

In Python this can also easily be done; first create the test object and obtain its chain code:

```
In :  a = zeros((5,5))
In :  a[1:4,1:4]=1
In :  c = shape(a,4)
```

Now create an array whose rows contain the cyclic shifts; these can be obtained using the roll method of numpy:

```
In :  N = len(c)
In :  z = zeros((N,N))
In :  for i in range(N):
...:      z[i,:] = np.roll(c,i)
...:
```

The array z can now be sorted in place along whichever axis we choose. For sorting the rows, the axis number is 1:

```
In :  z.sort(axis=1)
In :  print(z[0,:])
[ 0.  0.  1.  1.  2.  2.  3.  3.]
```

Let's try this with a slightly larger example: the test object from Figure 12.2:

```
>> c = shape(test,4);
>> N = length(c);
>> z = zeros(N)
>> for i = 1:N, z(i,:) = circshift(c',i)';end
>> zs = sortrows(z);
>> zs(1,:)

ans =

  0  0  0  1  0  1  1  1  2  1  2  2  3  3  3  2  3  3
```

This can be easily turned into a function. With Python we just repeat the commands above, starting with the new chain code array.

Shape Numbers

We now consider the second problem from above: defining a chain code that is independent of orientation of the object. For example, consider a simple "L" shape:

```
>> L = zeros(7,6); L(2:6,2:3) = 1; L(5:6,4:5) = 1

L =

     0     0     0     0     0     0
     0     1     1     0     0     0
     0     1     1     0     0     0
     0     1     1     0     0     0
     0     1     1     1     1     0
     0     1     1     1     1     0
     0     0     0     0     0     0
```

MATLAB/Octave

The normalized chain code can be found by the commands above to be

    0 0 0 1 2 2 1 1 1 2 3 3 3 3

Now suppose we rotate the shape so that it has a different orientation (we use the `rot90` function which rotates a matrix by 90 degrees):

```
>> L2 = rot90(L)

L2 =

     0     0     0     0     0     0     0
     0     0     0     0     1     1     0
     0     0     0     0     1     1     0
     0     1     1     1     1     1     0
     0     1     1     1     1     1     0
     0     0     0     0     0     0     0
```

MATLAB/Octave

The normalized chain code for this new orientation can be computed to be:

    0 0 0 0 1 1 1 2 3 3 2 2 2 3

Even when normalized, the chain codes are different.

To overcome this, take the *differences* of the chain code: for each two consecutive elements c_i and c_{i+1}, their difference is defined as

$$c_{i+1} - c_i \pmod 4.$$

(If we were dealing with 8-connected boundaries, then we would take differences mod 8). For an example, take the simple L shape and its rotation shown in Figure 12.6. The chain code for the first shape can be read off easily starting from the point shown; it is

    3 3 0 0 1 2 1 2.

To apply differences, subtract this list from a cyclic shift of itself:

$$
\begin{array}{ccccccccc}
  & 3 & 3 & 0 & 0 & 1 & 2 & 1 & 2 \\
- & 2 & 3 & 3 & 0 & 0 & 1 & 2 & 1 \\
\hline
= & 1 & 0 & 1 & 0 & 1 & 1 & 3 & 1 & (\mathrm{mod}\,4)
\end{array}
$$

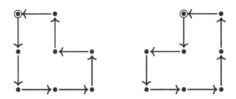

FIGURE 12.6: A simple L-shape

The normalized version of these differences is the *shape number* for this L shape and can be found to be

0 1 0 1 1 3 1 1.

The chain code for the rotated L can be found to be

3 2 3 0 0 1 1 2.

Even when normalized the result is not the same as the chain code for this shape in its original orientation. But if the differences are obtained:

$$
\begin{array}{r}
3\ 2\ 3\ 0\ 0\ 1\ 1\ 2 \\
-\ \ 2\ 3\ 2\ 3\ 0\ 0\ 1\ 1 \\
\hline
1\ 3\ 1\ 1\ 0\ 1\ 0\ 1
\end{array}
$$

then the normalized third row is

0 1 0 1 1 3 1 1.

which is exactly the same code obtained above.

This can be done easily in MATLAB or Octave: given a chain code c, simply normalize the difference with a cyclic shift:

```
>> c = shape(a)
c =

  3  3  0  0  1  1  2  2
>> d = mod(c - circshift(c',1)',4)
d =

  1  0  0  0  1  0  1  0
>> chain_norm(d)
ans =

  0  0  0  1  0  1  0  1
```

MATLAB/Octave

Let's try this with our L shape and its rotation.

```
>> c = shape(L,4);
>> d = mod(c - circshift(c',1)',4);
>> chain_norm(d)
ans =

  0  0  0  1  0  0  1  1  0  3  0  0  1  1

>> c2 = shape(L2,4);
>> d2 = mod(c2 - circshift(c2',1)',4);
>> chain_norm(d2)
ans =

  0  0  0  1  0  0  1  1  0  3  0  0  1  1
```

MATLAB/Octave

and both results are exactly the same. With Python, the following commands obtain the same outcome:

```
In :   c = shape(a,4)
In :   d = (c-np.roll(c,1))%4
In :   print(d)
[1 0 1 0 1 0 1 0]

In :
```

Python

12.3   Fourier Descriptors

The idea is this: suppose we walk around an object, but instead of writing down the directions, we write down the boundary coordinates. The final list of (x, y) coordinates can be turned into a list of complex numbers $z = x + yi$. The Fourier transform of this list of numbers is a *Fourier descriptor* of the object.

The beauty of a Fourier descriptor is that often only a few low frequency terms around the DC coefficient are enough to distinguish objects, or to classify them.

The chain code functions at the end of the chapter also produce the list of boundary pixels, so that, for example, with the L shape from the previous section:

```
>> [cc,bdy] = shape(L);
>> bdy'
ans =

  3  4  5  6  6  6  6  5  5  5  4  3  2  2
  2  2  2  2  3  4  5  5  4  3  3  3  3  2
```

MATLAB/Octave

Turning these into complex numbers is easy:

```
>> c = complex(bdy(:,1),bdy(:,2)); c'
ans =

 Columns 1 through 8:

  3 - 2i   4 - 2i   5 - 2i   6 - 2i   6 - 3i   6 - 4i   6 - 5i   5 - 5i

 Columns 9 through 14:

  5 - 4i   5 - 3i   4 - 3i   3 - 3i   2 - 3i   2 - 2i
```
`MATLAB/Octave`

The same array can be obtained with Python:

```
In :  cc,bdy = shape(L)
In :  c = bdy[:,0] + bdy[:,1]*1j
```
`Python`

These can be plotted:

```
>> plot(c,'.-',"markersize",10),axis([1,7,1,6]),axis equal
```
`MATLAB/Octave`

or with

```
In :  import matplotlib.pyplot as plt
In :  plt.plot(c.real,c.imag,'.-b',markersize=10)
In :  plt.axis([0,6,0,5])
In :  plt.show()
```
`Python`

and the result is shown in Figure 12.7. Supposing we take the Fourier transform, and from it extract only a few terms:

```
>> cf = fft(c)

cf =

   62.00000 + 43.00000i
   -9.76237 - 18.27796i
   -5.69003 -  0.64111i
    2.87152 -  2.43834i
    0.52068 +  0.82866i
    1.45451 -  0.95368i
   -0.16670 +  0.47640i
    0.00000 +  1.00000i
   -0.47640 +  0.16670i
    0.75862 +  0.27672i
   -0.82866 -  0.52068i
   -0.64200 -  0.93602i
    0.64111 +  5.69003i
   -8.68028 +  0.32927i
```
`MATLAB/Octave`

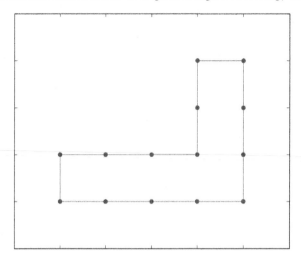

FIGURE 12.7: Boundary pixels

Note that as this transform is not shifted, the first term will be the DC coefficient, and low frequency terms occur after the DC coefficient, and at the end. What we can do is pick a value k, and keep k terms after the DC coefficient, and at the end:

```
>> k = 2
>> cf2 = cf;
>> N = len(cf);
>> cf2(k+2:N-k) = 0;
```
MATLAB/Octave

At this stage, 9 of the original 14 values have been removed. They can be plotted with lines and points, and to ensure the lines meet up the first values will be repeated as the last:

```
>> bdy2 = ifft(cf2);
>> bdy2 = [bdy2;bdy2(1,:)];
>> plot(bdy2,'.-',"markersize",10),axis([1,7,1,6]),axis equal
```
MATLAB/Octave

The result is shown in Figure 12.8(a).

Images (b) and (c) show the result by keeping a few more pairs of transform values. However even with $k = 2$ there is enough information to gauge the size and rough shape of the object; and with $k = 4$ (five transform values removed) the result is very close.

Recall that in an unshifted transform, the DC coefficient is at the start, and the high frequency values are toward the middle, as shown in Figure 12.9.

In this figure, the arrows give the direction of increasing frequency; the elements labeled X and Y are those of highest frequency. Around those central values, and excluding the DC coefficient, all other elements can be paired together with equal frequencies, as shown.

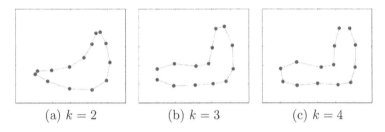

(a) $k = 2$ (b) $k = 3$ (c) $k = 4$

FIGURE 12.8: Fourier descriptors of an L shape

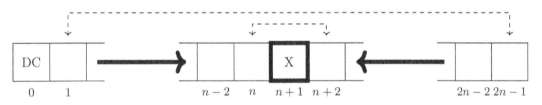

Number of elements is even, $2n$

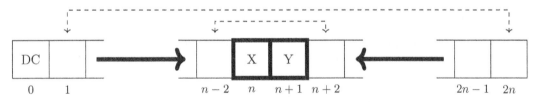

Number of elements is odd, $2n + 1$

FIGURE 12.9: Unshifted one-dimensional DFT

This means that when we exclude elements from the transform, they must be excluded in pairs. In other words, given a transform with N elements, all but $2k+1$ elements can be excluded as shown in Figure 12.10.

DC	k elements	$N - 2k - 1$ zeros	k elements

FIGURE 12.10: Excluding elements from the DFT

In order to approximate a shape, given $2k+1$ non-zero elements of a transform, there are several possibilities:

1. Apply the inverse transform to the complete N length vector (including the zeros), and plot the result.

2. Simply invert the smaller vector of $2k+1$ elements, and plot those values joined by lines.

3. Place any number of zeros between the first $k+1$ and last k values before inversion.

Consider a large shape with many boundary pixels:

```
>> z = zeros(256);
>> z(32:224,32:96) = 1; z(32:96,96:224) = 1; z(192:224,96:160) = 1;
>> [cc,bdy] = shape(z,4)
```

MATLAB/Octave

There are 896 pixels in the boundary. Now to transform them and delete most of them, first in MATLAB or Octave:

```
>> c = complex(bdy(:,1),bdy(:,2));
>> N = length(c)
N = 896

>> cf = fft(c);
>> k = 12;
>> cf2 = cf; cf2(k+2:N-k) = 0;
>> bdy2 = ifft(cf2);
>> cf3 = [cf(1:k+1);cf(N-k+1:N)];
>> bdy3 = ifft(cf3);
>> cf4 = [cf3(1:13);zeros(100,1);cf3(14:25)];
>> bdy4 = ifft(cf4);
```

MATLAB/Octave

or in Python:

```
In :   from numpy.fft import fft, ifft
In :   execfile('chaincode.py')
In :   z = np.zeros((256,256))
In :   z[31:224,31:96] = 1;z[31:96,95:223] = 1; z[192:224,95:160]=1
In :   cc,bdy = shape(z)
In :   c = bdy[:,0]+bdy[:,1]*1j
In :   k = 12; N = len(c)
In :   cf = fft(c); cf2 = np.copy(cf); cf2[k+1:N-k]=0; bdy2 = ifft(cf2);
In :   cf3 = np.hstack((cf[:k+1],cf[-k:])); bdy3 = ifft(cf3)
In :   cf4 = np.hstack((cf[:k+1],np.zeros((100)),cf[-k:])); bdy4 = ifft(cf4)
```
Python

At this stage `cf2` consists of 896 values, most of them zero, and so `bdy2` has 896 values. But `cf3` consists only of the first 13 and last 12 values of `cf`—in effect `cf3` is `cf2` with all the zero elements removed—and so both it and `bdy3` have only 25 values. Finally, `cf4` is created from `cf3` by putting in 100 zeros between the first $k+1$ and final k values. There is nothing special about 100—any value will do—but the larger the value, the more points will fill out the boundary.

So the plots can be created as

```
>> plot(bdy2,'.'),axis([0,256,0,256]),axis equal,axis nolabel
>> plot([bdy3;bdy3(1)],'.-',"markersize",5),axis([0,10000,0,10000]),...
   axis equal,axis nolabel
>> plot(bdy4,'.'),axis([0 2000 0 2000]),axis equal,axis nolabel
```
MATLAB/Octave

or as

```
In :   plt.plot(bdy2.real,bdy2.imag,'.'); plt.axis('equal')
In :   bdy3 = np.hstack(bdy3,bdy3[0])
In :   plt.plot(bdy3.real,bdy3.imag,'.-',markersize=5); plt.axis
   ([0,9000,0,9000])
In :   plt.plot(bdy4.real,bdy4.imag,'.'); plt.axis([0.0,2000.0,0.0,2000.0])
```
Python

and the results are shown in Figure 12.11.

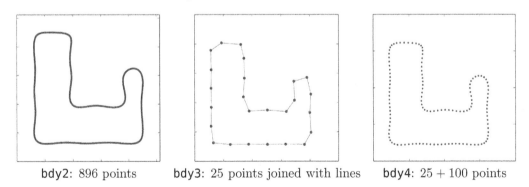

bdy2: 896 points   bdy3: 25 points joined with lines   bdy4: $25 + 100$ points

FIGURE 12.11: Fourier descriptors drawn differently

To give a slightly more substantial example, consider the three shapes in Figure 12.12. Each one can be plotted from a list of points; for example, the plane can be created in MATLAB or Octave with:

```
>> pl = dlmread('plane.txt');
>> c = complex(pl(:,1),pl(:,2));
>> plot(c,'.'),axis equal, axis nolabel
```
MATLAB/Octave

or in Python with

```
In :  pl = np.loadtxt('plane.txt')
In :  c = pl[:,0] + pl[:,1]*1j
In :  plt.plt.plot(c.real,c.imag,'.')
In :  plt.axis('equal')
In :  plt.tick_params(axis='both',labelbottom='off',labelleft='off')
In :  plt.show()
```
Python

where the `plt.tick_params` function turns off the tick marks on the axes. Each of the

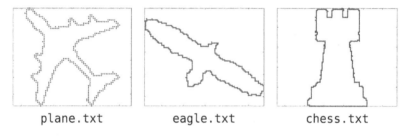

plane.txt   eagle.txt   chess.txt

FIGURE 12.12: Three shapes

three shapes has roughly the same number of boundary points: between 415 and 420.

For each of the values $k = 8, 20, 32$ set all but the first $k + 1$ and last k of the transform values to zero, and invert the result. So Figure 12.13 shows the Fourier descriptors with $k = 8$. Although the boundaries are quite blurred, and fine details are missing, the basic

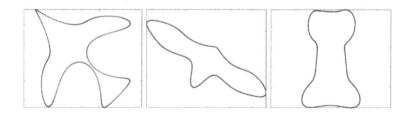

FIGURE 12.13: Using only 8 pairs of values from the Fourier transform

nature of the shapes, and their sizes and orientations are quite clear.

So Figure 12.14 shows the Fourier descriptors with $k = 20$. With more values used some fine detail begins to appear: the engines on the plane wings and the detail on the top of the chess piece. Even more detail is apparent with $k = 32$, as shown in Figure 12.15.

Although the boundaries are still not as sharp as the originals, a remarkable amount of detail is apparent for only 32 of the roughly 210 pairs of values in the transform.

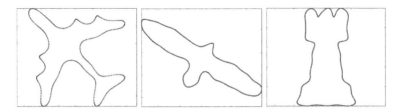

FIGURE 12.14: Using 20 pairs of values from the Fourier transform

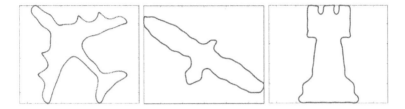

FIGURE 12.15: Using 32 pairs of values from the Fourier transform

Exercises

1. Find the chain codes, normalized chain codes, and shape numbers for each of the 4-connected shapes given in Figure 12.16.

2. Now repeat the above exercise for all possible reflections and rotations of those shapes.

3. Repeat the above exercises for the 8-connected shapes given in Figure 12.17.

4. Check your answers to the previous questions with the **shape** program.

5. Generate the shapes with the following 4-connected chain codes:

   (a) **3 3 3 0 0 0 0 1 1 2 2 1 2 2**

   (b) **3 3 3 0 3 0 0 1 0 1 1 2 2 1 2 2**

6. Generate the shapes with the following 8-connected chain codes:

   (a) **5 6 7 6 0 0 1 2 2 4 3 4**

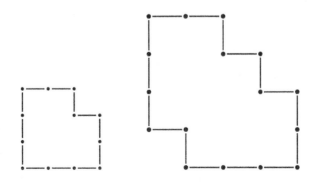

FIGURE 12.16: Shapes for Question 1

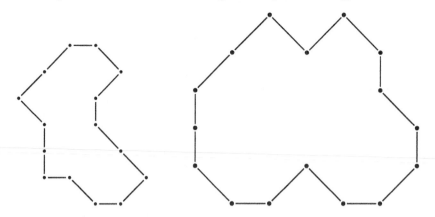

FIGURE 12.17: Shapes for Question 3

(b) **5 6 7 7 1 7 1 2 2 3 5 3 4**

7. Obtain the normalized chaincodes and shape numbers of all the shapes in the previous questions.

   Check your answers with **shape**.

8. Use your system to generate a T shape, and obtain its Fourier descriptors. How many terms are required to identify the

   (a) Symmetry of the object?

   (b) Size of the object?

   (c) Shape of the object?

9. Repeat the above question but use an X shape. Then again, but with an E shape.

10. Compare the results of the shapes in the previous two questions. How many terms are required to distinguish the

    (a) Symmetries

    (b) Sizes

    (c) Shapes

    of the three objects?

11. How does the size of an object affect its Fourier descriptors? Experiment with a 6×4 rectangle, and then with a 12×8 rectangle. Generalize your findings.

Chapter 13

Color Processing

For human beings, color provides one of the most important descriptors of the world around us. The human visual system is particularly attuned to two things: edges and color. We have mentioned that the human visual system is not particularly good at recognizing subtle changes in gray values. In this section we shall investigate color briefly, and then some methods of processing color images

13.1 What Is Color?

Color study consists of

1. The physical properties of light which give rise to color

2. The nature of the human eye and the ways in which it detects color

3. The nature of the human vision center in the brain, and the ways in which messages from the eye are perceived as color

Physical Aspects of Color

As we have seen in Chapter 1, visible light is part of the electromagnetic spectrum. The values for the wavelengths of blue, green, and red were set in 1931 by the CIE (Commission Internationale d'Eclairage), an organization responsible for color standards.

Perceptual Aspects of Color

The human visual system tends to perceive color as being made up of varying amounts of red, green, and blue. That is, human vision is particularly sensitive to these colors; this is a function of the cone cells in the retina of the eye. These values are called the *primary colors*. If we add together any two primary colors, we obtain the *secondary colors*:

$$\text{magenta (purple)} = \text{red} + \text{blue}$$
$$\text{cyan} = \text{green} + \text{blue}$$
$$\text{yellow} = \text{red} + \text{green}$$

The amounts of red, green, and blue that make up a given color can be determined by a *color matching experiment*. In such an experiment, people are asked to match a given color (a *color source*) with different amounts of the additive primaries red, green, and blue. Such an experiment was performed in 1931 by the CIE, and the results are shown in Figure 13.1. Note that for some wavelengths, various red, green, or blue values are *negative*. This is a

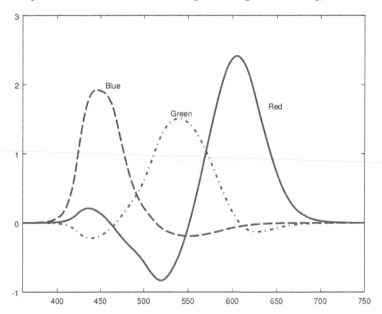

FIGURE 13.1: RGB color matching functions (CIE, 1931)

physical impossibility, but it can be interpreted by adding the primary beam to the color source, to maintain a color match.

To remove negative values from color information, the CIE introduced the XYZ color model. The values of X, Y, and Z can be obtained from the corresponding R, G, and B values by a linear transformation:

$$\begin{bmatrix} X \\ Y \\ Z \end{bmatrix} = \begin{bmatrix} 0.431 & 0.342 & 0.178 \\ 0.222 & 0.707 & 0.071 \\ 0.020 & 0.130 & 0.939 \end{bmatrix} \begin{bmatrix} R \\ G \\ B \end{bmatrix}$$

The inverse transformation is easily obtained by inverting the matrix:

$$\begin{bmatrix} R \\ G \\ B \end{bmatrix} = \begin{bmatrix} 3.063 & -1.393 & -0.476 \\ -0.969 & 1.876 & 0.042 \\ 0.068 & -0.229 & 1.069 \end{bmatrix} \begin{bmatrix} X \\ Y \\ Z \end{bmatrix}$$

The XYZ color matching functions corresponding to the R, G, B curves of Figure 13.1 are shown in Figure 13.2. The matrices given are not fixed; other matrices can be defined according to the definition of the color white. Different definitions of white will lead to different transformation matrices.

The CIE required that the Y component corresponded with *luminance*, or perceived brightness of the color. That is why the row corresponding to Y in the first matrix (that is, the second row) sums to 1, and also why the Y curve in Figure 13.2 is symmetric about the middle of the visible spectrum.

In general, the values of X, Y, and Z needed to form any particular color are called the *tristimulus values*. Values corresponding to particular colors can be obtained from published tables. In order to discuss color independent of brightness, the tristimulus values can be

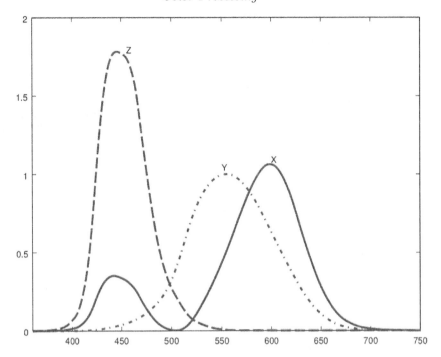

FIGURE 13.2: XYZ color matching functions (CIE, 1931)

normalized by dividing by $X + Y + Z$:

$$x = \frac{X}{X + Y + Z}$$
$$y = \frac{Y}{X + Y + Z}$$
$$z = \frac{Z}{X + Y + Z}$$

and so $x+y+z = 1$. Thus, a color can be specified by x and y alone, called the *chromaticity coordinates*. Given x, y, and Y, we can obtain the tristimulus values X and Z by working through the above equations backward:

$$X = \frac{x}{y}Y$$
$$Z = \frac{1 - x - y}{y}Y.$$

We can plot a chromaticity diagram, using the `ciexyz31.csv`[1] file of XYZ values. In MATLAB/Octave:

[1]This file can be obtained from the *Color & Vision Research Laboratories* web page `http://www.cvrl.org`.

```
>> nxyz = dlmread('ciexyz31.csv');
>> xyz = wxyz(:,2:4)';
>> xy = xyz'./(sum(xyz)'*[1 1 1]);
>> x = xy(:,1)';
>> y = xy(:,2)';
>> figure,plot([x x(1)],[y y(1)]),xlabel('x'),ylabel('y'),axis square
```

MATLAB/Octave

In Python:

```
In :  import matplotlib.pyplot as plt
In :  nxyz = np.loadtxt('ciexyz31_1.csv',delimiter=',')
In :  xyz = nxyz[:,1:]
In :  sums = xyz.sum(axis=1)
In :  newxyz = xyz/sums[:,np.newaxis]
In :  x = newxyz[0,:]; x1 = np.hstack((x,x[0]))
In :  y = newxyz[1,:]; y1 = np.hstack((y,y[0]))
In :  plt.plot(x1,y1)
```

Python

Here the matrix xyz consists of the second, third, and fourth columns of the data, and plot is a function that draws a polygon with vertices taken from the x and y vectors. The extra x(1) and y(1) ensures that the polygon joins up. The result is shown in Figure 13.3. The

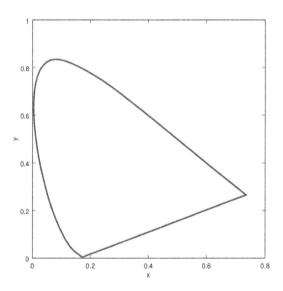

FIGURE 13.3: A chromaticity diagram

values of x and y that lie within the horseshoe shape in Figure 13.3 represent values that correspond to physically realizable colors. A good account of the XYZ model and associated color theory can be found in Foley et al. [11].

13.2 Color Models

A *color model* is a method for specifying colors in some standard way. It generally consists of a three-dimensional coordinate system and a subspace of that system in which each color is represented by a single point. We shall investigate three systems.

RGB

In this model, each color is represented as three values R, G, and B, indicating the amounts of red, green, and blue which make up the color. This model is used for displays on computer screens; a monitor has three independent electron "guns" for the red, green, and blue component of each color. We have met this model in Chapter 2.

Note also from Figure 13.1 that some colors require negative values of R, G or B. These colors are not realizable on a computer monitor or TV set, on which only positive values are possible. The colors corresponding to positive values form the *RGB gamut*; in general a color "gamut" consists of all the colors realizable with a particular color model. We can plot the RGB gamut on a chromaticity diagram, using the xy coordinates obtained above. To define the gamut, we shall create a $100 \times 100 \times 3$ array, and to each point (i, j) in the array, associate an XYZ triple defined by $(i/100, j/100, 1 - i/100 - j/100)$. We can then compute the corresponding RGB triple, and if any of the RGB values are negative, make the output value white. Programs to display the gamut, which is shown in Figure 13.4, are given at the end of the chapter

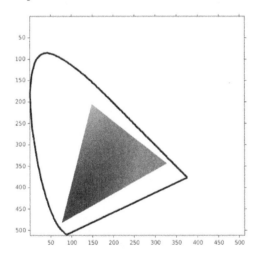

FIGURE 13.4: **SEE COLOR INSERT** The RGB gamut

HSV

HSV stands for hue, saturation, and value. These terms have the following meanings:

Hue: The "true color" attribute (red, green, blue, orange, yellow, and so on).

Saturation: The amount by which the color has been diluted with white. The more white in the color, the lower the saturation. So a deep red has high saturation, and a light red (a pinkish color) has low saturation.

Value: The degree of brightness: a well-lit color has high intensity; a dark color has low intensity.

This is a more intuitive method of describing colors, and as the intensity is independent of the color information, this is a very useful model for image processing. We can visualize this model as a cone, as shown in Figure 13.5.

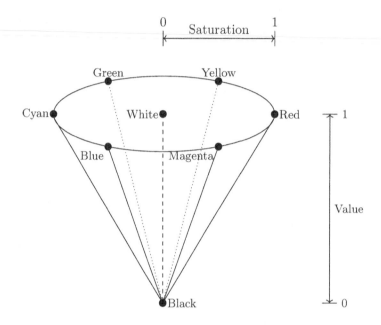

FIGURE 13.5: The color space HSV as a cone

Any point on the surface represents a purely saturated color. The saturation is thus given as the relative distance to the surface from the central axis of the structure. Hue is defined to be the angle measurement from a pre-determined axis, say red.

Conversion between RGB and HSV

Suppose a color is specified by its RGB values. If all three values are equal, then the color will be a grayscale; that is, an intensity of white. Such a color, containing just white, will thus have a saturation of zero. Conversely, if the RGB values are very different, we would expect the resulting color to have a high saturation. In particular, if one or two of the RGB values are zero, the saturation will be one, the highest possible value.

Hue is defined as the fraction around the circle starting from red, which thus has a hue of zero. Reading around the circle in Figure 13.5 produces the following hues:

Color	Hue
Red	0
Yellow	0.1667
Green	0.3333
Cyan	0.5
Blue	0.6667
Magenta	0.8333

Suppose we are given three R, G, B values, which we suppose to be between 0 and 1. So if they are between 0 and 255, we first divide each value by 255. We then define:

$$V = \max\{R, G, B\}$$
$$\delta = V - \min\{R, G, B\}$$
$$S = \frac{\delta}{V}$$

To obtain a value for hue, we consider several cases:

1. if $R = V$ then $H = \dfrac{1}{6} \dfrac{G - B}{\delta}$

2. if $G = V$ then $H = \dfrac{1}{6} \left(2 + \dfrac{B - R}{\delta}\right)$

3. if $B = V$ then $H = \dfrac{1}{6} \left(4 + \dfrac{R - G}{\delta}\right)$

If H ends up with a negative value, we add 1. In the particular case $(R, G, B) = (0, 0, 0)$, for which both $V = \delta = 0$, we define $(H, S, V) = (0, 0, 0)$.

For example, suppose $(R, G, B) = (0.2, 0.4, 0.6)$ We have

$$V = \max\{0.2, 0.4, 0.6\} = 0.6$$

$$\delta = V - \min\{0.2, 0.4, 0.6\} = 0.6 - 0.2 = 0.4$$

$$S = \frac{0.4}{0.6} = 0.6667$$

Since $B = G$ we have

$$H = \frac{1}{6} \left(4 + \frac{0.2 - 0.4}{0.4}\right) = 0.5833.$$

Conversion in this direction is implemented by the `rgb2hsv` function, and by the `rgb2hsv` method in the `skimage.color` module in Python. This is of course designed to be used on $m \times n \times 3$ arrays, but let's just experiment with our previous example:

```
>> rgb2hsv([0.2 0.4 0.6])
ans =

    0.5833    0.6667    0.6000
```
MATLAB/Octave

and

```
In :  sk.color.rgb2hsv(np.array([[[0.2,0.4,0.6]]]))
Out: array([[[ 0.58333333,  0.66666667,  0.6       ]]])
```
Python

and these are indeed the H, S, and V values we have just calculated.

To go the other way, we start by defining:

$$H' = \lfloor 6H \rfloor$$
$$F = 6H - H'$$
$$P = V(1 - S)$$
$$Q = V(1 - SF)$$
$$T = V(1 - S(1 - F))$$

Since H' is an integer between 0 and 5, we have six cases to consider:

H'	R	G	B
0	V	T	P
1	Q	V	P
2	P	V	T
3	P	Q	V
4	T	P	V
5	V	P	Q

Let's take the HSV values we computed above. We have:

$$H' = \lfloor 6(0.5833) \rfloor = 3$$
$$F = 6(0.5833) - 3 = 0.5$$
$$P = 0.6(1 - 0.6667) = 0.2$$
$$Q = 0.6(1 - (0.6667)(0.5)) = 0.4$$
$$T = 0.6(1 - 0.6667(1 - 0.5)) = 0.4$$

Since $H' = 3$ we have

$$(R, G, B) = (P, Q, V) = (0.2, 0.4, 0.6).$$

Conversion from HSV to RGB is implemented by the `hsv2rgb` function in MATLAB and Octave, and in the `skimage.color` module in Python.

YIQ

This color space is used for TV/video in the United States and other countries where NTSC is the video standard (Australia uses PAL). In this scheme, Y is the "luminance" (this corresponds roughly with intensity), and I and Q carry the color information. The conversion between RGB is straightforward:

$$\begin{bmatrix} Y \\ I \\ Q \end{bmatrix} = \begin{bmatrix} 0.299 & 0.587 & 0.114 \\ 0.596 & -0.274 & -0.322 \\ 0.211 & -0.523 & 0.312 \end{bmatrix} \begin{bmatrix} R \\ G \\ B \end{bmatrix}$$

and

$$\begin{bmatrix} R \\ G \\ B \end{bmatrix} = \begin{bmatrix} 1.000 & 0.956 & 0.621 \\ 1.000 & -0.272 & -0.647 \\ 1.000 & -1.106 & 1.703 \end{bmatrix} \begin{bmatrix} Y \\ I \\ Q \end{bmatrix}$$

The two conversion matrices are of course inverses of each other. Note the difference between Y and V:

$$Y = 0.299R + 0.587G + 0.114B$$
$$V = \max\{R, G, B\}$$

This reflects the fact that the human visual system assigns more intensity to the green component of an image than to the red and blue components. We note here that other transformations [13] $RGB \leftrightarrow HSV$ have

$$V = 0.333R + 0.333G + 0.333B$$

where the intensity is a simple average of the primary values. Note also that the Y of YIQ is different to the Y of XYZ, with the similarity that both represent luminance.

Since YIQ is a linear transformation of RGB, we can picture YIQ to be a parallelepiped (a rectangular box that has been skewed in each direction) for which the Y axis lies along the central $(0, 0, 0)$ to $(1, 1, 1)$ line of RGB. Figure 13.6 shows this.

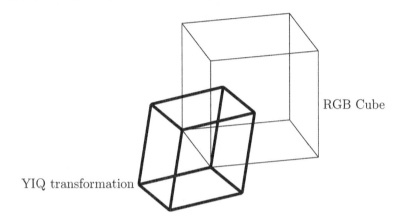

FIGURE 13.6: The RGB cube and its YIQ transformation

That the conversions are linear, and hence easy to do, makes this a good choice for color image processing. Conversion between RGB and YIQ is implemented with the MATLAB/Octave functions `rgb2ntsc` and `ntsc2rgb`. In Python, we can use the `rgb_to_yiq` and `yiq_to_rgb` functions in the standard `colorsys` module.

13.3 Manipulating Color Images

Since a color image requires three separate items of information for each pixel, a (true) color image of size $m \times n$ can be represented by an array of size $m \times n \times 3$: a three-dimensional array. We can think of such an array as a single entity consisting of three separate matrices aligned vertically. Figure 13.7 shows a diagram illustrating this idea. Suppose we read in an RGB image:

```
>> x = imread('twins');
>> size(x)

ans =

   256   256     3
```
MATLAB/Octave

or with

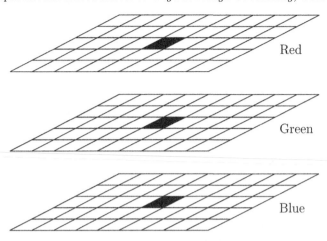

FIGURE 13.7: A three-dimensional array for an RGB image

```
In :   x = io.imread('twins.png')
In :   x.shape
Out:   (256, 256, 3)
```
Python

Each color component can be isolated by the colon operator:

MATLAB/Octave	Python	
x(:,:,1)	x[:,:,0]	The first, or red component
x(:,:,2)	x[:,:,1]	The second, or green component
x(:,:,3)	x[:,:,2]	The third, or blue component

These can all be viewed:

```
>> imshow(x)
>> for i = 1:3, figure(i+1),imshow(x(:,:,i)),end
```
MATLAB/Octave

or in Python with

```
In :   io.imshow(x)
In :   for i in range(3):
...:       plt.figure(i+2); io.imshow(x[:,:,i])
...:
```
Python

(Note that although arrays and lists in Python are indexed starting at 0, figures are automatically indexed starting at 1.)

These are all shown in Figure 13.8. Notice how the colors with particular hues show up with high intensities in their respective components. The orange clothing shows up as very light colored, as it is mostly red, in the red component. In the green component, that orange clothing appears dark in comparison with the yellow clothing. This is because yellow consists of more green than does orange. There are no particularly bright patches in the blue component, but the clothing now appears very dark, indicating the small amount of blue in each of the two colors.

We can convert to YIQ or HSV and view the components again:

| A color image | Red component | Green component | Blue component |

FIGURE 13.8: **SEE COLOR INSERT** An RGB color image and its components

```
>> xh = rgb2hsv(x);
>> for i = 1:3, figure(i+1),imshow(xh(:,:,i)),end
```
MATLAB/Octave

```
In :  xh = sk.color.rgb2hsv(x)
In :  for i in range(3):
...:      plt.figure(i); io.imshow(xh[:,:,i])
...:
```
Python

and these are shown in Figure 13.9. We can do precisely the same thing for the YIQ color

| Hue | Saturation | Value |

FIGURE 13.9: The HSV components

space:

```
>> xyiq = rgb2ntsc(x);
>> for i = 1:3, figure(i+1),imshow(xyiq(:,:,i)),end
```
MATLAB/Octave

In Python, the function to use is `rgb_to_yiq`, which is part of the `colorsys` module and which takes not just the image but the individual color components as inputs. To obtain the components quickly the `dsplit` command can be used to split the $256 \times 256 \times 3$ RGB array into a list of three separate arrays of size $256 \times 256 \times 1$. Preceding the `dsplit` function with an asterisk "unpacks" the elements of the list into individual elements, which can be used as inputs to `rgb_to_yiq`.

```
In :   xyiq = colorsys.rgb_to_yiq(*np.dsplit(x,3))
In :   for i in range(3):
...:        plt.figure(i); io.imshow(xyiq[i][:,:,0])
...:
```

`Python`

Note that the same result could be obtained simply by multiplying each column of the matrix x by the RGB to YIQ conversion matrix:

```
In :   rgb2yiq = np.array
       ([[.299,.587,0.114],[.596,-.275,-.321],[.212,-.528,.311]])
In :   xyiq = x.dot(rgb2yiq)
In :   for i in range(3):
...:        plt.figure(i); io.imshow(xyiq[:,:,i])
...:
```

`Python`

The Y, I, and Q images are shown in Figure 13.10. Notice that the Y component of YIQ gives

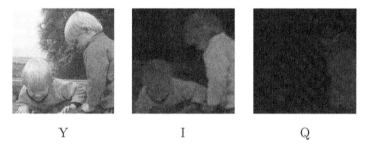

Y   I   Q

FIGURE 13.10: The YIQ components

a better grayscale version of the image than the value of HSV. The clothing, in particular, is quite washed out in Figure 13.9 (value), but shows better contrast in Figure 13.10 (Y).

We shall see below how to put three matrices, obtained by operations on the separate components, back into a single three dimensional array for display.

13.4   Pseudocoloring

This means assigning colors to a grayscale image in order to make certain aspects of the image more amenable for visual interpretation—for example, for medical images. There are different methods of pseudocoloring.

Intensity Slicing

In this method, the image is split into various gray level ranges, and a different color is assigned to each range. For example:

Gray level:	0–63	64–127	128–191	192–255
Color:	Blue	Magenta	Green	Red

This can be considered as a mapping, as shown in Figure 13.11.

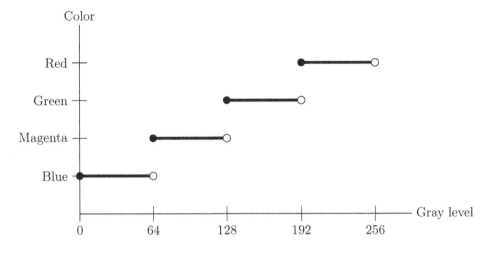

FIGURE 13.11: Intensity slicing as a mapping

Gray-Color Transformations

We have three functions $f_R(x)$, $f_G(x)$, $f_B(x)$ which assign red, green, and blue values to each gray level x. These values (with appropriate scaling, if necessary) are then used for display. Using an appropriate set of functions can enhance a grayscale image with impressive results. An example is shown in Figure 13.12.

The gray level x in the diagram is mapped onto red, green and blue values of approximately 0.8, 0.7, and 0.3, respectively.

In MATLAB or Octave, a simple way to view an image with added color is to use **imshow** with an extra **colormap** parameter. For example, consider the image **blocks.png**. We can add a color map with the **colormap** function; there are several existing color maps to choose from. Figure 13.13 shows the children's blocks image (from Figure 1.4) after color transformations. Color image (a) can be created with:

```
>> b = imread('blocks.png');
>> imshow(b,colormap(jet(256)))
```

MATLAB/Octave

or with:

```
In :  b = io.imread('blocks.png')
In :  io.imshow(b,cmap=plt.cm.jet)
```

Python

However, a bad choice of color map can ruin an image. Image (b) in Figure 13.13 is an example of this, where we apply the **vga** color map. This map in fact is not included in Octave, so here it is:

```
>> x = [1 1 1;1 0 0;1 1 0;0 1 0;0 1 1;0 0 1;1 0 1;0 0 0]
>> vga = [4*x(1,:);3 3 3;4*x(2:8,:);2*x(1:7,:)]/4
```

Octave

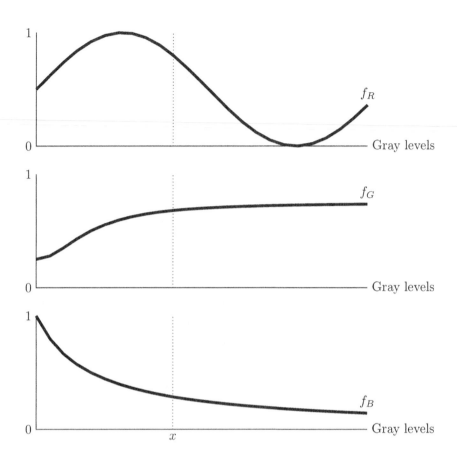

FIGURE 13.12: Mapping grays to colors

(a)   (b)

FIGURE 13.13: **SEE COLOR INSERT** Applying a color map to a grayscale image

Since this only has 16 rows, we need to reduce the number of grayscales in the image to 16. This can be done with the `grayslice` function:

```
>> b16 = grayslice(im2double(b),16);
>> figure,imshow(b16+1,colormap(vga))
```

MATLAB/Octave

In Python, the same effect can be obtained by first creating the array of VGA colors, and then turning that array into a colormap using the `vstack` function, which stacks arrays vertically:

```
In :  x = np.array([[1, 1, 1],[1, 0, 0],[1, 1, 0],[0, 1, 0],[0, 1, 1],[0,
    0, 1],[1,
0, 1],[0, 0, 0]])
In :  vga = np.vstack((4*x[0,:],np.array([[3, 3, 3]]),4*x[1:,:],2*x[:7,:]))
    /4.0
In :  b16 = b/16
In :  myvga = mpl.colors.ListedColormap(vga)
In :  io.imshow(b16,cmap=myvga)
```

Python

The result, although undeniably colorful, is not really an improvement on the original image. Each of MATLAB, Octave, and Python provide slightly different methods of listing the colormaps.

```
>> help graph3d
  Color maps.
    parula    - Blue-green-orange-yellow color map
    hsv       - Hue-saturation-value color map.
    hot       - Black-red-yellow-white color map.
    gray      - Linear gray-scale color map.
    bone      - Gray-scale with tinge of blue color map.
    copper    - Linear copper-tone color map.
    pink      - Pastel shades of pink color map.
    white     - All white color map.
    flag      - Alternating red, white, blue, and black color map.
    lines     - Color map with the line colors.
    colorcube - Enhanced color-cube color map.
    vga       - Windows colormap for 16 colors.
    jet       - Variant of HSV.
    prism     - Prism color map.
    cool      - Shades of cyan and magenta color map.
    autumn    - Shades of red and yellow color map.
    spring    - Shades of magenta and yellow color map.
    winter    - Shades of blue and green color map.
    summer    - Shades of green and yellow color map.
```

MATLAB

```
>> colormap('list')
ns =
{
  [1,1] = autumn
  [1,2] = bone
  [1,3] = cool
  [1,4] = copper
  [1,5] = flag
  [1,6] = gmap40
  [1,7] = gray
  [1,8] = hot
  [1,9] = hsv
  [1,10] = jet
  [1,11] = lines
  [1,12] = ocean
  [1,13] = pink
  [1,14] = prism
  [1,15] = rainbow
  [1,16] = spring
  [1,17] = summer
  [1,18] = white
  [1,19] = winter
}
```

`Octave`

In Python, the colormaps are given as dictionaries stored as `datad` in the `cm` module of `matplotlib.pyplot`:

```
In :  import matplotlib.pyplot as plt
In :  plt.cm.datad.viewkeys()
Out:  dict_keys(['Spectral', 'summer', 'coolwarm', 'pink_r', 'Set1', 'Set2'
      , 'Set3',
'brg_r', 'Dark2', 'hot', 'PuOr_r', 'afmhot_r', 'terrain_r', 'PuBuGn_r', '
    RdPu',
'gist_ncar_r', 'gist_yarg_r', 'Dark2_r', 'YlGnBu', 'RdYlBu', 'hot_r',
'gist_rainbow_r', 'gist_stern', 'gnuplot_r', 'cool_r', 'cool', 'gray', '
    copper_r',
'Greens_r', 'GnBu', 'gist_ncar', 'spring_r', 'gist_rainbow', 'RdYlBu_r',
'gist_heat_r', 'OrRd_r', 'CMRmap', 'bone', 'gist_stern_r', 'RdYlGn', '
    Pastel2_r',
'spring', 'terrain', 'YlOrRd_r', 'Set2_r', 'winter_r', 'PuBu', 'RdGy_r', '
    spectral',
'flag_r', 'jet_r', 'RdPu_r', 'Purples_r', 'gist_yarg', 'BuGn', 'Paired_r',
    'hsv_r',
'bwr', 'cubehelix', 'YlOrRd', 'Greens', 'PRGn', 'gist_heat', 'spectral_r',
    'Paired',
'hsv', 'Oranges_r', 'prism_r', 'Pastel2', 'Pastel1_r', 'Pastel1', 'gray_r',
     'PuRd_r',
'Spectral_r', 'gnuplot2_r', 'BuPu', 'YlGnBu_r', 'copper', 'gist_earth_r', '
    Set3_r',
'OrRd', 'PuBu_r', 'ocean_r', 'brg', 'gnuplot2', 'jet', 'bone_r', '
    gist_earth',
'Oranges', 'RdYlGn_r', 'PiYG', 'CMRmap_r', 'YlGn', 'binary_r', 'gist_gray_r
    ',
```

`Python`

```
'Accent', 'BuPu_r', 'gist_gray', 'flag', 'seismic_r', 'RdBu_r', 'BrBG', '
    Reds', 'BuGn_r', 'summer_r', 'GnBu_r', 'BrBG_r', 'Reds_r', 'RdGy', '
    PuRd', 'Accent_r', 'Blues', 'Grays', 'autumn', 'cubehelix_r', '
    nipy_spectral_r', 'PRGn_r', 'Grays_r', 'pink', 'binary', 'winter', '
    gnuplot', 'RdBu', 'prism', 'YlOrBr', 'coolwarm_r', 'rainbow_r', '
    rainbow', 'PiYG_r', 'YlGn_r', 'Blues_r', 'YlOrBr_r', 'seismic', '
    Purples', 'bwr_r', 'autumn_r', 'ocean', 'Set1_r', 'PuOr', 'PuBuGn', '
    nipy_spectral', 'afmhot'])
```
Python

There are help files for each of these color maps, so that in both MATLAB and Octave the command

```
>> help jet
```
MATLAB/Octave

will provide some information on the `jet` color map. Python doesn't give any information as such, but the colormaps can be viewed with the **viewitems** method:

```
In :  plt.cm.datad['jet'].viewitems()
Out:  dict_items([('blue', ((0.0, 0.5, 0.5), (0.11, 1, 1), (0.34, 1, 1),
      (0.65, 0, 0), (1, 0, 0))), ('green', ((0.0, 0, 0), (0.125, 0, 0),
      (0.375, 1, 1), (0.64, 1, 1), (0.91, 0, 0), (1, 0, 0))), ('red', ((0.0,
      0, 0), (0.35, 0, 0), (0.66, 1, 1), (0.89, 1, 1), (1, 0.5, 0.5)))])
```
Python

We can easily create our own color map: it must by a matrix with 3 columns, and each row consists of RGB values between 0.0 and 1.0. Suppose we wish to create a blue, magenta, green, red color map as shown in Figure 13.11. Using the RGB values:

Color	Red	Green	Blue
Blue	0	0	1
Magenta	1	0	1
Green	0	1	0
Red	1	0	0

we can create our color map with:

```
>> mycolormap = [0 0 1;1 0 1;0 1 0;1 0 0];
```
MATLAB/Octave

Before we apply it to the blocks image, we need to scale the image down so that there are only the four grayscales 0, 1, 2, and 3:

```
>> b4 = grayslice(im2double(b),4);
>> imshow(b4+1,mycolormap)
```
MATLAB/Octave

In Python, the method is slightly more complex:

```
In :  b = io.imread('blocks.png')
In :  b4 = b/64
In :  mycolormap = mpl.colors.ListedColormap([[0, 0, 1],[1, 0, 1],[0, 1,
      0],[1, 0, 0]])
In :  io.imshow(b4,cmap=mycolormap)
```
`Python`

and the result is shown in Figure 13.14.

FIGURE 13.14: SEE COLOR INSERT An image colored with a "handmade" color map

13.5   Processing of Color Images

There are two methods we can use:

1. We can process each R, G, B matrix separately.

2. We can transform the color space to one in which the intensity is separated from the color, and process the intensity component only.

Schemas for these are given in Figure 13.15.

We shall consider a number of different image processing tasks, and apply either of the above schema to color images.

Contrast Enhancement

We know that histogram equalization, in general, provides a simple method of contrast enhancement. So one way of enhancing the contrast in a color image would be to enhance each of the color layers separately. For example, take the image `color_sunset.png`, which is the original color image that we used before as a grayscale image in Chapter 4. Each layer can be processed separately, and the final result viewed. In MATLAB or Octave:

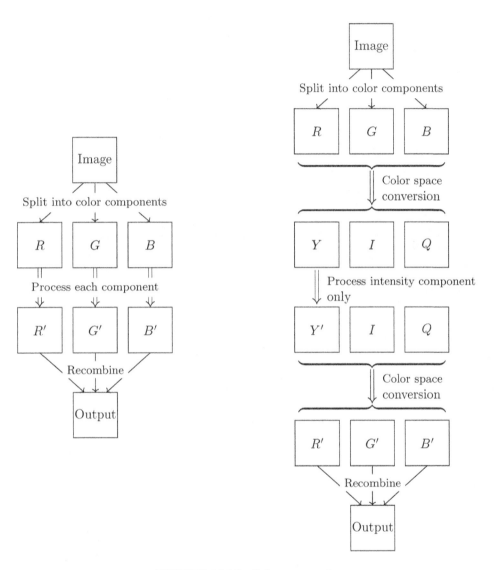

FIGURE 13.15: Color processing

```
>> s = imread('color_sunset.png');
>> s2 = zeros(size(s));
>> for i = 1:3, s2(:,:,i) = histeq(s(:,:,i)); end
>> imshow(s2)
```
MATLAB/Octave

or in Python:

```
In :   s = io.imread('color_sunset.png').astype('float64')
In :   s2 = np.zeros_like(s)
In :   for i in range(3):
...:       s2[:,:,i] = sk.exposure.equalize_hist(s[:,:,i])
...:
In :   io.imshow(s2)
```
Python

Both images are shown in Figure 13.16.

(a) The original image   (b) Attempt at equalization

FIGURE 13.16: **SEE COLOR INSERT** An attempt at color contrast enhancement

Note that although the result is certainly better contrasted, there are some curious new colors introduced: the orange sheen on the water has been replaced with a brownish-white and in general both water and sky are "washed out," and there is a light purple tint in the clouds.

A better result here may be obtained by decoupling the intensity information from the intensity component, for example, by converting from RGB to YIQ, and then processing the Y component only:

```
>> syiq = rgb2ntsc(s);
>> syiq(:,:,1) = histeq(syiq(:,:,1));
>> s2 = ntsc2rgb(syiq);
```
MATLAB/Octave

For simplicity, instead of using Y in TIQ for the intensity information, in Python choose instead the V from HSV, and import the **color** module first:

```
In :   import skimage.color as co
In :   sh = co.rgb2hsv(s)
In :   sh[:,:,2] = sk.exposure.equalize_hist(sh[:,:,2])
In :   s2 = co.hsv2rgb(sh)
```
Python

The result is shown in Figure 13.17

FIGURE 13.17: **SEE COLOR INSERT** A better attempt at color contrast enhancement

Note that the colors here are much more "true to life" than in the original attempt. One more example is of an emu; Figure 13.18 shows the original, and the enhancement with all RGB layers, or with an intensity layer only.

(a) The original image   (b) Processing each RGB   (c) Processing Y only

FIGURE 13.18: **SEE COLOR INSERT** Another example of color contrast enhancement

In this case, there is only minimal visible difference, mainly because the initial image is quite well contrasted already.

Spatial Filtering

It very much depends on the filter as to which schema we use. In general, if the filter will not dramatically change the values, it is probably safe to use either schema. For example, consider an image of a koala, `koala_color.png`, loaded and stored as a three-dimensional array `k`. Now try unsharp masking, first with each RGB component, and then with an intensity component only. In MATLAB and Octave, just use the unsharp filter given by `fspecial`.

```
>> u = fspecial('unsharp');
>> k2 = zeros(size(k));
>> for i = 1:3, k2(:,:,i) = imfilter(k(:,:,i),u); end
>> imshow(k2/255)
>> kyiq = rgb2ntsc(k);
>> kyiq(:,:,1) = imfilter(kyiq(:,:,1),u);
>> k3 = ntsc2rgb(kyiq);
>> figure, imshow(k3)
```

MATLAB/Octave

In Python the same filter can be used, but it needs to be entered first.

```
In :  u = np.array([[-1,-4,-1],[-4,26,-4],[-1,-4,-1]])/6.0
In :  k2 = np.zeros_like(k)
In :  for i in range(3):
...:      k2[:,:,i] = ndi.convolve(k[:,:,i],u)
...:
In :  kh = co.rgb2hsv(k)
In :  kh[:,:,2] = ndi.convolve(kh[:,:,2],u)
In :  k3 = co.hsv2rgb(kh)
```

Python

and the results are shown in Figure 13.19.

(a) The original image   (b) Processing each RGB   (c) Processing Y only

FIGURE 13.19: **SEE COLOR INSERT** Unsharp masking of a color image

As an example of a low pass filter consider a Kuwahara filter, first on each RGB component and then on the intensity component only. Note that the Kuwahara implementation is entirely unoptimized, so this is a very slow process!

```
>> c = imread('cat.png');
>> ck = zeros(size(c));
>> for i = 1:3, ck(:,:,i) = kuwahara(c(:,:,i)); end
>> imshow(ck)
>> cyiq = rgb2ntsc(c);
>> cyiq(:,:,1) = kuwahara(cyiq(:,:,1));
>> c3 = ntsc2rgb(cyiq);
>> figure, imshow(c3)
```

MATLAB/Octave

The Python approach is very similar, as it was for unsharp masking, but for intensity again use the V from HSV:

```python
In :   ck = np.zeros_like(c)
In :   for i in range(3):
...:       ck[:,:,i] = kuwahara(c[:,:,i])
...:
In :   ch = co.rgb2hsv(c)
In :   ch[:,:,2] = kuwahara(ch[:,:,2])
In :   c3 = co.hsv2rgb(ch)
```

Python

The results are shown in Figure 13.20, and as before there is not a huge observable difference between the outputs.

(a) The original image   (b) Processing each RGB   (c) Processing Y only

FIGURE 13.20: **SEE COLOR INSERT** Kuwahara filtering of a color image

Noise Reduction

Consider the twins image, with salt and pepper noise added:

```matlab
>> tn = imnoise(t,'salt_&_pepper',0.2);
>> imshow(tn)
>> figure,imshow(tn(:,:,1))
>> figure,imshow(tn(:,:,2))
>> figure,imshow(tn(:,:,3))
```

MATLAB/Octave

These are all shown in Figure 13.21. It would appear that median filtering should be applied

Salt and pepper noise   The red component   The green component   The blue component

FIGURE 13.21: **SEE COLOR INSERT** Noise on a color image

to each of the RGB components. This is easily done, using exactly the same commands as for Kuwahara filtering and histogram equalization, except for using `medfilt2` instead of `kuwahara` or `histeq` (or `equalize_hist`).

The result is shown in Figure 13.22(a). In this instance we cannot apply the median filter to the intensity component only, because the conversion from RGB to YIQ spreads the noise across all the YIQ components. If we remove the noise from Y only, then the noise has been only slightly diminished as shown in Figure 13.22(b). If the noise applies to only

(a) Denoising each RGB component   (b) Denoising intensity only

FIGURE 13.22: **SEE COLOR INSERT** Attempts at denoising a color image

one of the RGB components, then it would be appropriate to apply a denoising technique to this component only.

Also note that the method of noise removal must depend on the generation of noise. In the above example, we tacitly assumed that the noise was generated after the image had been acquired and stored as RGB components. But as noise can arise anywhere in the image acquisition process, it is quite reasonable to assume that noise might affect only the brightness of the image. In such a case, denoising an intensity component only will produce the best results.

Edge Detection

An edge image will be a binary image containing the edges of the input. We can go about obtaining an edge image in two ways:

1. We can take the intensity component only, and find its edges.

2. We can find the edges of each of the RGB components, and join the results.

To implement the first method in MATLAB or Octave, we start with the **rgb2gray** function, applied to an image of a Venetian canal:

```
>> v = imread('venice.png');
>> vg = rgb2gray(v);
>> ve = edge(vg);
>> imshow(ve)
```

                                                              MATLAB/Octave

In Python the **sobel** function can be used with a threshold value:

```
In :  v = io.imread('venice.png')
In :  vg = co.rgb2gray(v)
In :  ve1 = fl.sobel(vg)>0.11
```
Python

Recall that the **edge** function with no parameters implements Sobel detection. The result is shown in Figure 13.23(a). For the second method, we can join the results with the logical "or":

```
>> v2 = zeros(size(v));
>> for i = 1:3, v2(:,:,i) = edge(v(:,:,i)); end
>> ve2 = ve2 = v2(:,:,1) | v2(:,:,2) | v2(:,:,3);
>> figure,imshow(ve2)
```
MATLAB/Octave

The method is similar in Python, except that since **v2** is a numeric array, the results of the Sobel filtering will be cast to numeric values:

```
In :  v2 = np.zeros_like(v)
In :  for i in range(3):
...:      v2[:,:,i] = fl.sobel(v[:,:,i])>0.11
...:
In :  ve2 = (v2[:,:,0] + v2[:,:,1] + v2[:,:,2])>0
```
Python

and this is shown in Figure 13.23(b). The edge image **ve2** is a much more complete edge

ve1: Edges after **rgb2gray**   ve2: Edges of each RGB component

FIGURE 13.23: The edges of a color image

image. Frames of doors and windows are more complete, as are roof edges. However, the success of these methods will also depend on the parameters of the edge function chosen; for example, the threshold value used. In the examples shown, the **edge** function has been used with its default threshold.

The Retinex Algorithm

In the 1970s Edwin Land (of Polaroid camera fame) proposed a "retinex" theory of human vision [27]; the word being a mix of "retina" and "cortex." Retinex algorithms, of which there are many, have much in common with homomorphic filtering which was discussed in Section 7.9. Retinex algorithms are designed to improve visual contrast, such as in dark shadowed areas of an image, while maintaining color.

One particular algorithm, known as "center/surround retinex" [25], works by subtracting from the logarithm of the image the logarithm of the image filtered with a Gaussian filter:

$$R(x, y) = \log I(x, y) - \log \left(I(x, y) * F(x, y) \right).$$

The filtering is usually done using the Fourier transform, as one version uses a filter the same size as the image itself.

Figure 13.24(a) shows a street scene taken in low light (through a car window). It can be processed in MATLAB or Octave as:

```
>> s = im2double(rgb2gray(imread('street.png')));
>> s1 = mat2gray(s)+0.01;
>> f = fspecial('gaussian',size(s),40);
>> sf = abs(ifft2(fft2(s1).*fft2(f)));
>> sf(find(sf==0))=0.01;
>> s2 = log(s1) - log(sf);
>> imshow(mat2gray(sf))
```

MATLAB/Octave

or in Python as

```
In :   s = ut.img_as_float(co.rgb2gray(io.imread('street.png')))
In :   s1 = ex.rescale_intensity(s,out_range=(0,1))+0.01
In :   r,c = s.shape
In :   i,j = np.mgrid[-(r//2):(r+1)//2,-(c//2):(c+1)//2]
In :   f = np.exp(-(i**2+j**2)/40**2)
In :   sf = abs(ifft2(fft2(s1)*fft2(f)))
In :   sf[np.where(sf==0)]=0.01
In :   s2 = np.log(s1) - np.log(sf)
In :   io.imshow(s2)
```

Python

and the result is shown in Figure 13.24(b).

For a color image, there is the option of either processing each RGB component, or the intensity component only from either HSV or YIQ. Figure 13.25 shows a picture with a lot of dark shadows: on the far side of the ornamental lake it is hard to see any details. Figure 13.26 shows two outcomes of the center/surround retinex algorithm. This retinex algorithm is known as a "single-scale retinex" as it uses a single Gaussian filter; more sophisticated algorithms use multiple scales—different Gaussian filters—and can preserve color intensity and saturation.

Programs to implement this single-scale retinex are given below.

(a) A "dull" image   (b) Result of retinex processing

FIGURE 13.24: An application of center/surround retinex

FIGURE 13.25: **SEE COLOR INSERT** The botanic gardens, Melbourne, Australia

(a) Applying retinex to each of RGB   (b) Applying retinex to Y only

FIGURE 13.26: **SEE COLOR INSERT** Center/surround retinex on a color image

13.6 Programs

Here are programs for computing and displaying the RGB gamut, first in MATLAB or Octave:

```
function plotgamut()
  xyz2rgb = [3.063 -1.393 -0.476; -0.969 1.876 0.042;0.068 -0.229 1.069];
  ciexyz = csvread('ciexyz31.csv');
  ciexyz = [ciexyz;ciexyz(1,:)];
  W = ciexyz(:,1);
  X = ciexyz(:,2);
  Y = ciexyz(:,3);
  Z = ciexyz(:,4);
  S = X+Y+Z;

  [xx,yy] = meshgrid(1:512);
  z0 = zeros([size(xx),3]);
  z0(:,:,1)=xx/512;
  z0(:,:,2) = yy/512;
  z0(:,:,3) = 1-xx/512-yy/512;
  z0r = reshape(z0,prod(size(xx)),3)';
  rgb0 = xyz2rgb*z0r;
  mn = min(rgb0)<0;
  rgb1 = max(rgb0,[mn;mn;mn]);
  % rgb1(find(rgb1>1))=1;
  % mm = min(rgb0)>=0 & max(rgb0)<=1;
  % rgb1 = rgb0.*[mm;mm;mm];
  % rgb1(find(rgb1==0))=1;
  rgbs = reshape(rgb1',[size(xx),3]);
  figure,imshow(flipdim(rgbs,1))
  %figure,imshow(rgbs)
  hold on
  plot(floor(X./S*512),512-floor(Y./S*512),'-k','linewidth',3),axis on
  %plot(floor(X./S*512),floor(Y./S*512),'-k','linewidth',3),axis on
end
```
MATLAB/Octave

and second in Python:

```
def plotgamut():
  xyz2rgb = np.array([[3.063, -1.393, -0.476],[-0.969, 1.876,
      0.042],[0.068, -0.229, 1.069]])
  ciexyz = np.loadtxt('ciexyz31_1.csv',delimiter=',')
  ciexyz1 = np.vstack((ciexyz,ciexyz[0,:,np.newaxis].T))
  W = ciexyz1[:,0]
  X = ciexyz1[:,1]
  Y = ciexyz1[:,2]
  Z = ciexyz1[:,3]
  S = X+Y+Z

  xx,yy = np.mgrid[0:512,0:512].astype('float64')
  z0 = np.zeros((512,512,3))
```
Python

```python
z0[:,:,0] = xx/512;
z0[:,:,1] = yy/512;
z0[:,:,2] = 1.0-xx/512-yy/512;
z0r = np.reshape(z0,(xx.size,3),order='F').T
rgb0 = xyz2rgb.dot(z0r)
mn = (rgb0.min(axis=0)<0)*1.0
rgb1 = np.maximum(rgb0,np.tile(mn,(3,1)))
# rgb1(find(rgb1>1))=1;
# mm = min(rgb0)>=0 & max(rgb0)<=1;
# rgb1 = rgb0.*[mm;mm;mm];
# rgb1(find(rgb1==0))=1;
rgbs = np.reshape(rgb1.T,(xx.shape+(3,)),order='F');
rgbs[np.where(rgbs>1)]=1.0
#
#figure,imshow(rgbs)
plt.plot(np.ceil(X/S*512),512-np.ceil(Y/S*512))
plt.imshow(np.flipud(rgbs),aspect='equal',origin='upper')
plt.show()
#plot(floor(X./S*512),floor(Y./S*512),"-k","linewidth",3),axis on
```

Python

The retinex algorithm in MATLAB or Octave:

```matlab
% Single-scale retinex, adapted from
%    http://au.mathworks.com/MATLABcentral/fileexchange/26523
%
% Usage: ssretinex(image,hsize)
%
% where hsize is the spread of the Gaussian filter to be used
%
function out = ssretinex(X,hsize)
if length(size(X))==2
    imsize = size(X);
else
    imsize = size(X(:,:,1));
end
filt = fspecial('gaussian',imsize,hsize/sqrt(2));
if length(size(X))==2
    X1 = mat2gray(im2double(X))+0.05;
    Z = abs(ifft2(fft2(X1).*fft2(filt)));
    Z(find(Z==0))=0.05;
    L = log(Z);
    R = log(X1)-L;
    out = mat2gray(R);
else
    out = zeros(size(X));
    for i = 1:3
        X1 = mat2gray(im2double(X(:,:,i)))+0.01;
        Z = abs(ifft2(fft2(X1).*fft2(filt)));
        Z(find(Z==0))=0.01;
        L = log(Z);
        out(:,:,i) = mat2gray(log(X1)-L);
    end
end
```

MATLAB/Octave

```
    end
```

and here is retinex in Python:

```python
def ssretinex(im,hsize):
    from scipy.fftpack import fft2,ifft2
    from skimage.exposure import rescale_intensity
    r,c = im.shape[:2]
    i,j = np.mgrid[-(r//2):(r+1)//2,-(c//2):(c+1)//2]
    f = np.exp(-((i**2+j**2)/hsize**2))
    if len(im.shape)==2:
        im1 = rescale_intensity(im.astype(float),out_range=(0,1))+0.01
        imf = abs(ifft2(fft2(im1)*fft2(f)))
        imf[np.where(imf==0)] = 0.01
        return np.log(im1) - np.log(imf)
    else:
        out = np.zeros_like(im).astype(float)
        im2 = im.astype(float)
        for i in range(3):
            imr = rescale_intensity(im2[:,:,i],out_range=(0,1))+0.01
            imf = abs(ifft2(fft2(imr)*fft2(f)))
            imf[np.where(imf==0)] = 0.01
            out[:,:,i] = rescale_intensity(np.log(imr) - np.log(imf),
                out_range=(0,1))
        return out
```

Exercises

1. By hand, determine the saturation and intensity components of the following image, where the RGB values are as given:

$(0,1,1)$	$(1,2,3)$	$(7,7,7)$	$(5,1,2)$	$(1,1,7)$
$(2,1,2)$	$(1,7,7)$	$(2,0,2)$	$(3,3,2)$	$(5,5,0)$
$(4,4,4)$	$(4,6,7)$	$(4,5,6)$	$(1,5,7)$	$(3,6,7)$
$(3,0,3)$	$(5,2,2)$	$(1,1,1,$	$(6,6,0)$	$(2,2,2)$
$(1,2,1)$	$(0,4,4)$	$(3,1,6)$	$(3,3,3)$	$(2,4,6)$

2. Suppose the intensity component of an HSV image was thresholded to just two values. How would this affect the appearance of the image?

3. By hand, perform the conversions between RGB and HSV or YIQ, for the values:

R	G	B	H	S	V
0.5	0.5	0			
0	0.7	0.7			
0.5	0	0.5			
			0.33	0.5	1
			0.67	0.7	0.7
			0	0.2	0.8

R	G	B	Y	I	Q
0.3	0.3	0.7			
0.7	0.9	0			
0.8	0.8	0.7			
			1	0.3	0.3
			0.5	0.5	0.5
			0	1	1

You may need to normalize the RGB values.

4. Check your answers to the conversions in Question 3 by using the relevant functions in MATLAB, Octave, or Python.

5. Threshold the intensity component of a color image of your choice, and see if the result agrees with your guess from Question 2.

6. The image **emu.png** is an indexed color image. Experiment with applying different color maps to the index matrix of this image.

   Which color map seems to give the best results? Which color map seems to give the worst results?

7. View the image **street.png**. Experiment with histogram equalization on:

   (a) The intensity component of HSV

   (b) The intensity component of YIQ

   Which seems to produce the best result?

8. Create and view a random "patchwork quilt" with:

```
>> r = uint8(floor(256*rand(16,16,3)));
>> r = imresize(r,16);
>> imshow(r),pixval on
```
MATLAB/Octave

or

```
In :   r = np.random.randint(0,256,(16,16,3))
In :   r = tr.resize(r.astype('uint8'),(256,256),order=0)
In :   io.imshow(r)
```
Python

What RGB values produce (a) a light brown color? (b) a dark brown color?

Convert these brown values to HSV, and plot the hues on a circle.

9. Using the image **tropical_flower.png**, see if you can obtain an edge image from the intensity component alone, that is, as close as possible to the image **fe2** in Figure 13.23. What parameters to the **edge** function did you use? How close to **fe2** could you get?

10. Add Gaussian noise to an RGB color image x with

```
>> xn = imnoise(x,'gaussian');
```
MATLAB/Octave

or

```
In :  xn = skimage.util.random_noise(x)
```
Python

View your image, and attempt to remove the noise with

(a) Average filtering on each RGB component

(b) Wiener filtering on each RGB component

11. Take any color image of your choice and add salt and pepper noise to the intensity component. This can be done with

```
>> ty = rgb2ntsc(tw);
>> tn = imnoise(ty(:,:,1).'salt & pepper');
>> ty(:,:,1) = tn;
```
MATLAB/Octave

or

```
In :   rgb2yiq = np.array
          ([[.299,.587,0.114],[.596,-.275,-.321],[.212,-.528,.311]])
In :   yiq2rgb = np.linalg.inv(rgb2yiq)
In :   ty = ut.img_as_float(tw).dot(rgb2yiq)
In :   tn = ut.random_noise(ty[:,:,0],'s&p')
In :   ty[:,:,0] = tn
In :   t2 = ty.dot(yiq2rgb)
```
Python

Now convert back to RGB for display.

(a) Compare the appearance of this noise with salt and pepper noise applied to each RGB component as shown in Figure 13.21. Is there any observable difference?

(b) Denoise the image by applying a median filter to the intensity component.

(c) Now apply the median filter to each of the RGB components.

(d) Which one gives the best results?

(e) Experiment with larger amounts of noise.

(f) Experiment with Gaussian noise.

Chapter 14

Image Coding and Compression

14.1 Lossless and Lossy Compression

We have seen that image files can be very large. It is thus important for reasons both of storage and file transfer to make these file sizes smaller, if possible. In Section 1.9 we touched briefly on the topic of compression; in this section, we investigate some standard compression methods. It will be necessary to distinguish between two different classes of compression methods: *lossless compression*, where all the information is retained, and *lossy compression* where some information is lost.

Lossless compression is preferred for images of legal, scientific, or political significance, where loss of data, even of apparent insignificance, could have considerable consequences. Unfortunately this style tends not to lead to high compression ratios. However, lossless compression is used as part of many standard image formats.

14.2 Huffman Coding

The idea of Huffman coding is simple. Rather than using a fixed length code (8 bits) to represent the gray values in an image, we use a variable length code, with smaller length codes corresponding to more probable gray values.

A small example will make this clear. Suppose we have a 2-bit grayscale image with only four gray levels: 0, 1, 2, 3, with the probabilities 0.2, 0.4, 0.3, and 0.1, respectively. That is, 20% of pixels in the image have gray value 50; 40% have gray value 100, and so on. The following table shows fixed length and variable length codes for this image:

Gray Value	Probability	Fixed Code	Variable Code
0	0.2	00	000
1	0.4	01	1
2	0.3	10	01
3	0.1	11	001

Now consider how this image has been compressed. Each gray value has its own unique identifying code. The average number of bits per pixel can be easily calculated as the expected value (in a probabilistic sense):

$$(0.2 \times 3) + (0.4 \times 1) + (0.3 \times 2) + (0.1 \times 3) = 1.9.$$

Notice that the longest codewords are associated with the lowest probabilities. This average is indeed smaller than 2.

This can be made more precise by the notion of *entropy*, which is a measure of the amount of information. Specifically, the entropy H of an image is the theoretical minimum number of bits per pixel required to encode the image with no loss of information. It is defined by

$$H = -\sum_{i=0}^{L-1} p_i \log_2(p_i)$$

where the index i is taken over all grayscales of the image, and p_i is the probability of gray level i occurring in the image. Very good accounts of the basics of information theory and entropy are given by Roman [39] and Welsh [55]. In the example given above,

$$H = -\Big(0.2\log_2(0.2) + 0.4\log_2(0.4) + 0.3\log_2(0.3) + 0.1\log_2(0.1)\Big) = 1.8464.$$

This means that no matter what coding scheme is used, it will never use less than 1.8464 bits per pixel. On this basis, the Huffman coding scheme given above, giving an average number of bits per pixel much closer to this theoretical minimum than 2, provides a very good result.

To obtain the Huffman code for a given image we proceed as follows:

1. Determine the probabilities of each gray value in the image.

2. Form a binary tree by adding probabilities two at a time, always taking the two lowest available values.

3. Now assign 0 and 1 arbitrarily to each branch of the tree from its apex.

4. Read the codes from the top down.

To see how this works, consider the example of a 3-bit grayscale image (so the gray values are 0–7) with the following probabilities:

Gray Value	0	1	2	3	4	5	6	7
Probability	0.19	0.25	0.21	0.16	0.08	0.06	0.03	0.02

For these probabilities, the entropy can be calculated to be 2.6508. We can now combine probabilities two at a time as shown in Figure 14.1.

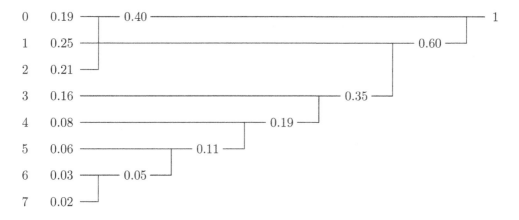

FIGURE 14.1: Forming the Huffman code tree

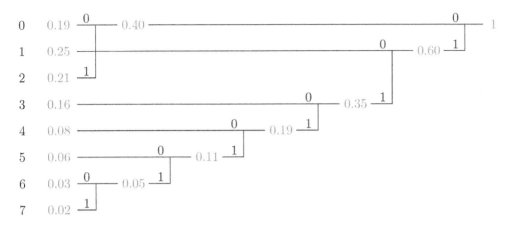

FIGURE 14.2: Assigning 0's and 1's to the branches

Note that if we have a choice of probabilities we choose arbitrarily. The second stage consists of arbitrarily assigning 0's and 1's to each branch of the tree just obtained. This is shown in Figure 14.2.

To obtain the codes for each gray value, start at the 1 on the top right, and work back toward the gray value in question, listing the numbers passed on the way. This produces:

Gray Value	Huffman Code
0	00
1	10
2	01
3	110
4	1110
5	11110
6	111110
h 7	111111

As above, we can evaluate the average number of bits per pixel as an expected value:

$$(0.19 \times 2) + (0,25 \times 2) + (0.21 \times 2) + (0.16 \times 3) +$$
$$(0.08 \times 4) + (0.06 \times 5) + (0.03 \times 6) + (0.02 \times 6) = 2.7$$

which is a significant improvement over 3 bits per pixel, and very close to the theoretical minimum of 2.6508 given by the entropy.

Huffman codes are *uniquely decodable*, in that a string can be decoded in only one way. For example, consider the string

$$1 \ 1 \ 0 \ 1 \ 1 \ 1 \ 0 \ 0 \ 0 \ 0 \ 0 \ 1 \ 0 \ 0 \ 1 \ 1 \ 1 \ 1 \ 1 \ 0$$

to be decoded with the Huffman code generated above. There is no code word 1, or 11, so we may take the first three bits 110 as being the code for gray value 3. Notice also that no other code word begins with this string. For the next few bits, 1110 is a code word; no other begins with this string, and no other smaller string is a codeword. So we can decode this string as gray level 4. Continuing in this way we obtain:

as the decoding for this string.

For more information about Huffman coding, and its limitations and generalizations, see [13, 37].

14.3   Run Length Encoding

Run length encoding (RLE) is based on a simple idea: to encode strings of zeros and ones by the number of repetitions in each string. RLE has become a standard in facsimile transmission. For a binary image, there are many different implementations of RLE; one method is to encode each line separately, starting with the number of 0's. So the following binary image:

```
0 1 1 0 0 0
0 0 1 1 1 0
1 1 1 0 0 1
0 1 1 1 1 0
0 0 0 1 1 1
1 0 0 0 1 1
```

would be encoded as

$(123)(231)(0321)(141)(33)(0132)$

Another method [49] is to encode each row as a list of pairs of numbers; the first number in each pair given the starting position of a run of 1's, and the second number its length. So the above binary image would have the encoding

$(22)(33)(1361)(24)(43)(1152)$

Grayscale images can be encoded by breaking them up into their *bit planes*; these were discussed in Chapter 3.

To give a simple example, consider the following 4-bit image and its binary representation:

```
10   7   8   9        1010  0111  1000  1001
11   8   7   6        1011  1000  0111  0110
 9   7   5   4   →    1001  0111  0101  0100
10  11   2   1        1010  1011  0010  0001
```

We may break it into bit planes as shown:

```
0 1 0 1      1 1 0 0      0 1 0 0      1 0 1 1
1 0 1 0      1 0 1 1      0 0 1 1      1 1 0 0
1 1 1 0      0 1 0 0      0 1 1 1      1 0 0 0
0 1 0 1      1 1 1 0      0 0 0 0      1 1 0 0
 0th plane    1st plane    2nd plane    3rd plane
```

and then each plane can be encoded separately using our chosen implementation of RLE.

However, there is a problem with bit planes, and that is that small changes of gray value may cause significant changes in bits. For example, the change from value 7 to 8 causes the change of all four bits, since we are changing the binary strings 0111 to 1000. The problem is of course exacerbated for 8-bit images. For RLE to be effective, we should hope that long runs of very similar gray values would result in very good compression rates for the code. But this may not be the case. A 4-bit image consisting of randomly distributed 7's and 8's would thus result in uncorrelated bit planes, and little effective compression.

To overcome this difficulty, we may encode the gray values with their binary *Gray codes*. A Gray code is an ordering of all binary strings of a given length so that there is only one bit change between a string and the next. So, a 4-bit Gray code is:

15	1	0	0	0
14	1	0	0	1
13	1	0	1	1
12	1	0	1	0
11	1	1	1	0
10	1	1	1	1
9	1	1	1	0
8	1	1	0	0
7	0	1	0	0
6	0	1	0	1
5	0	1	1	1
4	0	1	1	0
3	0	0	1	0
2	0	0	1	1
1	0	0	0	1
0	0	0	0	0

See [37] for discussion and detail. To see the advantages, consider the following 4-bit image with its binary and Gray code encodings:

```
8 8 7 8        1000  1000  0111  1000        1100  1100  0100  1100
8 7 8 7        1000  0111  1000  0111        1100  0100  1100  0100
7 7 8 7  ⟶     0111  0111  1000  0111        0100  0100  1100  0100
7 8 7 7        0111  1000  0111  0111        0100  1100  0100  0100
```

where the first binary array is the standard binary encoding, and the second array the Gray codes. The binary bit planes are:

```
0 0 1 0        0 0 1 0        0 0 1 0        1 1 0 1
0 1 0 1        0 1 0 1        0 1 0 1        1 0 1 0
1 1 0 1        1 1 0 1        1 1 0 1        0 0 1 0
1 0 1 1        1 0 1 1        1 0 1 1        0 1 0 0
 0th plane      1st plane      2nd plane      3rd plane
```

and the bit planes corresponding to the Gray codes are:

```
0 0 0 0        0 0 0 0        1 1 1 1        1 1 0 1
0 0 0 0        0 0 0 0        1 1 1 1        1 0 1 0
0 0 0 0        0 0 0 0        1 1 1 1        0 0 1 0
0 0 0 0        0 0 0 0        1 1 1 1        0 1 0 0
 0th plane      1st plane      2nd plane      3rd plane
```

Notice that the Gray code planes are highly correlated except for one bit plane, whereas all the binary bit planes are uncorrelated.

Implementing Run Length Coding

We can experiment with run length encoding by writing a simple function to implement it. We shall assume our data is binary, and in one single column. The first step is to find the places at which the data changes values, starting with the value 1, indicating that the first value is indeed different from the previous (non-existent!) value. We also need to append a 1 at the end, so that we can find the length of the final run. This can be done very easily by comparing the data with a shifted version of itself:

```
>> changes = [data ~= circshift(data,1);1];
>> changes(1) = 1;
```
MATLAB/Octave

In Python:

```
In :    changes = np.hstack(((data != np.roll(data,1))*1,[1]))
In :    changes[0] = 1
```
Python

Now the run-length code is simply the difference between the indices of the ones:

```
>> diffs = find(changes==1);
>> rle = diffs - circshift(diffs,1);
```
MATLAB/Octave

```
In :    diffs = np.nonzero(changes)
In :    rle = (diffs-np.roll(diffs,1)).tolist()[0]
```
Python

The only problem here is the first element, which should be changed to zero if the initial value of **data** is 1, and removed if the first element of data is zero:

```
>> if data(1)==1, rle(1) = 0; else rle = rle(2:end); end
```
MATLAB/Octave

```
In :    if data[0] == 1:
...:        rle[0] = 0
...:    else:
...:        rle = rle[:,1:]
...:
```
Python

For example, if **data** was the list

1 1 1 0 0 0 0 1 1 1

then the value of **changes** would consist of

1 0 0 1 0 0 0 1 0 1

giving the values of **diffs** as

```
0 3 7 9
```

and the initial `rle` as

```
-9 3 4 2
```

Since the first value of **data** is 1, then the first value **-9** is changed to **0** giving the run-length code

```
0 3 4 2
```

If we experiment with a binary image, for example `circles2.png` then:

```
>> c = imread('circles2.png');
>> data = c(:);
```
MATLAB/Octave

Repeating the above commands for this new **data**, and producing a new `rle`, then by using **whos rle** it can be found that `rle` consists of 1207 values of type **double**, taking up **9656** bytes of storage. If we change the data type to **uint16** the number of bytes decreases to 2414:

```
>> rle2 = uint16(rle)
>> whos c rle rle2
```

Attr	Name	Size	Bytes	Class
	c	256x256	65536	logical
	rle	1207x1	9656	double
	rle2	1207x1	2414	uint16

MATLAB/Octave

(Note that the outputs for MATLAB and Octave from **whos** are slightly different). In this case, RLE provides a remarkable compression rate.

In Python:

```
In :  c = np.uint8((io.imread('circles2.png')>0)*1)
In :  c.nbytes
Out:  65536

In :  data = c.flatten()
```
Python

Again the above commands can be applied, to produce `rle`.

14.4   Dictionary Coding: LZW Compression

The principle idea behind all dictionary compression schemes (of which there are a great many) is that there is a dictionary of symbols and of groups of symbols. As the algorithm works its way through the input data, when it encounters a group of symbols not in the dictionary, it adds that group—with an appropriate code—to the dictionary. This means that when that group of symbols is encountered again, there is a dictionary entry for it.

The initial scheme was published in 1978 by Abraham Lempel and Jacob Ziv (and is known as LZ78); the LZW scheme is a variant devised by Terry Welch in 1984. It has been a hugely popular scheme because of its simplicity. It achieved some notoriety by being used in the GIF image format; the algorithm was thus patented to the Unisys Corporation, who attempted to enforce its licensing. The commercial aspect of the GIF format and its algorithms was one of the main drivers behind the development of a patent-free alternative: the result was the PNG format. The patent however expired in 2003, and the algorithm can now be used freely and without restriction.

The LZW algorithm is best explained by an example. To keep matters as simple as possible, data will consist only of capital letters, and so the test phrase "A banana bandanna ban" will appear as "ABANANABANDANNABAN". The initial dictionary consists of the 26 letters of the alphabet, and their codes 0 to 25:

Symbol:	A	B	C	D	$\cdots$	W	X	Y	Z
Code:	00	01	02	03	$\cdots$	22	23	24	25

The algorithm keeps current letters from the data in a buffer, which is initially empty. At each step, the next letter from the data is added to the buffer. When a group of letters is reached that is *not* in the dictionary, then the buffer except for the most recent addition is added to the dictionary with the next unused code value, and the buffer is changed to hold only the last letter.

For example, suppose we collect these letters one by one into the buffer:

Letter	Buffer
A	A
N	AN
N	ANN
A	ANNA

Suppose that the first three groups of letters A, AN, ANN were already in the dictionary, but the last group ANNA was not. Then the output would be the code for ANN, and the buffer would be emptied except for the last letter A.

The uncompressed text, as codes is:

```
00 01 00 12 00 12 00 01 00 12 03 00 12 12 00 01 00 12
```

Here is the algorithm applied to the test text:

Data	Buffer	In dict?	New dict and code	Output	Comments
A	A	yes			Do nothing at this stage
B	AB	No	AB = 26	00	Output the code for A, reduce the buffer to B
A	BA	No	BA = 27	01	Output the code for B, reduce the buffer to A
N	AN	No	AN = 28	00	Output the code for A, reduce the buffer to N
A	NA	no	NA = 29	12	Output the code for N, reduce the buffer to A
N	AN	yes			Do nothing at this stage
A	ANA	no	ANA = 30	28	Output the code for AN, reduce the buffer to A
B	AB	yes			Do nothing at this stage
A	ABA	No	ABA = 31	26	Output the code for AB, reduce the buffer to A
N	AN	yes			Do nothing at this stage
D	AND	no	AND = 32	28	Output the code for AN, reduce the buffer to D
A	DA	no	DA = 33	03	Output the code for D, reduce the buffer to A
N	AN	yes			Do nothing at this stage
N	ANN	no	ANN = 34	28	Output the code for AN, reduce the buffer to N
A	NA	yes			Do nothing at this stage
B	NAB	no	NAB = 35	29	Output the code for NA, reduce the buffer to B
A	BA	yes			Do nothing at this stage
N	BAN	no	BAN = 36	27	Output the code for BA, reduce the buffer to N
#	N#	no		12	End of data reached, so stop

The compressed output, as code, would be

```
00 01 00 12 28 26 28 03 28 29 27 12
```

So the algorithm has reduced 18 code values to just 11. Clearly this example is highly contrived to have good results for a small input; but, in fact, for large data we could expect in most cases to achieve a reasonable compression rate.

One of the many elegant aspects of LZW compression is that it is self-uncompressing: given the initial dictionary, the compressed output contains all the information needed to build up the dictionary. That is, there is no need to include the extra dictionary values with the code. Basically, we do exactly the same as for compression: add the current letter to the buffer, check whether it's in the dictionary. Sometimes the output code might involve several symbols, so we treat each one separately:

Code	Data	Buffer	In dict?	New dict and code	Comments
00	A	A	yes		
01	B	AB	no	AB = 26	Reduce buffer to B
00	A	BA	no	BA = 27	Reduce buffer to A
12	N	AN	no	AN = 28	Reduce buffer to N
28	A	NA	no	NA = 29	Reduce buffer to A
	N	AN	yes		
26	A	ANA	no	ANA = 30	Reduce buffer to A
	B	AB	yes		
28	A	ABA	no	ABA = 31	Reduce buffer to A
	N	AN	yes		
03	D	AND	no	AND = 32	Reduce buffer to D
28	A	DA	no	DA = 33	Reduce buffer to A
	N	AN	yes		
29	N	ANN	no	ANN = 34	Reduce buffer to N
	A	NA	yes		
27	B	NAB	no	NAB = 35	Reduce buffer to B
	A	BA	yes		
12	N	Last code value so stop			

The original uncompressed data can be seen in the second column.

Programs for LZW compression are given at the end of the chapter. For example, suppose the programs are applied to the resized version of the cameraman image:

```
>> c = imresize(imread('cameraman.png'),0.5);
>> [code,dict] = imLZW(c);
```
MATLAB/Octave

To compare the total number of bits in the image and the LZW code, note that the number of bits required for an integer n is given by

$$\lceil \log_2(n) \rceil$$

This is the *ceiling* (next integer above) of the logarithm of n to base 2. So, the total number of bits required for all the code values can be computed as:

```
>> sum(ceil(log2(code)))
ans = 91928
```
MATLAB/Octave

Then we can calculate the total number of bits of all the pixels in c, first with 8 bits per pixel, and next with the number of bits per integer value:

```
>> 128*128*8
ans = 131072
>> c(find(c==0))=1;
>> sum(ceil(log2(c(:))))
ans =  113341
```
MATLAB/Octave

In Python, the values are slightly different, owing to the differences in the resizing algorithm and the conversion to data type `uint8`:

```
In :   c = ut.img_as_ubyte(tr.rescale(io.imread('cameraman.png'),0.5))
In :   code = imLZW(c)
In :   sum(np.ceil(np.log2(code)))
Out: 85286.0
```

Python

Total bits in the image:

```
In :   np.ceil(np.log2(c.flatten().tolist())).sum()
Out:   114243.0
```

Python

There is clearly some compression.

The version of LZW discussed here is the simplest version; there are variants that are faster or with more optimal compression rates. For more detail, the texts by Salomon [41, 42] or by Sayood [43] are excellent references and introductions not only to LZW but also to other compression algorithms.

14.5   The JPEG Algorithm

Lossy compression trades some acceptable data loss for greater rates of compression. Of the many compression methods available, the algorithm developed by the Joint Photographic Experts Group (JPEG) has become one of the most popular. It uses *transform coding*, where the coding is done not on the pixel values themselves, but on a transform.

The heart of this algorithm is the *discrete cosine transform* (DCT), which is defined similarly to the Fourier transform. Although it can be applied to an array of any size, in the JPEG algorithm it is applied only to 8×8 blocks. If $f(j, k)$ is one such block, then the forward (two-dimensional) DCT is defined as:

$$F(u,v) = \frac{C(u)C(v)}{4} \sum_{j=0}^{7} \sum_{k=0}^{7} f(j,k) \cos\left(\frac{(2j+1)u\pi}{16}\right) \cos\left(\frac{(2k+1)v\pi}{16}\right)$$

and the corresponding inverse DCT as

$$f(i,j) = \sum_{u=0}^{7} \sum_{v=0}^{7} f(u,v)C(u)c(v) \cos\left(\frac{(2j+1)u\pi}{16}\right) \cos\left(\frac{(2k+1)v\pi}{16}\right)$$

where $C(w)$ is defined as

$$C(w) = \begin{cases} \frac{1}{\sqrt{2}} & \text{if } w = 0 \\ 0 & \text{otherwise.} \end{cases}$$

The DCT has a number of properties that make it particularly suitable for compression:

1. It is real-valued, so there is no need to manipulate complex numbers

2. It has a high information packing ability, in that it packs large amounts of information into a small number of coefficients

3. It can be implemented very efficiently in hardware

4. Like the FFT, there is a "fast" version of the transform which maximizes efficiency

5. The basis values are independent of the data [13]

Like the 2-dimensional FFT, the 2-dimensional DCT (given above), is separable, and so can be calculated by a sequence of 1-dimensional DCTs; first we apply the 1-D DCT to the rows, then the 1-D DCT to the columns of the result.

To see an example of the information packing capabilities, we consider the simple linear sequence given by

```
>> a = [10:15:115]

a =

    10    25    40    55    70    85    100    115
```
MATLAB/Octave

We shall apply both the FFT and DCT to this sequence, remove half the result, and invert the remainder. First the FFT:

```
>> fa = fft(a);
>> fa(5:8) = 0;
>> round(abs(ifft(fa)))

ans =

    49    41    56    57    71    70    85    90
```
MATLAB/Octave

Now the DCT (using the `dct` and `idct` functions):

```
>> da = dct(a);
>> da(5:8) = 0;
>> round(idct(da))

ans =

    11    23    41    56    69    84    102    114
```
MATLAB/Octave

In Python, this can be handled by first importing a DCT function:

```
In :    from scipy.fftpack import dct, idct, fft, ifft
In :    a = np.array(range(10,120,15)).astype('float64')
In :    fa = fft(a); fa[4:] = 0
In :    da = dct(a,norm='ortho'); da[4:] = 0
In :    print np.round(np.abs(ifft(fa)))
In :    print np.round(np.abs(idct(da,norm='ortho')))
```
Python

Python's default is not to perform automatically scaling with the DCT, so we need to include the parameter `norm='ortho'` to ensure that the appropriate scaling takes place.

Notice how much closer the DCT result is to the original, in spite of the loss of information from the transform. This example is illustrated in Figure 14.3.

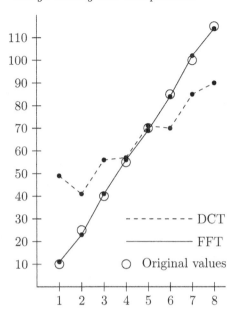

FIGURE 14.3: Comparison of the FFT and the DCT

The JPEG baseline compression scheme is applied as follows:

1. The image is divided up into 8×8 blocks; each block is transformed and compressed separately.

2. For a given block, the values are "shifted" by subtracting 128 each value.

3. The DCT is applied to this shifted block.

4. The DCT values are "normalized" by dividing through by a "normalization matrix" Q.

   It is this normalization that provides the compression, by making most of the elements of the block zero.

5. This matrix is formed into a vector by reading off all non-zero values from the top left in a zig-zag fashion as shown in Figure 14.4.

6. The first coefficients of each vector, which will be the largest elements in each vector, and which are known as the *DC coefficients*, are encoded by listing the *difference* between each value and the values from the previous block. This helps keep all values (except for the first) small.

7. These are then compressed using run length encoding.

8. All other values (known as the *AC coefficients*) are compressed using a Huffman coding.

The amount of lossy compression can be changed by scaling the normalization matrix Q in Step 4 above.

To decompress, the above steps are applied in reverse; the Huffman encoding and run length encoding can be decoded with no loss of information. Having done that, then:

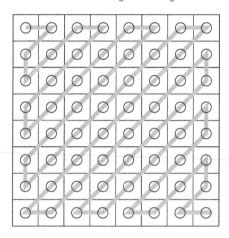

FIGURE 14.4: Reading the elements from a DCT block

1. The vector is read back into an 8×8 matrix.

2. The matrix is multiplied by the normalization matrix.

3. The inverse DCT is applied to the result.

4. The result is shifted back by 128 to obtain the original image block.

The normalization matrix, which has been used by the JPEG group, is

$$
\begin{bmatrix}
16 & 11 & 10 & 16 & 24 & 40 & 51 & 61 \\
12 & 12 & 14 & 19 & 26 & 58 & 60 & 55 \\
14 & 13 & 16 & 24 & 40 & 57 & 69 & 56 \\
14 & 17 & 22 & 29 & 51 & 87 & 80 & 62 \\
18 & 22 & 37 & 56 & 68 & 109 & 103 & 77 \\
24 & 35 & 55 & 64 & 81 & 104 & 113 & 92 \\
49 & 64 & 78 & 87 & 103 & 121 & 120 & 101 \\
72 & 92 & 95 & 98 & 112 & 100 & 103 & 99
\end{bmatrix}
$$

We can experiment with the DCT and quantization using standard functions, and with an 8×8 block from an image. First, define the block and subtract 128 from it:

```
>> c = imread('caribou.png');
>> x = 151; y = 90;
>> block = c(x:x+7,y:y+7);
>> b = block-128;
```

MATLAB/Octave

```
In :   c = io.imread('caribou.png')
In :   x,y = 150,90
In :   block = c[x:x+8,y:y+8]
In :   b = block.astype('float64')-128
```

Python

The block of values thus produced is

```
-41   -33   -36   -55   -69   -71   -71   -73
-54   -57   -60   -69   -74   -74   -77   -71
-64   -70   -71   -73   -70   -63   -62   -63
-71   -65   -60   -62   -54   -39   -30   -24
-33   -19   -11   -14    -9     6    17    12
  0    11    18    11    12    20    23    15
  9     7    -3   -10     9    28    26     4
 -6    -9   -15   -18     0    16    12    14
```

Now we apply the DCT, and divide by the normalization matrix:

```
>> q = dlmread('normalization.txt');
>> bd = dct2(b);
>> bq = round(bd./q);
>> bq(find(abs(bq)<1))=0
```

MATLAB/Octave

Note that Python does not have a **dct2** function, but it is easy to write one, using the separability of the transform:

```
In :  def dct2(x):
...:       return dct(dct(x,norm='ortho').T,norm='ortho').T
...:
In :  q = np.loadtxt('normalization.txt')
In :  bd = dct2(b)
In :  bq = np.round(bd/q)
In :  bq[np.where(abs(bq)<1)]=0
```

Python

The value of this final array is

```
-14   -3    2    0   -1    0    0    0
-20    4    0   -1    0    0    0    0
  0    4    0    0    0    0    0    0
  7    1   -1    0    0    0    0    0
  0   -1    0    0    0    0    0    0
 -1    0    0    0    0    0    0    0
  0    0    0    0    0    0    0    0
  0    0    0    0    0    0    0    0
```

At this stage we have turned our block from above into a block containing mostly zeros. If we were to output a vector from this block, it would be:

```
-14 -3 -20 0 4 2 0 0 4 7 0 1 0 -1 -1 0 0 0 -1 0 -1 EOB
```

where **EOB** signifies the end of the block. So, by this stage we have reduced an 8×8 block to a vector of length 21, containing only small values.

To uncompress, we take the vector and reorder its elements into the matrix **bq** above. The steps to recover the block are first multiplying by the normalization matrix, applying the inverse DCT, and finally adding 128:

```
>> bq2 = bq.*q
>> bd2 = idct2(bq2)
>> block2 = round(bd2+128)
```
MATLAB/Octave

In Python, as before, we have to provide a function for the inverse DCT:

```
In :  def idct2(x):
...:      return idct(idct(x,norm='ortho').T,norm='ortho').T
...:
In :  bq2 = bq*q
In :  bd2 = idct2(bq2)
In :  block2 = np.round(bd2+128)
```
Python

The value of `block2` is:

```
 80   89   88   74   62   59   58   53
 75   81   77   63   54   56   59   58
 57   62   58   48   47   58   68   70
 60   66   68   66   73   90  102  104
 98  106  111  113  123  137  143  139
130  135  137  137  143  153  152  144
131  133  129  126  133  144  144  136
125  124  118  114  124  141  147  143
```

It can be seen that these values are very close, if not quite exactly the same, as the values in the original block. The differences between original and reconstructed values are:

```
 7    6    4   -1   -3   -2   -1    2
-1  -10   -9   -4    0   -2   -8   -1
 7   -4   -1    7   11    7   -2   -5
-3   -3    0    0    1   -1   -4    0
-3    3    6    1   -4   -3    2    1
-2    4    9    2   -3   -5   -1   -1
 6    2   -4   -8    4   12   10   -4
-3   -5   -5   -4    4    3   -7   -1
```

The algorithm works best on regions of low frequency; in such cases, the original block can be reconstructed with only very small errors.

We can experiment with JPEG compression by experimenting with scaling the normalization matrix. In MATLAB or Octave, we can create anonymous functions:

```
>> jpg_in = @(x,n) round(dct2(double(x)-128)./(q*n));
>> jpg_out = @(x,n) round(idct2(x.*(q*n))+128);
```
MATLAB/Octave

and in Python we can create ordinary functions:

```
In :  def jpg_in(x,n):
...:      bd = dct2(np.float64(x)-128)
...:      return np.round(bd/(q*n))
...:
In :  def jpg_out(x,n):
...:      return np.round(idct2(x*q*n)+128)
...:
```

The functions in MATLAB/Octave can be applied to an image with the `blockproc` function. Recall that this applies a function to each block in the image; the size of the blocks being given as parameters to `blockproc`.

Applied to the matrix of the caribou image:

```
>> cj1 = blockproc(c,[8,8],jpg_in,1);
>> length(find(cj1==0))
ans =
      51937
```

Python does not have a `blockproc` equivalent, but in fact we can use Python's programming facility to do the job just as well:

```
In :  c = c = io.imread('caribou.png').astype('float64')
In :  rs,cs = c.shape
In :  cj1 = np.zeros_like(c)
In :  for i in range(0,rs,8):
...:      for j in range(0,cs,8):
...:          cj1[i:i+8,j:j+8] = jpg_in(c[i:i+8,j:j+8],1)
...:
In :  len(np.where(abs(cj1)<1)[0])
Out:  51937
```

So we have applied the compression scheme to each 8×8 block of the image, up to the point of dividing by the quantization matrix and round off. The point of the second command is to show how much information has been lost at this stage. The original image contained 65,536 different items of information (the pixel values); each one between 0 and 128. But now we have only $65,536 - 51,940 = 13,596$ items of information, and their maxima and minima are:

```
>> [max(cj1(:)),min(cj1(:))]
ans =
    60   -45
```

and in Python:

```
In :  cj1.max(),cj1.min()
Out:  (60.0, -45.0)
```

So, the information is not only far less in number, but also in range. Now we can go backward:

```
>> c1 = blockproc(cj1,[8,8],jpg_out,1);
>> c1 = uint8(c1);
```

MATLAB/Octave

or

```
In :   c1 = np.zeros_like(c)
In :   for i in range(0,rs,8):
...:        for j in range(0,cs,8):
...:            c1[i:i+8,j:j+8] = jpg_out(cj1[i:i+8,j:j+8],1)
...:
```

Python

The original image and result **c2** are shown in Figure 14.5. There is no apparent difference

FIGURE 14.5: An image before and after JPEG compression and decompression

between the images—they look identical. However, they are not identical:

```
>> diff = double(c(:))-double(c1(:));
>> [max(diff) min(diff)]
ans =

    31  -26
```

MATLAB/Octave

or

```
In :   diff.max(),diff.min()
Out: (31.0, -26.0)
```

Python

and we can see the difference (scaled for viewing):

```
>> imshow(mat2gray(double(c)-double(c1)))
```

MATLAB/Octave

or in Python with

```
In :   io.imshow(diff)
```

Python

This difference is shown in Figure 14.6. We can experiment with different levels of quan-

FIGURE 14.6: The difference between an image and its JPEG decompression

tization. Suppose we take our extra parameter n to be 2. This has the effect of doubling each value in the normalization matrix, so we should set more of the DCT values to zero:

```
>> cj2 = blockproc(c,[8,8],jpg_in,2);
>> length(find(cj2==0))

ans =

      56726
```

MATLAB/Octave

This means that only 8807 values are non-zero. We can also find that the maximum and minimum values of cj are 30 and −22. We can now decompress, and display the result and the difference with the same commands as above. The results are shown in Figure 14.7. Now we shall try a scale of 5:

FIGURE 14.7: JPEG compression with a scale factor of 2

```
In :   cj5 = np.zeros_like(c)
In :   for i in range(0,rs,8):
...:       for j in range(0,cs,8):
...:           cj5[i:i+8,j:j+8] = jpg_in(c[i:i+8,j:j+8],5)
...:
In :   len(np.where(abs(cj5)<1)[0])
Out:   61511
```

Python

and the result and difference image are shown in Figure 14.8. At this stage we may be

FIGURE 14.8: JPEG compression with a scale factor of 5

losing some fine detail, but the image is still remarkably good. We also note that as the scale factor increases, the range of values after the quantization decreases. For the matrix `cj` above, the maximum and minimum are 12 and -9.

Finally, with a scale factor of 10, the maximum and minimum are 6 and -4, and we have only 1852 items of information. The results are shown in Figure 14.9. This image

FIGURE 14.9: JPEG compression with a scale factor of 10

certainly shows considerable degradation; however, the animal is still quite clear.

We can see the results of JPEG compression and decompression by looking at a closeup of the image; we shall investigate that portion of the animal around the top of the head:

```
>> imshow(imresize(c(68-31:68+32,56-31:56+32),4))
```

MATLAB/Octave

which is shown in Figure 14.10. The same areas, with the scales of 1 and 2, are shown in

FIGURE 14.10: An image closeup

Figure 14.11. And the closeups, with the scales of 5 and 10, are shown in Figure 14.12.

FIGURE 14.11: Closeups after the scale factors of 1 and 2

Notice that "blockiness" becomes more apparent as the scale factor increases. This is due to the working of the algorithm; that each 8×8 block is processed independently of the others. This tends to produce discontinuities in the output, and is one of the disadvantages of the JPEG algorithm for high levels of compression.

In Chapter 15, we shall see that some of these disadvantages can be overcome by using wavelets for compression.

We have seen how changing the compression rate may affect the output. But the JPEG algorithm is particularly designed for storage. For example, suppose we take the original block from the caribou image, but divide it by double the quantization matrix. After reordering, the output vector will be

```
7 -1 -10 0 2 1 0 0 2 4 0 1 0 -1 EOB
```

This vector is further encoded using Huffman coding. To do this, each element of the vector (except for the first), that is, each AC value, is defined to be in a particular *category* depending on its absolute value. In general, the value 0 is given category 0, and category k

FIGURE 14.12: Closeups after the scale factors of 5 and 10

contains all elements x whose absolute value satisfies

$$2^k \leq |x| \leq 2^{k+1} - 1.$$

The categories are also used for the differences of the first (DC) value. The first few categories are:

Range	DC Category	AC Category
0	0	Not applicable
$-1, 1$	1	1
$-3, -2, 2, 3$	2	2
$-7, \ldots, -4, 4, \ldots, 7$	3	3
$-15, \ldots, -8, 8, \ldots, 15$	4	4

To encode the vector, the category is used with all non-zero terms, along with the number of preceding zeros. For example, for the vector above:

Values:	7	-1	10	0	2	1	0	0	2	4	0	1	0	-1
Category:	1	4		2	1				2	3		1		1
Preceding zeros:	0	0		1	0				2	0		1		1

For each non-zero AC value then, a binary vector is produced containing the Huffman code for its particular category and run of preceding zeros, followed by a sign bit (0 for negative, 1 for positive), and then for category k a run of $k - 1$ bits indicating the position within that category. The Huffman code table is provided as part of the JPEG baseline standard; for the first few categories and runs, they are:

Run	Category	Code
0	0	1010
0	1	00
0	2	01
0	3	100
0	4	1011
1	1	1100
1	2	111001
1	3	1111001
1	4	111110110
2	1	11011
2	2	11111000
2	3	1111110111

A full table is given by Gonzalez and Woods [13]. For the non-zero AC values in the above vector, the output binary strings will be created from:

Run	Category	Code	Sign	Position
0	1	00	0	0
0	4	1011	1	011
1	2	111001	1	0
0	1	00	1	1
2	2	11111000	1	0
0	3	100	1	00
1	1	1100	1	0
1	1	1100	0	0

The output string of bits for this block will consist of the code for the difference of the DC coefficient, followed by:

0000/10111011/11100110/0011/1111100010/100100/110010/110000

where the lines just show the individual strings. Assuming 8 bits for the first code, the entire block has been encoded with only 60 bits,

$$60/64 \approx 0.94$$

bit per pixel: a compression rate of over 8.5.

For a detailed account of the JPEG algorithm, see Pennebaker [56].

14.6 Programs

Start with some programs in MATLAB/Octave for LZW compression of uppercase text:

```
% Local Variables:
% mode: Octave
% End:
function [out,dict] = LZW(data)
%
% A very simple function for LZW compression of strings of uppercase
    letters.
%
    L = length(data);
    dict = mat2cell(char(65:90),1,ones(26,1));
    buffer = "";
    out = [];
    n = 27;
    for i = 1:L,
        newbuffer = strcat(buffer, data(i));
        if ~ismember(newbuffer,dict),
            dict{n} = newbuffer;
            disp(cstrcat(newbuffer,'_',num2str(n)))
            [x,m] = ismember(buffer,dict);
            out = [out,m];
```

MATLAB/Octave

```
              n = n+1;
              buffer = data(i);
          else,
              buffer = newbuffer;
          end
      end
      [x,m] = ismember(data(end),dict);
      out = [out,m];
  end
```

and for decompression of the resulting code:

```
% Local Variables:
% mode: Octave
% End:
function out = unLZW(code)
%
% A very simple function for LZW decompression of strings of uppercase
    letters.
%
  L = length(code);
  dict = mat2cell(char(65:90),1,ones(26,1));
  buffer = "";
  out = dict{code(1)};
  n = 27;
  buffer = dict{code(1)};
  for i = 2:L
    c = code(i);
    if c <= length(dict)
      entry = dict{c};
    elseif c == length(dict)+1
      entry = strcat(buffer,buffer(1));
    else
      error('Bad_compressed_code');
    end
    out = strcat(out, entry);
    dict{n} = strcat(buffer,entry(1));
    n = n+1;
    buffer = entry;
  end
end
```

Finally, the LZW program slightly adjusted to manage images of type `unit8`. Note that this program does no error checking, so that if the image is not of type `unit8` the program will produce rubbish.

```
function [out,dict] = imLZW(im)
%
% A very simple function for LZW compression of an image
%
    data = dec2hex(im(:),2);
    L = length(im(:));
```

```
        dict = mat2cell(dec2hex([0:255],2),ones(256,1),2);
        buffer = "";
        out = uint16([]);
        n = 257;
        for i = 1:L,
            newbuffer = strcat(buffer,data(i,:));
            if ~ismember(newbuffer,dict),
                dict{n} = newbuffer;
                [x,m] = ismember(buffer,dict);
                out = [out,m];
                n = n+1;
                buffer = data(i,:);
            else,
                buffer = newbuffer;
            end
        end
        out = [out,im(end)];
    end
```

MATLAB/Octave

For Python, here is the program file which contains all three functions: compression of uppercase text and of images, and decompressing to text.

```
# Adapted from http://rosettacode.org/wiki/LZW_compression#Python to be
    super-simple


def LZW(data):
    """Compress␣an␣uppercase␣string␣to␣a␣list␣of␣output␣code␣values."""

    # Build the dictionary.
    dictionary = [chr(i) for i in range(65,91)]

    buffer = ""
    result = []
    for c in data:
        newbuffer = buffer + c
        if newbuffer in dictionary:
            buffer = newbuffer
        else:
            result += [dictionary.index(buffer)]
            # Add newbuffer to the dictionary.
            dictionary += [newbuffer]
            print newbuffer,dictionary.index(newbuffer)
            buffer = c

    # Output the code for buffer.
    if buffer:
        result += [dictionary.index(buffer)]
    return result


def unLZW(code):
    """Decompress␣a␣list␣of␣output␣codes␣to␣an␣uppercase␣string."""
```

Python

```
    # Build the dictionary.
    dictionary = [chr(i) for i in range(65,91)]

    buffer = dictionary[code.pop(0)]
    result = buffer
    for n in code:
        if n < len(dictionary):
            entry = dictionary[n]
        elif n == len(dictionary):
            entry = buffer + buffer[0]
        else:
            raise ValueError('Bad_compressed_code:_%s' % n)
        result += entry

        # Add buffer+entry[0] to the dictionary.
        print buffer+entry[0],len(dictionary)
        dictionary += [buffer + entry[0]]

        buffer = entry
    return result

def imLZW(im):
    """Compress_an_uppercase_string_to_a_list_of_output_code_values."""

    data = [format(i,'02x') for i in im.flatten().tolist()]

    # Build the dictionary.
    dictionary = [format(i,'02x') for i in range(256)]

    buffer = ""
    result = []
    for c in data:
        newbuffer = buffer + c
        if newbuffer in dictionary:
            buffer = newbuffer
        else:
            result += [dictionary.index(buffer)]
            # Add newbuffer to the dictionary.
            dictionary += [newbuffer]
            buffer = c

    # Output the code for buffer.
    if buffer:
        result += [dictionary.index(buffer)]
    return result
```

Python

Exercises

1. Construct a Huffman code for each of the probability tables given:

Grayscale		0	1	2	3	4	5	6	7
Probability	(a)	.07	.11	.08	.04	.5	.05	.06	.09
	(b)	.13	.12	.13	.13	.12	.12	.12	.13
	(c)	.09	.13	.15	.1	.14	.12	.11	.16

   In each case, determine the average bits/pixel given by your code.

2. From your results of the previous question, what do you think are the conditions of the probability distribution which give rise to a high compression rate using Huffman coding?

3. Download a Huffman code program (or write one of your own), and test it on the examples given.

   Then try it on different images. What sort of compression rates do you obtain? What sorts of images seem to have the best compression rates?

   Note: Both Octave and MATLAB have Huffman routines in their Communications toolboxes; a Python routine can be found at `http://rosettacode.org/wiki/Huffman_coding#Python`.

4. Encode each of the following binary images using run length encoding:

```
        1 0 0 1 1 1           1 0 1 0 0 0
        0 1 0 1 1 1           0 0 1 1 0 1
   (a)  1 0 0 1 1 1      (b)  1 1 0 0 0 0
        0 1 1 1 0 1           0 0 0 0 1 1
        1 0 1 0 1 1           1 1 1 1 0 0
        0 1 1 1 1 0           1 1 1 0 0 0
```

5. Using run length encoding, encode each of the following 4-bit images:

```
        1  1  3  3  1  1           0  0  0  6 12 12  1  9
        1  7 10 10  7  1           1  1  1  6 12 11  9 13
   (a)  6 13 15 15 13  6      (b)  2  2  2  6 11  9 13 13
        6 13 15 15 13  6           8 10 15 15  7  5  5  5
        1  7 10 10  7  1          14  8 10 15  7  4  4  4
        1  1  3  3  1  1          14 14  5 10  7  3  3  3
```

6. Check your answers to the previous two questions with the techniques discussed in this chapter. You can isolate the bit planes by using the technique discussed in Section 3.3.

7. Encode the preceding images using the 4-bit Gray code, and apply run length encoding to the bit planes of the result.

   Compare the results obtained using Gray codes, and standard binary codes.

8. Write a function for restoring a binary image from a run length code. Test it on the images and codes from the previous questions.

9. The following are the run-length encodings for a 4 × 4 4-bit image from most to least important bit-planes:

```
3  1  2  2  1  4  1  2
1  2  1  2  1  2  1  2  1  3
2  1  2  1  2  2  1  5
0  3  1  3  2  3  1  2  1
```

Construct the image.

10. (a) Given the following 4-bit image:

```
0  4  4  4  4   4   6   7
0  4  5  5  5   4   6   7
1  4  5  5  5   4   6   7
1  4  5  5  5   4   6   7
1  4  4  4  4   4   6   7
2  2  8  8  8  10  10  11
2  2  9  9  9  12  13  13
3  3  9  9  9  15  14  14
```

transform it to a 3-bit image by removing the least most significant bit plane. Construct a Huffman code on the result and determine the average number of bits/pixel used by the code.

(b) Now apply Huffman coding to the original image and determine the average number of bits/pixel used by the code.

(c) Which of the two codes gives the best rate of compression?

11. By hand, apply LZW encoding to the phrases

   (a) HOWNOWBROWNCOW
   (b) REALLYREALCEREALDEAL

and decode the results.

12. Experiment with LZW coding of images. Which image has the best compression rate and which image has the worst?

13. Write a program to perform LZW uncompression of an image code, and apply it to the codes from the previous question.

14. Apply JPEG compression to an 8×8 block consisting of (a) all the same value, (b) the left half one value, and the right half another, (c) random values uniformly distributed in the 0–255 range.

Compare the length of the code vector in each case, and the results of decompression.

15. Open the image `engineer.png`.

Using the JPEG compression commands described above, attempt to compress this image using greater and greater compression rates. What is the largest quantization scale factor for which the image is still recognizable? How many zeros are in the DCT block matrix?

16. Apply the given JPEG Huffman codes to the vector:

```
-14 -3 -7 0 4 2 0 0 4 3 0 1 0 -1 -1 0 0 0-1 0 -1 EOB
```

and determine the compression rate.

Chapter 15

Wavelets

15.1 Waves and Wavelets

We have seen in Chapters 7 and 14 that the discrete Fourier transform and its cousin, the discrete cosine transform, are of enormous use in image processing and image compression. But as we saw with the Fourier transform, this power does not come without cost. In particular, as the Fourier transform is based on the idea of an image being decomposed into periodic sine or cosine functions, we have to assume some sort of periodicity in the image to make the theory fit the practice. This periodicity was obtained by assuming that the image "bent round" to meet itself at the top and bottom, and at the left and right sides. But this means that we have introduced unnecessary discontinuities in the image, and these may affect the transform and our use of it.

The idea of wavelets is to keep the wave idea, but drop the periodicity. So we may consider a *wavelet* to be a little part of a wave: a wave which is only non-zero in a small region. Figure 15.1 illustrates the idea. Figure 15.1(a) is just the graph of

$$y = \sin(x)$$

for $-10 \le x \le 10$, and Figure 15.1(b) is the graph of a scaled version of

$$y = (1 - x^2)e^{-x^2/2}$$

over the same interval. This second function, in the more general form

$$y = \frac{2}{\sqrt{3\sigma}\pi^{1/4}} \left(1 - \frac{x^2}{\sigma^2}\right) e^{-x^2/(2\sigma^2)}$$

is known as the *Mexican hat wavelet*.

Suppose we are given a wavelet. What can we do with it? Well, if

$$f = w(x)$$

is the function that defines our wavelet, we can:

Dilate it by applying a scaling factor to x: $f(2x)$ would "squash" the wavelet; $f(x/2)$ would expand it

Translate it by adding or subtracting an appropriate value from x: $f(x-2)$ would shift the wavelet 2 to the right; $f(x+3)$ would shift the wavelet 3 to the left

Change its height by simply multiplying the function by a constant.

Of course, we can do any or all at once:

$$6f(x/2 - 3), \quad \frac{1}{3}f(16x + 17), \quad f(x/128 - 33.5), \ldots$$

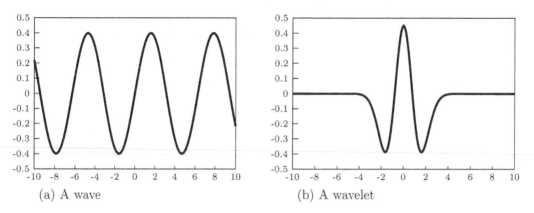

(a) A wave   (b) A wavelet

FIGURE 15.1: Comparing a wave and a wavelet

Let

$$w(x) = \sin(x)e^{-x^2}$$

be the function shown in Figure 15.1(b). Figure 15.2 shows some dilations and shifts of this wavelet.

Given our knowledge of the working of the Fourier transform, it should come as no surprise that given a suitable starting wavelet $w(x)$, we can express any function $f(x)$ as a sum of wavelets of the form

$$aw(bx + c)$$

Moving into two dimensions, we can apply wavelets to images in the same way we applied sine and cosines with the Fourier transform.

Using wavelets has provided a new class of very powerful image processing algorithms: wavelets can be used for noise reduction, edge detection, and compression. The use of wavelets has superseded the use of the DCT for image compression in the JPEG2000 image compression algorithm.

A possible problem with wavelets is that the theory, which has been very well researched, can be approached from many different directions. There is thus no "natural" method of presenting wavelets: it depends on your background, and on the use to which they will be put. Also, much writing on wavelets tends to dive into the theory very quickly. In this chapter, though, we shall just look at the very simplest examples of the use of wavelets and their application to images.

A Simple Wavelet Transform

To obtain a feeling for how a wavelet transform works and behaves, we shall look at a very simple example. All wavelet transforms work by taking weighted averages of input values and providing any other necessary information to be able to recover the original input.

For our example, we shall perform just two operations: adding two values and subtracting. For example, suppose we are given two numbers 14 and 22. Their sum and difference are

$$14 + 22 = 36$$
$$14 - 22 = -8$$

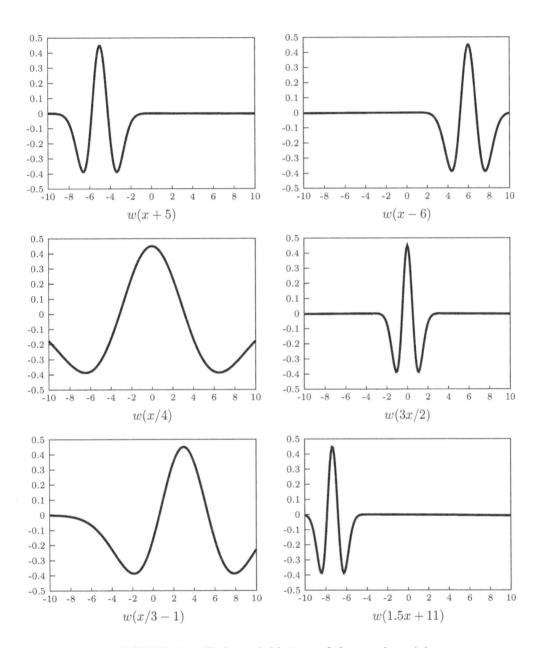

FIGURE 15.2: Shifts and dilations of the wavelet $w(x)$

To recover the original two values from the sum and difference, we take the sum and difference again, and divide by two:

$$(36 - (-8))/2 = 22$$
$$(36 + (-8))/2 = 14$$

In general, if a and b are our two numbers, we have their sum and difference

$$s = a + b$$
$$d = a - b$$

from which the two original numbers can be recovered with

$$a = (s + d)/2$$
$$b = (s - d)/2$$

The forward operations (adding, subtracting) and the backward operations (adding and dividing by two, subtracting and dividing by two) differ only in the extra division. This function can be extended to any vector with an even number of elements. For example, suppose

$$v = [71, \quad 67, \quad 24, \quad 26, \quad 36, \quad 32, \quad 14, \quad 18].$$

Then define s to be the sum of the elements in pairs:

$$s = [71 + 67, \quad 24 + 26, \quad 36 + 32, \quad 14 + 18]$$
$$= [138, \quad 50, \quad 68, \quad 32]$$

and d to be the difference of pairs:

$$d = [71 - 67, \quad 24 - 26, \quad 36 - 32, \quad 14 - 18]$$
$$= [4, \quad -2, \quad 4, \quad -4].$$

The concatenation of these two vectors:

$$W = [138, \quad 50, \quad 68, \quad 32, \quad 4, \quad -2, \quad 4, \quad -4]$$

is the *discrete wavelet transform at 1 scale* of the original vector.

Recall that in general adding values (such as when using an averaging filter) produces a low pass, blurred output. Conversely, subtracting (such as when using an edge detection filter) produces a high pass output. In general, a wavelet transform will produce a mixture of low and high pass results. These can be separated or combined.

The preceding operation on the vector v could have been performed by a matrix product:

$$
\begin{bmatrix} 138 \\ 50 \\ 68 \\ 32 \\ 4 \\ -2 \\ 4 \\ -4 \end{bmatrix}
=
\begin{bmatrix}
1 & 1 & 0 & 0 & 0 & 0 & 0 & 0 \\
0 & 0 & 1 & 1 & 0 & 0 & 0 & 0 \\
0 & 0 & 0 & 0 & 1 & 1 & 0 & 0 \\
0 & 0 & 0 & 0 & 0 & 0 & 1 & 1 \\
1 & -1 & 0 & 0 & 0 & 0 & 0 & 0 \\
0 & 0 & 1 & -1 & 0 & 0 & 0 & 0 \\
0 & 0 & 0 & 0 & 1 & -1 & 0 & 0 \\
0 & 0 & 0 & 0 & 0 & 0 & 1 & -1
\end{bmatrix}
\begin{bmatrix} 71 \\ 67 \\ 24 \\ 26 \\ 36 \\ 32 \\ 14 \\ 18 \end{bmatrix}
$$

We can continue by performing this same adding and differencing of the first four elements of the result; this will involve the concatenation of two vectors of two elements each:

$$s_1 = [138 + 50, \quad 68 + 32]$$
$$= [188, \quad 100]$$
$$d_1 = [138 - 50, \quad 68 - 32]$$
$$= [88, \quad 36]$$

Replacing the first four elements of w above with this new s_1 and d_1 produces

$$W_2 = [188, \quad 100, \quad 88, \quad 36, \quad 4, \quad -2, \quad 4, \quad -4]$$

which is the *discrete wavelet transform at 2 scales* of the original vector. We can go one more step, replacing the first two elements of w_2 with their sum and difference:

$$W_3 = [188 + 100, \quad 188 - 100, \quad 88, \quad 36, \quad 4, \quad -2, \quad 4, \quad -4]$$
$$= [288, \quad 88, \quad 88, \quad 36, \quad 4, \quad -2, \quad 4, \quad -4]$$

and this is the *discrete wavelet transform at 3 scales* of the original vector.

To recover the original vector, we simply add and subtract, dividing by two each time; first using only the first two elements, then the first four, and finally the lot:

$$\left[\frac{288 + 88}{2}, \quad \frac{288 - 88}{2}, \quad 88, \quad 36, \quad 4, \quad -2, \quad 4, \quad -4 \right]$$
$$= [188, \quad 100, \quad 88, \quad 36, \quad 4, \quad -2, \quad 4, \quad -4]$$

$$\left[\frac{[188, \quad 100] + [88, \quad 36]}{2}, \quad \frac{[188, \quad 100] - [88, \quad 36]}{2}, \quad 4, \quad -2, \quad 4, \quad -4 \right]$$
$$= [138, \quad 50, \quad 68, \quad 32, \quad 4, \quad -2, \quad 4, \quad -4]$$

$$\left[\frac{[138, \quad 50, \quad 68, \quad 32] + [4, \quad -2, \quad 4, \quad -4]}{2}, \right.$$
$$\left. \frac{[138, \quad 50, \quad 68, \quad 32] - [4, \quad -2, \quad 4, \quad -4]}{2} \right]$$
$$= [71, \quad 67, \quad 24, \quad 26, \quad 36, \quad 32, \quad 14, \quad 18]$$

At each stage, the sums produce a "lower resolution" version of the original vector. Wavelet transforms produce a mix of lower resolutions of the input, and the extra information required for inversion.

We notice that the differences may be small if the input values are close together. This leads to an idea for compression: we apply a threshold by setting to zero all values in the transform that are less than a predetermined value. Suppose we take a threshold of 0, thus removing all negative values, so that after thresholding W_3 becomes:

$$W_3' = [288, \quad 88, \quad 88, \quad 36, \quad 4, \quad 0, \quad 4, \quad 0]$$

If we now use this as the starting place for our adding and subtracting, we end up with

$$v' = [71, \quad 67, \quad 25, \quad 25, \quad 36, \quad 32, \quad 16, \quad 16].$$

Notice how close this is to the original vector, in spite of the loss of some information from the transform.

15.2   A Simple Wavelet: The Haar Wavelet

The Haar wavelet has been around for a long time, and has been used with images as the *Haar transform*. Only recently has the Haar transform been viewed as a simple wavelet transform. In fact, the Haar wavelet is the simplest possible wavelet, and for that reason is a good starting place.

The Haar wavelet is defined by the function

$$\psi(x) = \begin{cases} 1 & \text{if } 0 < x < 1/2 \\ -1 & \text{if } 1/2 \le x < 1 \\ 0 & \text{otherwise} \end{cases}$$

and is shown in Figure 15.3. As with our wavelet function $w(x)$, we can compress and

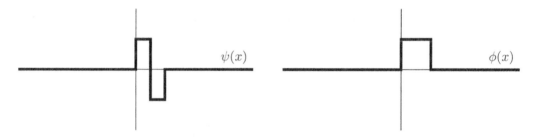

FIGURE 15.3: The Haar wavelet and pulse function

expand this wavelet horizontally or vertically, and shift it.

Applying the Haar Wavelet

Given the Haar wavelet, what do we do with it? That is, how do we use the Haar wavelet in a wavelet transform? We haven't yet defined what a wavelet transform is, or what it might look like. Without going into too many details, a wavelet transform can be defined in much the same way as the DFT or the DCT: as a sum of function values multiplied by wavelet values. In comparison, for the DFT we multiplied function values by complex exponentials, and in the DCT we multiplied our function values by cosines.

The discrete wavelet transform (DWT) can be written [13] as a sum of input values multiplied by the values from a particular class of functions. And as with the Fourier transform, the discrete wavelet transform can be written as a matrix multiplication. We shall show how this is done.

Notice that the Haar wavelet can be written in terms of the simpler *pulse function*:

$$\phi(x) = \begin{cases} 1 & \text{if } 0 \le x < 1 \\ 0 & \text{otherwise} \end{cases}$$

using the relation

$$\psi(x) = \phi(2x) - \phi(2x - 1). \tag{15.1}$$

We can see that this is true by noting that $\phi(2x)$ is equal to 1 for $0 \le x \le 1/2$, and that $\phi(2x - 1)$ is equal to 1 for $1/2 \le x < 1$.

The pulse function also satisfies the equation

$$\phi\left(\frac{x}{2}\right) = \phi(x) - \phi(x-1). \tag{15.2}$$

In the theory of wavelets, the "starting wavelet," in this case our function $\psi(x)$, is called the *mother wavelet*, and the corresponding function $\phi(x)$ is called the *scaling function* (and sometimes called a *father wavelet*).

We can also write our scaling expression 15.2 as

$$\phi(x) = \phi(2x) + \phi(2x-1). \tag{15.3}$$

Equations 15.2 and 15.3 are important because they tell us how the wavelets are rescaled at different resolutions. Equation 15.3 is a very important equation: it is called the *dilation equation*, as it relates the scaling function to dilated versions of itself. Generalizations of this equation, as we shall see below, have led to other wavelets than the Haar wavelet. Equation 15.1 is the *wavelet equation* for the Haar wavelet. Note that the dilation and wavelet equations have the same right-hand side, except for a change of sign. These can be generalized to produce

$$\phi(x) = \cdots + h_{-2}\phi(2x+2) + h_{-1}\phi(2x+1) + h_0\phi(2x) + h_1\phi(2x-1)$$
$$+ h_2\phi(2x-2) + h_3\phi(2x-3) + \cdots \tag{15.4}$$
$$\psi(x) = \cdots - h_{-2}\phi(2x-3) + h_{-1}\phi(2x-2) - h_0\phi(2x-1) + h_1\phi(2x)$$
$$- h_2\phi(2x+1) + h_3\phi(2x+2) - \cdots \tag{15.5}$$

where the values h_i are called the *filter coefficients* or the *taps* of the wavelet. A wavelet is completely specified by its taps.

We thus have, for the Haar wavelet:

$$\phi(x) = h_0\phi(2x) + h_1\phi(2x-1) \tag{15.6}$$
$$\psi(x) = h_1\phi(2x) - h_0\phi(2x-1) \tag{15.7}$$

where $h_0 = h_1 = 1$. It is actually these h_i values that we use in a calculation of the DWT. We can put them into a DWT matrix:

$$H_{2^n} = \left[\begin{array}{cccccccccc}
1 & 1 & 0 & 0 & 0 & 0 & \cdots & 0 & 0\ 0 & 0 \\
0 & 0 & 1 & 1 & 0 & 0 & \cdots & 0 & 0\ 0 & 0 \\
0 & 0 & 0 & 0 & 1 & 1 & \cdots & 0 & 0\ 0 & 0 \\
\vdots & \vdots & \vdots & \vdots & \vdots & \vdots & & \vdots & \vdots\ \vdots & \vdots \\
0 & 0 & 0 & 0 & 0 & 0 & \cdots & 1 & 1\ 0 & 0 \\
0 & 0 & 0 & 0 & 0 & 0 & \cdots & 0 & 0\ 1 & 1 \\
1 & -1 & 0 & 0 & 0 & 0 & \cdots & 0 & 0\ 0 & 0 \\
0 & 0 & 1 & -1 & 0 & 0 & \cdots & 0 & 0\ 0 & 0 \\
0 & 0 & 0 & 0 & 1 & -1 & \cdots & 0 & 0\ 0 & 0 \\
\vdots & \vdots & \vdots & \vdots & \vdots & \vdots & & \vdots & \vdots\ \vdots & \vdots \\
0 & 0 & 0 & 0 & 0 & 0 & \cdots & 1 & -1\ 0 & 0 \\
0 & 0 & 0 & 0 & 0 & 0 & \cdots & 0 & 0\ 1 & -1
\end{array}\right]$$

If v is the vector we investigated earlier, then $vH_8 = d_1$. We can obtain d_2 and d_3 by multiplying the first values of the vector by the appropriately sized H matrix.

If we compare the results W_2 and W_3, the transforms at 2 and 3 scales, respectively, with the input, we find that the matrices corresponding to these transforms are:

$$H_{8,2} = \begin{bmatrix} 1 & 1 & 1 & 1 & 0 & 0 & 0 & 0 \\ 0 & 0 & 0 & 0 & 1 & 1 & 1 & 1 \\ 1 & 1 & -1 & -1 & 0 & 0 & 0 & 0 \\ 0 & 0 & 0 & 0 & 1 & 1 & -1 & -1 \\ 1 & -1 & 0 & 0 & 0 & 0 & 0 & 0 \\ 0 & 0 & 1 & -1 & 0 & 0 & 0 & 0 \\ 0 & 0 & 0 & 0 & 1 & -1 & 0 & 0 \\ 0 & 0 & 0 & 0 & 0 & 0 & 1 & -1 \end{bmatrix}$$

and

$$H_{8,3} = \begin{bmatrix} 1 & 1 & 1 & 1 & 1 & 1 & 1 & 1 \\ 0 & 0 & 0 & 0 & -1 & -1 & -1 & -1 \\ 1 & 1 & -1 & -1 & 0 & 0 & 0 & 0 \\ 0 & 0 & 0 & 0 & 1 & 1 & -1 & -1 \\ 1 & -1 & 0 & 0 & 0 & 0 & 0 & 0 \\ 0 & 0 & 1 & -1 & 0 & 0 & 0 & 0 \\ 0 & 0 & 0 & 0 & 1 & -1 & 0 & 0 \\ 0 & 0 & 0 & 0 & 0 & 0 & 1 & -1 \end{bmatrix}$$

In order to enter such matrices easily, a useful operation is the *Kronecker product* of matrices, denoted $A \otimes B$ and which is defined as the block matrix for which the block at position (i, j) is $a_{ij}B$. For example:

$$\begin{bmatrix} 2 & 3 \\ -1 & 4 \end{bmatrix} \otimes \begin{bmatrix} -2 & 5 \\ 1 & 2 \end{bmatrix} = \begin{bmatrix} 2\begin{bmatrix} -2 & 5 \\ 1 & 2 \end{bmatrix} & 3\begin{bmatrix} -2 & 5 \\ 1 & 2 \end{bmatrix} \\ (-1)\begin{bmatrix} -2 & 5 \\ 1 & 2 \end{bmatrix} & 4\begin{bmatrix} -2 & 5 \\ 1 & 2 \end{bmatrix} \end{bmatrix} = \begin{bmatrix} -4 & 10 & -6 & 15 \\ 2 & 4 & 3 & 6 \\ 2 & -5 & -8 & 20 \\ -1 & 2 & 4 & 8 \end{bmatrix}$$

With this new product, the matrix H can be defined as

$$H_{2^n} = \begin{bmatrix} I_{2^{n-1}} \otimes [1 \quad 1] \\ \hline I_{2^{n-1}} \otimes [1 \quad -1] \end{bmatrix}$$

In MATLAB and Octave, the Kronecker product is implemented with `kron`:

```
>> v = [71 67 24 26 36 32 14 18];
>> H = [kron(eye(4),[1 1]); kron(eye(4),[1 -1])];
>> w = (H*v')'
w =

   138    50    68    32     4    -2     4    -4
```

and in Python by `np.kron`:

```
In:    from numpy import kron, array, eye, vstack
In :   v = array([71, 67, 24, 26, 36, 32, 14, 18])
In :   H = vstack([kron(eye(4),[1,1]),kron(eye(4),[1,-1])])
In :   w = (H.dot(v.T)).T
In :   print w
[ 138.    50.    68.    32.     4.    -2.     4.    -4.]
```

Given

$$H_1 = \begin{bmatrix} 1 & 1 \\ 1 & -1 \end{bmatrix}$$

as the DWT matrix for 2 variables at 1 scale, a sequence of matrices can be created as:

$$H_{n+1} = \left[\begin{array}{c} H_n \otimes \begin{bmatrix} 1 & 1 \end{bmatrix} \\ \hline I_{2^n} \otimes \begin{bmatrix} 1 & -1 \end{bmatrix} \end{array} \right]$$

Then H_n is the matrix for a DWT of 2^n variables at n scales—the maximum number of scales possible.

Although we have spoken of the DWT formed by adding and subtracting as using the Haar wavelet, this is not precisely true. In order to be a wavelet proper, the DWT matrices must be *orthogonal*; that is

$$HH^T = I.$$

and the H matrices constructed so far are not quite; for example, consider the 4×4 2-scale matrix H. Then:

$$HH^T = \begin{bmatrix} 1 & 1 & 1 & 1 \\ 1 & 1 & -1 & -1 \\ 1 & -1 & 0 & 0 \\ 0 & 0 & 1 & -1 \end{bmatrix} \begin{bmatrix} 1 & 1 & 1 & 0 \\ 1 & 1 & -1 & 0 \\ 1 & -1 & 0 & 1 \\ 1 & -1 & 0 & -1 \end{bmatrix} = \begin{bmatrix} 4 & 0 & 0 & 0 \\ 0 & 4 & 0 & 0 \\ 0 & 0 & 2 & 0 \\ 0 & 0 & 0 & 2 \end{bmatrix}.$$

In order to make H orthogonal, each row must be divided by the square root of the diagonal elements in this last matrix:

$$H' = \begin{bmatrix} 1/2 & 1/2 & 1/2 & 1/2 \\ 1/2 & 1/2 & -1/2 & -1/2 \\ 1/\sqrt{2} & -1/\sqrt{2} & 0 & 0 \\ 0 & 0 & 1/\sqrt{2} & -1/\sqrt{2} \end{bmatrix}$$

As an example, here is the 3-scale DWT using the Haar wavelet applied to the original vector. In MATLAB or Octave:

```
>> H = [1 1;1 -1];
>> for i = 1:2, H = [kron(H,[1 1]);kron(eye(size(H)),[1,-1])]; end
```
MATLAB/Octave

So far this produces the 8×8 matrix for a 3-scale transform. Now we must divide the rows by the square roots of the diagonal elements of HH^T:

```
>> D = diag(diag(H*H'))
>> H = sqrt(inv(D))*H
>> format bank
>> w = (H*v')
w =

   101.82    31.11    44.00    18.00    2.83    -1.41    2.83    -2.83
```
MATLAB/Octave

And in Python:

```
In :  H = array([[1,1],[1,-1]])
In :  for i in range(2):
...:      H = vstack([kron(H,[1,1]),kron(eye(H.shape[0]),[1,-1])])
...:
In :  D = H.dot(H.T)
In :  H = np.linalg.inv(np.sqrt(D)).dot(H)
In :  w = (H.dot(v.T)).T
In :  print np.around(w,2)
[ 101.82   31.11   44.    18.     2.83   -1.41    2.83   -2.83]
```

Python

Any wavelet transform can be computed by matrix multiplications. But as we have seen in our discussion of the DFT, this can be very inefficient. Happily many wavelet transforms can be quickly computed by a fast algorithm similar in style to the average/difference method we saw earlier. These methods are called *lifting methods* [24].

If we go back to our averaging and differencing example, we can see that the averaging part of the transform corresponds to low pass filtering, in that we are "coarsening" or "blurring" our input. Similarly, the differencing part of the transform corresponds to a high pass filter, of the sort we looked at in our discussion of edges in Chapter 9. Thus, a wavelet transform contains within it both high and low pass filtering of our input, and we can consider a wavelet transform entirely in terms of filters. This approach will be discussed in Section 15.4.

Two-Dimensional Wavelets

The two-dimensional wavelet transform is *separable*, which means we can apply a one-dimensional wavelet transform to an image in the same way as for the DFT: we apply a one-dimensional DWT to all the columns, and then one-dimensional DWTs to all the rows of the result. This is called the *standard decomposition*, and is illustrated in Figure 15.4.

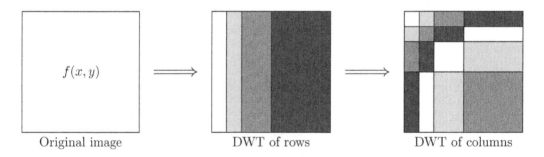

$f(x, y)$ $\Longrightarrow$ $\Longrightarrow$

Original image   DWT of rows   DWT of columns

FIGURE 15.4: The standard decomposition of the two-dimensional DWT

We can also apply a wavelet transform differently. Suppose we apply a wavelet transform (say, using the Haar wavelet) to an image by columns and then rows, but using our transform at 1 scale only. This will produce a result in four quarters: the top left will be a half-sized version of the image; the other quarters will be high pass filtered images. These quarters will contain horizontal, vertical, and diagonal edges of the image. We then apply a 1-scale DWT to the top left quarter, creating smaller images, and so on. This is called the *nonstandard decomposition*, and is illustrated in Figure 15.5.

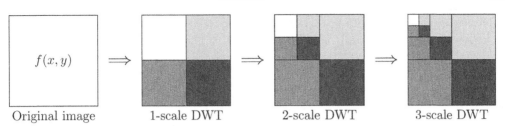

FIGURE 15.5: The non-standard decomposition of the two-dimensional DWT

15.3 Wavelets and Images

Each of MATLAB, Octave, and Python has some wavelet functionality, either built in or available through an added toolbox or package. However, the standard packages all seem to work a little differently. Happily for us, there is an open-source wavelet package with bindings for each of MATLAB, Octave, and Python from Rice University, obtainable from

http://dsp.rice.edu/software/rice-wavelet-toolbox

Assuming that the toolbox has been downloaded and installed, we can try it, using the same vector as above:

```
>> [h,g] = daubcqf(2)
h =
    0.70711    0.70711

g =
   -0.70711    0.70711
```
MATLAB/Octave

Here h and g are the low pass and high pass filter coefficients for the forward transform. The daubcqf function produces the filter coefficients for a class of wavelets called *Daubechies wavelets*, of which the Haar wavelet is the simplest. Now we can apply the DWT to our vector:

```
>> format bank
>> w = mdwt(v,h,1)
w =

   97.58   35.36   48.08   22.63    2.83   -1.41    2.83   -2.83
```
MATLAB/Octave

This is not quite the same as the first vector W from above, but is off only by a single scaling factor:

```
>> w*sqrt(2)
ans =

   138.00    50.00    68.00    32.00     4.00    -2.00     4.00    -4.00
```
MATLAB/Octave

which is the same result we obtained earlier by adding and subtracting. The transforms at 2 and 3 scales can be given simply by adjusting the final parameter:

```
>> w2 = mdwt(v,h,2)
w2 =

  94.00   50.00   44.00   18.00    2.83   -1.41    2.83   -2.83

>> w3 = mdwt(v,h,3)
w3 =

 101.82   31.11   44.00   18.00    2.83   -1.41    2.83   -2.83
```
MATLAB/Octave

To go backward, use the inverse transform:

```
>> format
>> midwt(w3,h,3)
ans =

  71.000   67.000   24.000   26.000   36.000   32.000   14.000   18.000
```
MATLAB/Octave

Note that we can supply our own filter values:

```
>> h = [1 1]
>> mdwt(v,h,1)
ans =

   138     50     68     32      4     -2      4     -4
>> mdwt(v,h,2)
ans =

   188    100     88     36      4     -2      4     -4
>> mdwt(v,h,3)
ans =

   288     88     88     36      4     -2      4     -4
```
MATLAB/Octave

and these are indeed the values we obtained earlier simply by adding and subtracting.

The Python equivalent commands require the loading of the rwt library. Then:

```
In : h,g = rwt.daubcqf(2)
In : print rwt.dwt(v,h,1)[0]
[ 97.5807  35.3553  48.0833  22.6274  2.8284  -1.4142  2.8284  -2.8284]

In : print rwt.dwt(v,h,3)[0]
[ 101.8234  31.1127  44. 18.  2.8284  -1.4142  2.8284 -2.8284]

In : h = np.array([1,1]).astype('float')
In : vw = rwt.dwt(v,h,3)[0]
In : print vw[0]
[ 288.  88.  88.  36.  4.  -2.  4.  -4.]

In : hi = np.array([0.5,0.5]).astype('float')
In : print rwt.idwt(vw[0],hi,3)[0]
[ 71.  67.  24.  26.  36.  32.  14.  18.]
```
> Python

Let's try an image. We shall apply the Haar wavelet to an image of size 256×256, first at one scale. For display, it will be necessary to scale parts of the image for viewing.

```
>> c = imread('cameraman.png');
>> h = daubcqf(2)
>> cw1 = mdwt(double(c),h,1);
```
> MATLAB/Octave

Now at this stage some adjustment needs to be done to display the transform. Because the range of values in the transform `cw1` is large:

```
>> [max(cw1(:)),min(cw1(:))]
ans =

    501.00  -209.50
```
> MATLAB/Octave

some method will be needed to adjust those values for display. In MATLAB or Octave, since `cw1` is an array of data type `double`, elements outside the range $0.0 - -1.0$ will be displayed as black or white. Adjustment can be done in several ways; first by simply using `mat2gray`, second by a log function, as was done for the Fourier transform:

```
>> imshow(cw1)
>> figure, imshow(mat2gray(cw1))
>> cwlog = log(1+abs(cw1))
>> figure, imshow(mat2gray(cwlog))
```
> MATLAB/Octave

In Python, this is more easily done as images are automatically scaled for display:

```
In : h,g = rwt.daubcqf(2)
In : cw1 = rwt.dwt(castype('float'),h,1)[0]
In : cwlog = np.log(1+abs(cw1))
```
> Python

and both `cw1` and `cwlog` can be displayed simply with `io.imshow`. All three images are shown in Figure 15.6. Now consider the same transform, but at 3 scales:

(a) No adjustment   (b) With `mat2gray`   (c) With log and `mat2gray`

FIGURE 15.6: Different displays of a 1-scale DWT applied to an image

```
>> cw3 = mdwt(double(c),h,3);
```
MATLAB/Octave

or with

```
In :  cw1 = rwt.dwt(castype('float'),h,3)[0]
```
Python

and the result with the same adjustments as above is shown in Figure 15.7.

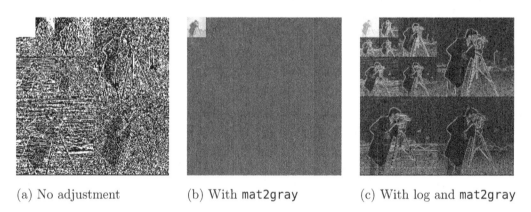

(a) No adjustment   (b) With `mat2gray`   (c) With log and `mat2gray`

FIGURE 15.7: Different displays of a 3-scale DWT applied to an image

This wavelet transform can be seen to use the non-standard decomposition. To see what a standard decomposition looks like, we can multiply the rows, and then the columns, by the appropriate Haar matrix:

```
>> H = [1 1;1 -1];
>> for i = 1:2, H = [kron(H,[1 1]);kron(eye(size(H)),[1,-1])]; end
```
MATLAB/Octave

or in Python as

```python
In : H = np.array([[1,1],[1,-1]])
In : for i in range(2):
...:     H = np.vstack([np.kron(H,[1,1]),np.kron(np.eye(H.shape[0]),[1,-1])
    ])
...:
```
Python

So far this is the 3-scale 8×8 matrix, which now needs to be scaled up to be 256×256:

```matlab
>> K1 = kron(eye(32),H(1,:));
>> K2 = kron(eye(32),H(2,:));
>> K3 = kron(eye(32),H(3:4,:));
>> K4 = kron(eye(32),H(5:8,:));
>> H = [K1;K2;K3;K4];
```
MATLAB/Octave

or

```python
In : K1 = np.kron(np.eye(32),H[0,:])
In : K2 = np.kron(np.eye(32),H[1,:])
In : K3 = np.kron(np.eye(32),H[2:4,:])
In : K4 = np.kron(np.eye(32),H[4:8,:])
In : H = np.vstack([K1,K2,K3,K4])
```
Python

The next step is to scale the rows so that the matrix is orthogonal:

```matlab
>> D = diag(diag(H*H'));
>> H = sqrt(inv(D))*H;
```
MATLAB/Octave

or

```python
In : D = H.dot(H.T)
In : np.sqrt(np.linalg.inv(D)).dot(H)
```
Python

This last H is the matrix we want, which implements a Haar 3-scale wavelet transform on vectors of length 256. It can be applied to the image:

```matlab
>> cw = (H*(H*double(c))')';
```
MATLAB/Octave

or with

```python
In : cw = H.dot(H.dot(c.astype('float')).T).T
```
Python

and displayed with the log scaling, as shown in Figure 15.8. With the log scaling, it is clear that in the standard decomposition, the filtered images in the transform are squashed, rather than all retaining their shape as in the non-standard decomposition.

FIGURE 15.8: A 3-scale standard decomposition DWT applied to an image

15.4   The Daubechies Wavelets

We have seen that the dilation and wavelet equations for the Haar wavelet are particular examples of the general equations 15.4 and 15.5. However, it is not at all obvious that there are any other solutions. One of the many contributions of Ingrid Daubechies to the theory of wavelets was to define an entire class of wavelets which may be considered solutions to these equations.

The Daubechies-4 wavelet has a scaling function $\phi(x)$ and wavelet function $\psi(x)$ which satisfy the equations:

$$\phi(x) = h_0\phi(2x) + h_1\phi(2x - 1) + h_2\phi(2x - 2) + h_1\phi(2x - 2) \tag{15.8}$$

$$\psi(x) = h_0\phi(2x - 1) - h_1\phi(2x) + h_2\phi(2x + 1) - h_3\phi(2x + 2) \tag{15.9}$$

where the values of the filter coefficients are:

$$h_0 = \frac{1 + \sqrt{3}}{4\sqrt{2}} \approx 0.48296$$

$$h_1 = \frac{3 + \sqrt{3}}{4\sqrt{2}} \approx 0.83652$$

$$h_2 = \frac{3 - \sqrt{3}}{4\sqrt{2}} \approx 0.22414$$

$$h_3 = \frac{1 - \sqrt{3}}{4\sqrt{2}} \approx -0.12941$$

These can be obtained with the **daubcqf** function of **rwt**:

```
>> [h,g] = daubcqf(4)
h =

    0.48296    0.83652    0.22414    -0.12941


g =

    0.12941    0.22414    -0.83652    0.48296
```

MATLAB/Octave

The vectors contain the same values, but in different orders and with different signs. As for the Haar wavelet, we can apply the Daubechies-4 wavelet by a matrix multiplication; the matrix for a 1-scale DWT on a vector of length 8 is

$$
\begin{bmatrix}
h_0 & h_1 & h_2 & h_3 & 0 & 0 & 0 & 0 \\
0 & 0 & h_0 & h_1 & h_2 & h_3 & 0 & 0 \\
0 & 0 & 0 & 0 & h_0 & h_1 & h_2 & h_3 \\
h_2 & h_3 & 0 & 0 & 0 & 0 & h_0 & h_1 \\
h_3 & -h_2 & h_1 & -h_0 & 0 & 0 & 0 & 0 \\
0 & 0 & h_3 & -h_2 & h_1 & -h_0 & 0 & 0 \\
0 & 0 & 0 & 0 & h_3 & -h_2 & h_1 & -h_0 \\
h_1 & -h_0 & 0 & 0 & 0 & 0 & h_3 & -h_2
\end{bmatrix}
$$

Notice that the filter coefficients overlap between rows, which is not the case for the Haar matrix H_{2^n}. This means that the use of the Daubechies-4 wavelet will have smoother results than using the Haar wavelet. The form of the matrix above is similar to circular convolution with the one-dimensional filters

$$
\begin{bmatrix} h_0 & h_1 & h_2 & h_3 \end{bmatrix}
$$

and

$$
\begin{bmatrix} h_3 & -h_2 & h_1 & -h_0 \end{bmatrix} .
$$

The discrete wavelet transform can indeed be approached in terms of filtering; the above two filters are then known as the low pass and high pass filters, respectively. Steps for performing a 1-scale wavelet transform are given by Umbaugh [52]:

1. Convolve the image rows with the low pass filter.

2. Convolve the columns of the result of Step 1 with the low pass filter, and rescale this to half its size by subsampling.

3. Convolve the result of Step 1 with the high pass filter, and again subsample to obtain an image of half the size.

4. Convolve the original image rows with the high pass filter.

5. Convolve the columns of the result of Step 4 with the low pass filter, and rescale this to half its size by subsampling.

6. Convolve the result of Step 4 with the high pass filter, and again subsample to obtain an image of half the size.

At the end of these steps there are four images, each half the size of the original. They are:

1. the "low pass/low pass" image (LL); result of Step 2.

2. the "low pass/high pass" image (LH); result of Step 3.

3. the "high pass/low pass" image (HL); result of Step 5.

4. the "high pass/high pass" image (HH); result of Step 6.

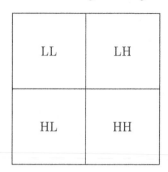

FIGURE 15.9: The 1-scale wavelet transform in terms of filters

These can then be placed into a single image grid as shown in Figure 15.9.

The filter coefficients of a wavelet are such that the transform may be inverted precisely to recover the original image. Using filters, this is done by taking each subimage, zero-interleaving to produce an image of double the size, and convolving with the inverse low pass and high pass filters. Finally, the results of all the filterings are added. For the Daubechies-4 wavelet, the inverse low pass and high pass filters are:

$$\begin{bmatrix} h_2 & h_1 & h_0 & h_3 \end{bmatrix}$$

and

$$\begin{bmatrix} h_3 & -h_0 & h_1 & -h_2 \end{bmatrix}$$

respectively.

A further approach to the discrete wavelet transform may be considered as a generalization to filtering; it is called *lifting*, and is discussed by Jensen and la Cour-Harbo [24]. Lifting starts with a sequence $s_j[n], n = 0 \ldots 2^j - 1$ and produces two sequences $s_{j-1}[n], n = 0 \ldots 2^{j-1} - 1$ and $d_{j-1}[n], n = 0 \ldots 2^{j-1} - 1$, each half the length of the original. The lifting scheme for the Haar wavelet may be described as

$$d_{j-1}[n] = s_j[2n + 1] - s_j[2n]$$
$$s_{j-1}[n] = s_j[2n] + d_{j-1}[n]/2,$$

and a lifting scheme for the Daubechies-4 wavelet is:

$$s_{j-1}^{(1)}[n] = s_j[2n] + \sqrt{3}s_j]2n + 1]$$
$$d_{j-1}^{(1)}[n] = s_j[2n + 1] - \frac{1}{4}\sqrt{3}s_{j-1}^{(1)}[n] - \frac{1}{4}(\sqrt{3} - 2)s_{j-1}^{(1)}[n - 1]$$
$$s_{j-1}^{(2)} = s_{j-1}^{(1)}[n] - d_{j-1}^{(1)}[n + 1]$$
$$s_{j-1}[n] = \frac{\sqrt{3} - 1}{\sqrt{2}}s_{j-1}^{(2)}[n]$$
$$d_{j-1}[n] = \frac{\sqrt{3} + 1}{\sqrt{2}}d_{j-1}^{(1)}[n].$$

Both the Haar and Daubechies-4 schemes can be reversed; see [24] for details. A single lifting, as shown above, produces a 1-scale wavelet transform. To produce transforms of higher scales, lifting can be applied to s_{j-1} to produce s_{j-2} and d_{j-2}, and then to s_{j-2}, and so on. In each lifting scheme, the sequence s_{j-1} may be considered the subsampled low pass version of s_j, and the sequence d_{j-1} may be considered the subsampled high pass version of s_j. Thus, a single lifting scheme provides, in effect, both the low pass and high pass filter results. Applying a lifting to an image, first to all the rows, and then to all the columns of the result, will produce a 1-scale wavelet transform.

We might hope that since the scaling and wavelet functions $\phi(x)$ and $\psi(x)$ have such simple forms for the Haar wavelet, as we saw in Figure 15.3, the same would be true for the Daubechies-4 wavelet. We can plot a wavelet by inverting a "basis vector": a vector consisting of all zeros except for a single one. Following [24] we shall choose a vector whose sixth element is one. We shall experiment first with the Haar wavelet:

```
>> [h,g] = daubcqf(2);
>> t = zeros(1,512); t(6) = 1;
>> td = miwt(t,h,9);
>> plot([1:512]/512,td,"linewidth",2)
```
MATLAB/Octave

The result is shown in Figure 15.10. Other basis vectors produce similar results, but with

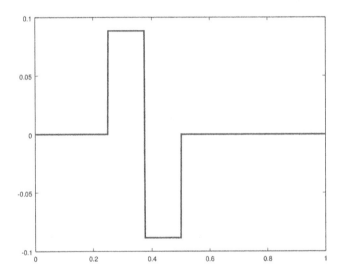

FIGURE 15.10: A plot of the Haar wavelet

the wavelet scaled or translated. The vector we have chosen seems to produce the best plot. For the Daubechies-4 wavelet, we repeat the above four commands, but starting with

```
>> [h,g] = daubcqf(4);
```
MATLAB/Octave

In Python, the plot can be generated with

```
In : t = np.array([0]*512); t[5]=1
In : td = rwt.idwt(t,h,9)[0]
In : plt.plot(np.arange(512),td)
```
Python

The result is shown in Figure 15.11. This is a very spiky and unusual graph! It is in fact a fractal, and unlike the Haar wavelet, the function cannot be described in simple terms.

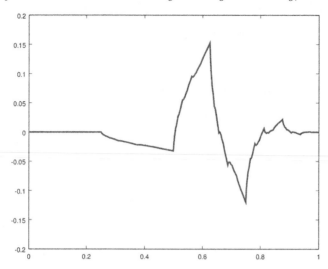

FIGURE 15.11: A plot of the Daubechies-4 wavelet

15.5   Image Compression Using Wavelets

Wavelets provide some of the most powerful tools known for image compression. As we stated earlier, they have replaced the use of the DCT in the JPEG2000 algorithm. In this section we shall play around with investigating what information we can remove from the DWT of an image, and still retain most of the original information.

Thresholding and Quantization

The idea is to take the DWT of an image, and for a given value d, set all values x in the DWT for which $|x| \leq d$ to zero. This is at the basis of the JPEG2000 compression [51]. We shall try this using the Daubechies-4 wavelets, and $d = 10$.

```
>> c = imread('caribou.png');
>> imshow(c);
>> cw = mdwt(double(c),h,8);
>> length(find(abs(cw)<=10))

ans =

      47427
>> cw(find(abs(cw)<=10)) = 0;
>> cw = round(cw);
>> ci=midwt(cw,h,8);
>> imshow(mat2gray(ci))
```
MATLAB/Octave

We are thus removing nearly three quarters of the information from the DWT. We further simplify the result by quantizing; in this case we just use the **round** function to turn fractions into integers. The original caribou image and the result of the above commands are shown in Figure 15.12. The result is indistinguishable from the original image. We

FIGURE 15.12: An image after DWT thresholding and inversion

can use different wavelets, by using either **daub(2)** or **daub(6)**, to obtain the Haar filter coefficients or the filter coefficients of the Daubechies-6 wavelet. The results of thresholding (again with $d = 10$) are shown in Figures 15.13(a) and (b). We could clearly obtain higher

(a) Using the Haar wavelet   (b) Using the Daubechies-6 wavelet

FIGURE 15.13: Use of different wavelets

compression rates by removing more information; this can be obtained by increasing the threshold value d. We will try with $d = 30$ and $d = 50$, and using the Daubechies-4 wavelet:

```
In :   cw = mdwt(c.astype(double),h,8)
In :   len(np.where(abs(cw)<=30)[0])
Out:   61093
In :   cw[np.where(abs(cw)<=30)] = 0
In :   cw = round(cw);
In :   ci = rwt.idwt(cw,h,8);
In :   io.imshow(ci)
In :   cw = rwt.dwt(c.astype(double),h,8)
In :   length(find(abs(cw)<=50))
Out:   63411
In :   cw[np.where(abs(cw)<=50)] = 0
In :   cw = np.round(cw);
In :   ci = rwt.idwt(cw,h,8);
In :   io.imshow(ci)
```

Python

The results are shown in Figure 15.14. Notice that with a threshold value of 50 we are

FIGURE 15.14: Results of wavelet compression with $d = 30$ and $d = 50$

throwing away 63,411 items of information, leaving only 2125 non-zero values. However, the result is extremely good when compared with the results of JPEG compression as shown in Figure 14.9 in Chapter 14.

As we did in Chapter 14, we shall investigate the appearance of the image with a closeup. Figure 15.15 shows the closeup after thresholding at $d = 10$ using both the Haar wavelets and the Daubechies-6 wavelets. Figure 15.16 shows the closeup after thresholding at $d = 30$ and $d = 50$ using the Daubechies-6 wavelets. Even though we can see that we are losing some fine detail, the result is far superior to the result obtained by using the JPEG compression algorithm.

Extraction

Here we cut off a portion of the DWT, setting all values to zero, and invert the rest. This is not a standard method of compression, but it does gives reasonably good results. It also has the advantage of being computationally easier than thresholding and quantization. For example, suppose we keep just the upper left 100×100 elements of the transform:

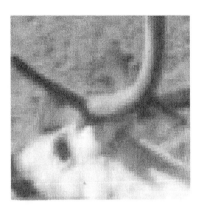

FIGURE 15.15: Closeups of wavelet compression using $d = 10$

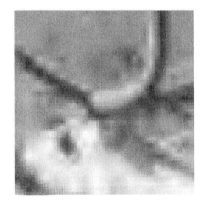

FIGURE 15.16: Closeups of wavelet compression using $d = 30$ and $d = 50$

```
>> [h,g] = daubcqf(4);
>> cw = mdwt(double(c),h,8);
>> temp = zeros(size(c));
>> temp(1:100,1:100) = cw(1:100,1:100);
>> ci = midwt(temp,h,8);
>> imshow(mat2gray(ci))
```

MATLAB/Octave

and the result is shown in Figure 15.17. Notice that whereas we took only 10,000 elements of

FIGURE 15.17: An image after DWT extraction and inversion

the 65,536 elements of the transform—less than one-sixth—the output is remarkably good. Figure 15.18 shows the results after two more extractions. On the left, we used the top left

FIGURE 15.18: Compression by DWT extraction

50×50 elements of the DWT; on the right only 20×20 elements. The left-hand figure is getting quite blurry: but as we have used only 2500 of the possible 65,536 elements of the transform, this is to be expected. On the right, with only 400 elements used, the figure is basically unrecognizable. Even so, we can still distinguish size, shapes, and grayscales.

15.6 High Pass Filtering Using Wavelets

If we look at a wavelet transform, then apart from the rescaled image in the top left, the rest is high frequency information. We would expect then, that if we were to eliminate the top left corner, by setting all the transform values to zero, the result after inversion would be a high pass filtered version of the original image.

We will perform a 2-scale decomposition (with Daubechies-4) on the caribou image, and remove the image from the top left.

```
>> cw = mdwt(double(c),h,2);
>> cw(1:64,1:64) = 0;
>> ci = midwt(cw,h,2);
>> imshow(mat2gray(ci))
```
MATLAB/Octave

This is shown on the left in Figure 15.19. We can do the same thing but using a 3-scale decomposition; this will mean that the corner image will be half the size.

```
In :  cw = rwt.dwt(c.astype(float),h,3)[0]
In :  cw[0:64,0:64] = 0;
In :  ci = rwt.idwt(cw,h,3);
In :  io.imshow(ci)
```
Python

This is shown on the right in Figure 15.19. These images can be thresholded to produce

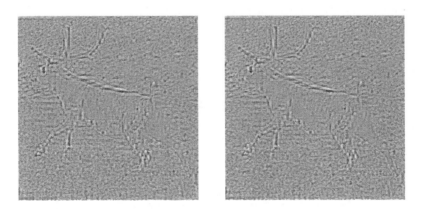

FIGURE 15.19: High pass filtering with wavelets

edge images.

We can be more specific in the part of the transform we eliminate or keep, and we can extract the horizontal or vertical edges only.

15.7   Denoising Using Wavelets

Since noise can be considered a high frequency component, some form of low pass filtering should remove it. As we did for compression, we can remove such components by thresholding. To compare with the results obtained in Chapter 8, we shall use a gray version of the seagull image, converted to type **double**.

```
>> f = im2double(rgb2gray(imread('gull.png')));
>> fg = imnoise(f,'gaussian');
>> imshow(fg)
```
MATLAB/Octave

Now we can attempt some noise removal. We shall assume that all the filter coefficients are those of the Daubechies-4 wavelet.

```
>> fw = mdwt(fg,h,4);
>> length(find(abs(fw)<=0.2))
ans =

        159192
```
MATLAB/Octave

Note that the image is of size $400 \times 400 = 160,000$, so we are looking at setting over 94% of the transform values to zero.

```
>> fw(find(abs(fw)<=0.2))=0;
>> fi = midwt(fw,h,4);
>> figure,imshow(mat2gray(fi))
```
MATLAB/Octave

The noisy image and the result after wavelet filtering are shown in Figure 15.20. At this

FIGURE 15.20: Denoising using wavelets

level of thresholding, not much of the noise has been removed, although there has been some slight reduction. Thresholding can of course be done at higher levels:

```
In :   f = ut.img_as_float(io.imread('fledgling.png'))
In :   fg = ut.random_noise(f,'gaussian')
In :   fw = rwt.dwt(fg,h,4)[0]
In :   fw[np.where(abs(fw)<=0.4)] = 0
In :   fi = rwt.idwt(fw,h,4)
In :   io.imshow(fi[0])
```
Python

Figure 15.21 shows the results using (a) a threshold of 0.3 and (b) a threshold of 0.5. The

(a)   (b)

FIGURE 15.21: Denoising using different thresholds

results compare very favorably to those we investigated in Chapter 8; however, we can see that too high a threshold results in unacceptable blurring of the image.

Exercises

1. For each of the sequences

   (a) 20 38 6 25 30 29 21 32
   (b) 30 11 0 38 0 15 22 32
   (c) 7 38 19 11 24 14 32 14

   apply, by hand, the simple wavelet transform discussed in Section 15.1 at 1, 2, and 3 scales.

2. Check your answers using the filter coefficients h = [1,1].

3. Work backward from the transforms and obtain the original sequence.

4. Repeat the above calculations in your system using the Haar wavelet transform; obtaining the filter coefficients with the command daubcqf(2).

5. For an 8-element vector v, determine its DWT at 1 scale using the Daubechies-4 wavelet. For the filter values h and g, determine the correlations with

```
>> v = double(v)
>> imfilter(v,h,'circular')
>> imfilter(v,g,'circular')
```

MATLAB/Octave

or

```
In :  v = np.array(v).astype(float)
In :  ndi.correlate(v,h,mode='wrap')
In :  ndi.correlate(v,g,mode='wrap')
```

Python

Can you see how the wavelet is related to the results of the correlations? Write a program to perform the DWT at 1 scale using the Daubechies-4 wavelet by filtering. Then generalize your program to higher scales.

6. Obtain a 256×256 grayscale image. Obtain the DWT at 1, 2, and 3 scales using daubcqf(2), daubcqf(4), and daubcqf(6).

   Are there any observable differences in the outputs of the transform for the different wavelets?

7. Write a function showdwt that will scale the result of a two-dimensional wavelet transform for easier viewing by scaling the individual square components of the DWT separately.

8. Choose any 256×256 grayscale image:

   Perform the forward DWT at 8 scales using the Daubechies-4 wavelet. For each of the following values of d:

   (a) 10,   (b) 25,   (c) 50,   (d) 60

   - Make all elements of the transform whose absolute values are less than d equal to zero.
   - Determine the number of zeros in the result.
   - Quantize the result by using the floor or round functions.
   - Invert the result and display.

   At what values of d do you notice degradation in the result?

9. Repeat the above question for some very large values of d: 100, 200, 300,...

   What size of d renders the image completely unrecognizable? How many zeros are in the transform?

10. Repeat the above two questions with an image of a face: take a selfie, or a photograph of somebody else, and reduce it to a square image whose rows and columns are a power of 2.

11. Choose any of your 256×256 grayscale images. Add Gaussian noise with mean 0, and with standard variations 0.01, 0.02, 0.05, and 0.1. For each noisy image, attempt to denoise it with wavelets.

    How reasonable a result can you obtain for the noise with the higher standard deviations?

Chapter 16

Special Effects

We have seen a great many different methods of changing the appearance of an image, and usually our need for the particular change was motivated by a problem: we may need to remove something from an image (noise, for example); we may need to make the image "look better"; we may need to calculate some aspect of the image (size, position, number of components, etc.)

But there is another totally different aspect of image processing, that of adding some sort of special effect to an image. We may want to do this for effect alone, or simply for fun. There are a great many different effects possible, and in this chapter we shall look at a few of them. This is a sort of gray area of image processing; we are encroaching toward the area of computer graphics. However, our algorithms involve the changing either of the value or the position of pixels, and we can achieve many effects with very simple means.

16.1 Polar Coordinates

Many effects have a "radial" nature, their appearance "radiates" outward from the center of the image. In order to achieve these effects, we need to transform our image from Cartesian coordinates to polar coordinates, and back again.

The origin. Suppose we have an image of size rows × cols. If the dimensions are odd, we can choose as our origin the very center pixel. If the dimensions are even, we choose as origin a pixel at the top left of the bottom right quadrant. so in a 7×9 image, our origin would be at pixel $(4, 5)$, and in a 6×8 image, our origin would be at position $(4, 4)$. We can express the origin formally by

$$x_0 = \lceil (r + 1)/2 \rceil$$
$$y_0 = \lceil (c + 1)/2 \rceil$$

where r and c are the numbers of rows and columns, respectively, and $\lceil x \rceil$ is the *ceiling function*, which returns the smallest integer not less than x. Of course, if one dimension is odd and the other even, the origin will be at the center of the odd dimension, and in the first place of the second half of the even dimension.

Thus, we can find coordinates of the origin by either

```
>> ox = ceil((rows+1)/2)
>> oy = ceil((cols+1)/2)
```

MATLAB/Octave

or

```
In :   ox = (rows+1)//2
In :   oy = (cols+1)//2
```

Python

Polar coordinates. To find the polar coordinates, the usual formulas can be used:

$$r = \sqrt{x^2 + y^2}$$
$$\theta = \tan^{-1}(y/x),$$

which can be implemented for an array by using the the `meshgrid` function from MATLAB and Octave, or the `mgrid` function of Python, where the indices are offset by the values of the origin:

```
>> [y,x] = meshgrid([1:cols]-oy,[1:rows]-ox);
>> r = sqrt(x.^2+y.^2);
>> theta = atan2(y,x);
```

MATLAB/Octave

```
In :   y,x = np.mgrid[-oy:cols-oy,-ox:rows-ox]
In :   r = np.sqrt(x**2+y**2)
In :   theta = np.arctan2(y,x)
```

Python

We can easily go back, using the fact that the (x, y) coordinates corresponding to the polar coordinates (r, θ) are

$$x = r \cos \theta$$
$$y = r \sin \theta.$$

Thus:

```
>> x2 = round(r.*cos(theta))+ox
>> y2 = round(r.*sin(theta))+oy
```

MATLAB/Octave

or

```
In :   x2 = np.round(r*np.sin(theta))+ox
In :   y2 = np.round(r*np.cos(theta))+oy
```

Python

and these arrays will contain the same indices as the original arrays.

It will be convenient to have a **polarmesh** function which provides r and θ values for a given array size, as well as a **polar2im** function which applies polar values to an image. Such functions are given at the end of the chapter.

An Example: Radial Pixelization

We have seen in Chapter 3 that an image can be "pixelated"; that is, be shown in large blocks of low resolution, by use of the `imresize` function. We can achieve the same effect, however, by using the `mod` function. In general, $\text{mod}(x, n)$ is the remainder when x is divided by n. For example:

```
>> x=1:12
x =
      1     2     3     4     5     6     7     8     9    10    11    12

>> mod(x,4)
ans =
      1     2     3     0     1     2     3     0     1     2     3     0
```
MATLAB/Octave

In Python, the `%` operator provides modulus functionality:

```
In :  x = np.array(range(1,13))
In :  print np.vstack((x,x%4))
[[ 1  2  3  4  5  6  7  8  9 10 11 12]
 [ 1  2  3  0  1  2  3  0  1  2  3  0]]
```
Python

So if x is divisible by 4, then there is no remainder, and so the modulus function returns 0. Now if we *subtract* the mod values form the original, we obtain repetition, which in an image would produce a pixelated effect:

```
>> x-mod(x,4)
ans =
      0     0     0     4     4     4     4     8     8     8     8    12
```
MATLAB/Octave

```
print x-x%4
[ 0  0  0  4  4  4  4  8  8  8  8 12]
```
Python

So a subtraction of moduli from the radius and argument values should produce a radial pixelated effect.

We can try it with the flowers image, which for simplicity we shall turn into a grayscale image.

```
>> f = imread('iris.png');
>> fg = rgb2gray(f);
>> [r,theta] = polarmesh(fg);
```
MATLAB/Octave

In Python, the commands would be

```
In :  f = io.imread('iris.png')
In :  fg = co.rgb2gray(f)
In :  r,theta = polarmesh(fg)
```
Python

At this stage we have the polar values of the mesh, so now we can perform the pixelation:

```
>> r2 = r-mod(r,5);
>> theta2 = theta-mod(theta,5*pi/180);
```
MATLAB/Octave

or

```
In :   r2 = r - r%5
In :   theta2 = theta - theta%(5*np.pi/180)
```
Python

Since the angles are given as radians, the modulus is equivalent to 5 degrees.

Now we transfer back into Cartesian coordinates:

```
>> x2 = r2.*cos(theta2);
>> y2 = r2.*sin(theta2);
```
MATLAB/Octave

or

```
In :   x2 = r2*np.cos(theta2)
In :   y2 = r2*np.sin(theta2)
```
Python

To obtain an image for display, we have to adjust the indices x2 and y2 by adding ox and oy to them, rounding off, and making sure that the row indices are between 1 and rows, and that the column indices are between 1 and cols. This last clipping will be made easier in MATLAB or Octave by a very small function:

```
function out = clip(x,y,z)
   out = x;
   out(find(out<y))=y;
   out(find(out>z))=z;
```
MATLAB/Octave

Now for the adjustment:

```
>> xx = round(x2)+ox;
>> yy = round(y2)+oy;
>> xx = clip(xx,1,rows);
>> yy = clip(yy,1cols);
```
MATLAB/Octave

In Python again a clipping function will be useful:

```
In :   def clip(x,y,z):
...:         x[np.where(x<y)]=y
...:         x[np.where(x>z)]=z
...:         return x
...:
```
Python

and the adjustment is very similar to that in MATLAB/Octave:

```
In :   xx = np.round(x2)+ox
In :   yy = np.round(y2)+oy
In :   xx = clip(xx,0,rows-1).astype(int)
In :   yy = clip(yy,0,cols-1).astype(int)
```
Python

Now we can use these new indices **xx** and **yy** to obtain an image from the original flower image matrix **fg**. This can be most simply done by using the **sub2ind** function, which basically finds places in an array indexed by given vectors.

```
>> f2 = fg(sub2ind([rows,cols],xx,yy));
```

MATLAB/Octave

In Python the **ravel** method produces one long vector of the array; the **xx** and **yy** vectors can be used as indices into **f**, the result of which is reshaped for display:

```
In :   f2 = np.reshape(f[xx.ravel(),yy.ravel()],(rows,cols))
```

Python

And in fact all the last commands can be bundled into **polar2im** functions, which are given at the end of the chapter, so that

```
In :   f2 = polar2im(im,r2,theta2);
```

MATLAB/Octave

(and the same command in Python) is all that is needed.

The initial image and the radiated pixelated result are shown in Figure 16.1.

The original **fg**   The pixelated **f2**

FIGURE 16.1: Radial pixelation

16.2 Ripple Effects

We will investigate two different ripple effects: "bathroom glass" ripples, which give the effect of an image seen through wavy glass, such as is found in bathrooms, and "pond ripples," which approximate a reflection in the surface of a pond.

We have seen that subtracting moduli produces a pixelated effect. A bathroom glass effect can be obtained by *adding* moduli:

```
>> x=1:12;
>> x+mod(x,4)
ans =
    2    4    6    4    6    8   10    8   10   12   14   12
```
MATLAB/Octave

or

```
In :   x = np.array(range(1,13))
In :   print np.vstack((x,x+x%4))
[[ 1  2  3  4  5  6  7  8  9 10 11 12]
 [ 2  4  6  4  6  8 10  8 10 12 14 12]]
```
Python

Notice that although the values increase from left to right, they do in small runs, which seem to overlap. On an image, the appearance will be of a ripple across the image. We shall experiment first using Cartesian coordinates:

```
>> [y,x] = meshgrid(1:cols,1:rows);
>> y2 = y+mod(y,32);
>> y2 = clip(y2,1,cols);
```
MATLAB/Octave

Here we have just added moduli across the columns, then adjusted the result to ensure that the values stay with the column range. Now we can use **sub2ind** to create a new image:

```
>> ripple1 = fg(sub2ind([rows cols],x,y2));
>> imshow(ripple1)
```
MATLAB/Octave

We can of course do the same thing down the rows:

```
>> x2 = x+mod(x,32);
>> x2 = clip(x2,1,rows);
>> ripple2 = fg(sub2ind([rows cols],x2,y));
>> imshow(ripple2)
```
MATLAB/Octave

or even both at once:

```
>> ripple3 = fg(sub2ind([rows cols],x2,y2));
>> imshow(ripple3)
```
MATLAB/Octave

In Python, the vertical ripples can be obtained with

```
In :   y,x = np.mgrid[0:cols,0:rows]
In :   x2 = clip(x+x%32,0,rows-1)
In :   ripple1 = np.reshape(fg[x2.ravel(),y.ravel()],(rows,cols)).T
```
Python

horizontal ripples with

```
In :  y2 = clip(y+y%32,0,cols-1)
In :  ripple2 = np.reshape(fg[x.ravel(),y2.ravel()],(rows,cols)).T
```
Python

and both with

```
In :  ripple3 = np.reshape(fg[x2.ravel(),y2.ravel()],(rows,cols)).T
```
Python

To obtain a radial ripple, corresponding to ripples on a pond, we create the polar coordinate matrices r and `theta` with `polarmesh`, adjust the radii. and finally use `polar2im` to obtain the output image:

```
>> r2=r+mod(r,30);
>> ripple4 = uint8(polar2im(fg,r2,theta));
```
MATLAB/Octave

All results are shown in Figure 16.2.

(a) `ripple1`

(b) `ripple2`

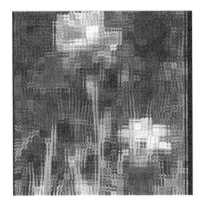

(c) `ripple3`

(d) Radial ripple: `ripple4`

FIGURE 16.2: "Bathroom glass" rippling on an image

To obtain a pond ripple effect, we move the pixels around by using a sine wave. For Cartesian effects:

```
>> [y,x] = meshgrid(1:cols,1:rows);
>> y2 = clip(round(y+7*sin(x/5)),1,cols);
>> x2 = clip(round(x+7*sin(y/5)),1,rows);
```
MATLAB/Octave

or with

```
In :  y,x = np.mgrid[0:rows,0:cols]
In :
```
Python

As above, we ensure that the values of x2 and y2 fit within the row and column bounds. Then:

```
>> ripple5 = fg(sub2ind([rows cols],x,y2));
>> ripple6 = fg(sub2ind([rows cols],x2,y));
>> ripple7 = fg(sub2ind([rows cols],x2,y2));
```
MATLAB/Octave

For a radial sine wave ripple, we adjust r first:

```
>> r2 = r + 7*sin(r/5);
>> ripple8 = polar2im(fg,r2,theta);
```
MATLAB/Octave

After that we create x2 and y2 as above, and turn them back into indices xx and yy with the appropriate bounds. Then

```
>> ripple8 = fg(sub2ind([rows cols],xx,yy));
```
MATLAB/Octave

These new ripples are shown in Figure 16.3. The effects can be varied by changing the parameters to the sine functions in the commands above.

16.3   General Distortion Effects

The rippling effects in the previous section are examples of more general effects called *distortion effects*. Here it is not so much the *value* of the pixels that are changed as their *position*. All distortion effects work in much the same way, which we shall describe in a sequence of steps. Suppose we have an effect that involves Cartesian coordinates only.

1. Start with an image matrix $p(i,j)$, of size $m \times n$.

2. Produce two new indexing arrays (the same size as the matrix): $x(i,j)$ and $y(i,j)$.

3. Create a new image $q(i,j)$ by

$$q(i,j) = p(x(i,j), y(i,j)).$$

(a) ripple5

(b) ripple6

(c) ripple7

(d) ripple8

FIGURE 16.3: "Pond rippling" on an image

We have seen above that the arrays $x(i,j)$ and $y(i,j)$ may be created by a procedure that produces values which may be fractional or be outside the bounds of the image matrix. So we will need to ensure that the indexing arrays contain integers only, and that for all i and j, that $1 \le x(i,j) \le m$, and $1 \le y(i,j) \le n$.

If we have an effect that requires polar coordinates, we add a few steps to the procedure above:

1. Start with an image matrix $p(i,j)$, of size $m \times n$.

2. Create the arrays $r(i,j)$ and $\theta(i,j)$ containing the polar coordinates of each position (i,j).

3. Produce new arrays $s(i,j)$ and $\phi(i,j)$ by adjusting the values of r and θ.

4. Write these new arrays out to Cartesian indexing arrays (the same size as the matrix):

$$x(i,j) = s(i,j)\cos(\phi(i,j)) + o_x$$
$$y(i,j) = s(i,j)\sin(\phi(i,j)) + o_y$$

   where (o_x, o_y) are the coordinates of the polar origin.

5. Create a new image $q(i,j)$ by

$$q(i,j) = p(x(i,j), y(i,j)).$$

As above, we need to ensure that the indexing arrays contain only integers, and have the appropriate bounds.

We have seen above how to implement various distortion effects in MATLAB, so just by changing the functions used to distort the indexing arrays or the polar coordinates, we can achieve many different effects.

Fisheye.   The "fisheye lens" effect can be obtained very easily with polar coordinates, with

$$r \leftarrow r^2/R$$

where

$$R = \max(r).$$

So with MATLAB and Octave, using the `polarmesh` and `polar2im` functions, the commands are:

```
>> [r,theta] = polarmesh(fg);
>> R = max(r(:));
>> s = r.^2/R;
>> f2 = uint8(polar2im(fg,s,theta));
```
MATLAB/Octave

And in Python they are very similar:

```
In :  r,theta = polarmesh(fg)
In :  s = r**2/r.max()
In :  f2 = polar2im(fg,s,theta)
```
Python

The result is shown in Figure 16.4.

Twirl. A "twirl" or "swirl" effect, as the top of a cup of coffee being swirled with a spoon, is also very easily done. Again, it is a polar effect. A twirl effect can be obtained by

$$s(i,j) = \theta(i,j) + r(i,j)/K$$

where K can be changed to adjust the amount of twirl. A small value provides more twirl, a large value less. So given the arrays r and theta, and a value K, the relevant commands are:

```
>> phi = theta+(r/K);
>> twirl1 = polar2im(fg,r,phi);
```

MATLAB/Octave

Figure 16.4 shows the results of twirls with different values of K.

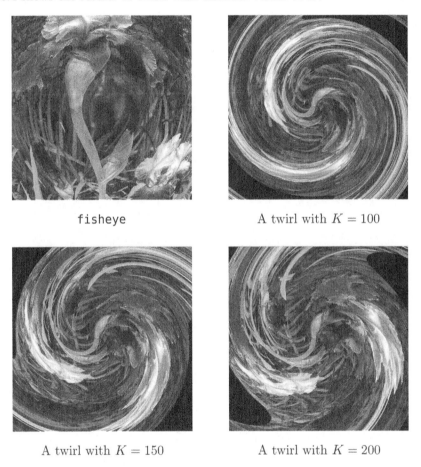

fisheye

A twirl with $K = 100$

A twirl with $K = 150$

A twirl with $K = 200$

FIGURE 16.4: Fisheye and twirls

Jitter. The "jitter" effect is more easily implemented than described. It is like a radial bathroom glass effect, but instead of manipulating the r matrix to obtain a ripple effect, we manipulate the theta matrix:

$$\phi = \theta + \mod (\theta, 8\pi/180) - 4\pi/180.$$

```
>> phi = theta+mod(theta,8*pi/180)-4*pi/180;
```
MATLAB/Octave

```
In :   phi = theta+(theta%(8*np.pi/180))-4*np.pi/180
```
Python

The result is shown in Figure 16.5.

Circular slice. This is another effect obtained by manipulating the angle. Its implementation is similar to that of the jitter effect:

$$\phi = \theta + \quad \mod (r, 6) * \pi/180$$

We include the extra $\pi/180$ factor to convert to radians. The result, which has a rippling appearance, is shown in Figure 16.5.

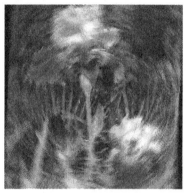   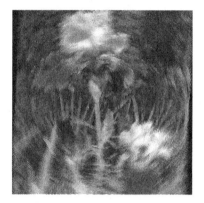

Jitter   Circular slice

FIGURE 16.5: Jitter and circular slice

Square slice This is a Cartesian effect. It works by cutting the image into small squares and perturbing them a little bit. It uses the "sign" function (sometimes called the *signum* function and abbreviated as sgn), which is given in MATLAB and Octave as `sign` and in Python as `np.sign`, and which is defined as:

$$\operatorname{sign}(x) = \begin{cases} 1 & \text{if } x > 0, \\ 0 & \text{if } x = 0, \\ -1 & \text{if } x < 0. \end{cases}$$

We start with

```
>> [y,x] = meshgrid(1:cols,1:rows);
```
MATLAB/Octave

and then implement the slice distortion with

```
>> K=8;
>> Q=10;
>> x2=round(x+K*sign(cos(y/Q)));
>> y2=round(y+K*sign(cos(x/Q)));
```

Then the final image can be computed with

```
>> f2 = fg(sub2ind(size(fg),clip(x2,1,rows),clip(y2,1,cols)));
```

The values K and Q influence the size of the squares and the amount of perturbation. A large value of K produces a very perturbed image; a large value of Q produces large squares. After rounding, fixing the bounds and writing to a new image, results are shown in Figure 16.6.

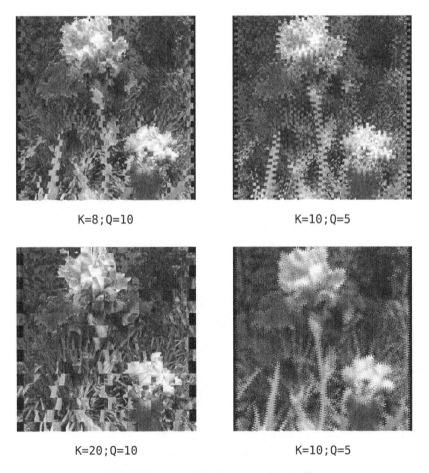

K=8;Q=10   K=10;Q=5

K=20;Q=10   K=10;Q=5

FIGURE 16.6: The "square slice" effect

Fuzzy effect. This is obtained by replacing each pixel with a randomly selected pixel from its neighborhood. To do this, we simply add random values to the index matrices x and y. If we are choosing values from a 7×7 neighborhood, then we randomize the indices with integer values chosen from $-3, -2, -1, 0, 1, 2, 3$. We can achieve this in MATLAB or Octave with

```
>> x2 = x+randi([-3,3],rows,cols);
>> y2 = y+randi([-3,3],rows,cols);
```

MATLAB/Octave

and in Python with

```
In :   x2 = x + np.random.randint(-3,3,size=(rows,cols))
In :   y2 = y + np.random.randint(-3,3,size=(rows,cols))
```

Python

The MATLAB/Octave `randi` function given here produces a matrix of size rows×cols where each element is a random integer in the range given by the initial vector. To obtain the image:

```
>> f2 = fg(sub2ind(size(fg),clip(x2,1,rows),clip(y2,1,cols)));
```

MATLAB/Octave

or

```
In :   xx = clip(x2,0,rows-1).astype(int)
In :   yy = clip(y2,0,cols-1).astype(int)
In :   f2 = np.reshape(f[xx.ravel(),yy.ravel()],(rows,cols)).T
```

Python

Results for different sized neighborhoods are shown in Figure 16.7. For a large neighborhood,

5×5 7×7

FIGURE 16.7: The fuzzy effect using different neighborhoods

the result begins to take on a frosted glass effect.

16.4   Pixel Effects

In a sense, *all* image effects are "pixel effects," because we are dealing with pixels. However, in the previous sections, we did not change the values of the grayscales explicitly; we merely copied them to different positions in the output arrays. However, many effects take the pixel values and apply some sort of processing routines to them.

Oil painting.   This is a very popular effect, and is included in most image manipulation and photo handling software. In conception it is very simple; it works by means of a non-linear filter; the output of the filter is the most common pixel value in the filter mask.

The most commonly occurring value in a neighborhood is the statistical *mode*, and is implemented in MATLAB and Octave by the function `mode`. In Python, it is implemented by the `mode` function in the `scipy.stats` module. It can be applied to the iris image as:

```
>> f2 = colfilt(fg,[9,9],'sliding',@mode);
```
MATLAB/Octave

with the `sliding` parameter indicating that the function is to be applied to each possible 9×9 neighborhood.

In Python, the `mode` function has to be adjusted slightly so as to return a single number, rather than both the number and its frequency:

```
In :  def mymode(x):
...:       return sc.stats.mode(x,axis=None)[0][0]
...:
In :  f2 = ndi.generic_filter(fg,mymode,size=(9,9))
```
Python

The result is shown in Figure 16.8(a). We can obtain a more "painterly" effect by choosing a larger block size in `nlfilter`. All of these functions can be slow!

(a) 9×9 \qquad\qquad (b) 15×15

FIGURE 16.8: The oil painting effect with different filter sizes

Solarization.   This is a photographic effect that is obtained by applying diffuse light to a developing photograph, and then continuing with the development. The result, which seems somehow magical, is that the image can be partly positive and partly negative.[1] This effect has been used by many photographers, but can be a fairly hit or miss affair in the photographic darkroom. However, it is very easy to create solarization effects digitally.

We have seen examples of solarization functions in Chapter 4; now we explore how they can be applied.

A simple solarization effect can be obtained by taking the complement of all pixels in an image $p(i, j)$ whose grayscales are less than 128:

$$\text{sol}(i, j) = \left\{ \begin{array}{ll} p(i, j) & \text{if } p(i, j) > 128, \\ 255 - p(i, j) & \text{if } p(i, j) \le 128. \end{array} \right.$$

This can be very easily implemented:

```
>> u = fg>128;
>> sol = u.*fg + (1-u).*(255-fg);
>> imshow(uint8(sol)),figure,imshow(mat2gray(sol))
```

MATLAB/Octave

We display the result twice; the second time the image is shown with better contrast.

In Python, the `rgb2gray` function returns a floating point array, so the image complement is obtained with `1-fg`:

```
In :   u = fg>0.5
In :   sol = u*fg + (1-u)*(1-fg)
In :   io.imshow(sol,vmin=0,vmax=1)
In :   io.imshow(sol)
```

Python

Note that Python automatically adjusts the display so that the lowest element in the array is mapped to black and the largest element to white. To prevent that adjustment the `vmin` and `vmax` parameters must be used; these supply minimum and maximum values to `imshow`. Both images are shown in Figure 16.9. We can make more interesting solarization effects

FIGURE 16.9: The solarization effect

by varying the complementation according to the value of the row or column. All we need to do is to change the matrix `u` above:

---

[1]A good account of solarization can be found at `http://www.cchem.berkeley.edu/wljeme/SOUTLINE.html`.

```
>> u1 = double(fg > 255*x/(2*rows));
>> u2=double(fg > 255*y/(2*cols));
```

MATLAB/Octave

or with

```
In :  u1 = fg > np.float32(x)/(2.0*rows)
In :  u2 = fg > np.float32(y)/(2.0*cols)
```

Python

The results are shown in Figure 16.10.

sol1   sol2

FIGURE 16.10: More solarization effects

16.5 Color Images

Most of the above effects look far more dramatic when applied to color images. For all the distortion effects, application to a color image means applying the effect to each of the RGB components separately. Recall that distortion effects do not change the values of the pixels so much as their position. Thus, in order to maintain the colors, we need to shift all the RGB components.

Here, for example, are twirl and circular ripple applied to the colored image, and assuming that all the index arrays x, y, r, and theta have already been computed:

```
>> f = imread('iris.png');
>> K = 150;
>> phi = theta + r/K;
>> r2 = r + mod(r,30);
>> twirl = zeros(size(f));
>> ripple = zeros(size(f));
>> for i = 1:3, twirl(:,:,i) = polar2im(f(:,:,i),r,phi); end
>> for i = 1:3, ripple(:,:,i) = polar2im(f(:,:,i),r2,theta); end
>> imshow(uint8(twirl)), figure, imshow(uint8(ripple))
```

MATLAB/Octave

The method in Python is very similar:

```
In :   f = io.imread('iris.png')
In :   K = 150; phi = theta + r/K
In :   r2 = r + (r%30)
In :   twirl = np.zeros_like(f)
In :   ripple = np.zeros_like(f)
In :   for i in range(3):
...:       twirl[:,:,i] = polar2im(f[:,:,i],r,phi)
...:       ripple[:,:,i] = polar2im(f[:,:,i],r2,theta)
...:
```

Python

The results are shown in Figure 16.11.

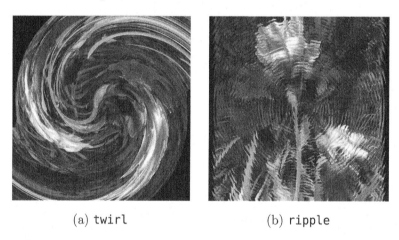

(a) `twirl`   (b) `ripple`

FIGURE 16.11: **SEE COLOR INSERT** Effects on a color image

Oil painting is again done by applying the effect to each of the color components separately.

Exercises

1. Experiment with changing the parameters to the effects described in this chapter. For example, see if you can use the pond ripple effect to obtain pictures like this:

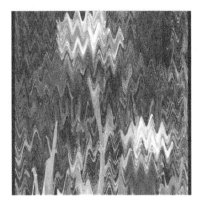

2. Write some functions to implement all the effects given. Include some parameters to change the output.

3. Extend your functions so that they can be applied to color images.

4. If you can get hold of Holzmann's book, [18][2] many other effects are described. For example, the following pictures show the result of a random "tile effect," where the image is broken up into squares that are then shifted by a random amount.

See if you can write a function to implement this effect, and then extend your function to work with color images.

---

[2]It is now available online, at http://spinroot.com/pico/.

Appendix A

Introduction to MATLAB and Octave

A.1 Introduction

MATLAB is a data analysis and visualization tool that has been designed with powerful support for matrices and matrix operations. As well as this, MATLAB has excellent graphics capabilities, and its own powerful programming language. One of the reasons that MATLAB has become such an important tool is through the use of sets of MATLAB programs designed to support a particular task. These sets of programs are called *toolboxes*, and the particular toolbox of interest to us is the *image processing toolbox*.

Concurrently with MATLAB, other software projects have grown up with similar features and design. One of the most powerful currently is GNU Octave,[1] initially written at the University of Wisconsin-Madison by John Eaton, and since extended and enhanced by many others. Eaton is still the official maintainer, and it is now mature software, having had its first official release in 1992. The Octave developers place great emphasis on its interoperability with MATLAB: the goal is that MATLAB programs and scripts—unless they rely on specific toolboxes—should run with no change on Octave.

For that reason, we are able to treat MATLAB and Octave as being "the same," and for the elementary purposes of this text they are. There are a number of advanced tools both in MATLAB and the image processing toolbox which currently have no equivalent in Octave. On the other hand, some of Octave's commands and functions are more flexible than MATLAB's.

Rather than give a description of all of the capabilities of each system, we shall restrict ourselves to just those aspects concerned with the handling of images. We shall introduce functions, commands, and techniques as required. A *function* is a keyword that accepts various parameters, and produces some sort of output: for example, a matrix, a string, a graph, or a figure. Examples of such functions are `sin`, `imread`, `imclose`. There are *many* functions in MATLAB, and as we shall see, it is very easy (and sometimes necessary) to write our own. A *command* is a particular use of a function. Examples of commands might be

```
>> sin(pi/3)
>> c = imread('cameraman.tif');
>> a = imclose(b);
```

MATLAB/Octave

As we shall see, we can combine functions and commands, or put multiple commnds on a single input line.

---

[1] https://www.gnu.org/software/octave/

The standard data type of MATLAB and Octave is the matrix—all data are considered to be matrices of some sort. Images, of course, are matrices whose elements are the gray values (or possibly the RGB values) of its pixels. Single values are considered to be 1×1 matrices, while a string is merely a $1 \times n$ matrix of characters; n being the string's length.

In this chapter we will look at the more generic commands, and discuss images in further chapters.

When you start up MATLAB or Octave, you have a blank window called the "Command Window" in which you enter commands. The initial MATLAB window is shown in Figure A.1, and the initial Octave window is shown in Figure A.2. Given the vast number of functions, and the different parameters they can take, a command line style interface is in fact much more efficient than a complex sequence of pull-down menus.

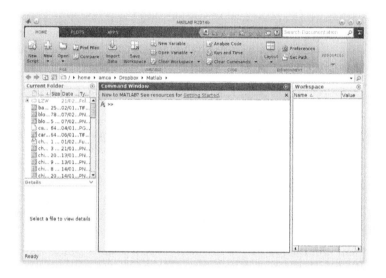

FIGURE A.1: The MATLAB command window ready for action

The default MATLAB prompt consists of two right arrows:

```
>>
```
MATLAB/Octave

and Octave can be configured to have the same prompt.

A.2   Basic Use

If you have never used MATLAB or Octave before, we will experiment with some simple calculations. We first note that each is *command line driven*; all commands are entered by typing them after the prompt symbol. Let's start off with a mathematical classic:

```
>> 2+2
```
MATLAB/Octave

FIGURE A.2: The Octave command window ready for action

What this means is that you type in

```
2+2
```

at the prompt, and then press your ‖Enter‖ key. This sends the command to the system kernel. What you should now see is

```
ans =

    4
```

MATLAB/Octave

Good, huh?

Either system of course can be used as a calculator; each understands the standard arithmetic operations of addition (as we have just seen), subtraction, multiplication, division, and exponentiation. Try these:

```
>> 3*4

>> 7-3

>> 11/7

>> 2^5
```

MATLAB/Octave

The results should not surprise you. Note that for the output of **11/7** the result was only given to a few decimal places. In fact, all calculations are performed internally to *double precision*. However, the default display format is to use only eight decimal places. We can change this by using the **format** function. For example:

```
>> format long
>> 11/7

ans =

    1.57142857142857
```

Entering the command `format` by itself returns to the default format.

All of the elementary mathematical functions are built in:

```
>> sqrt(2)

ans =

    1.4142

>> sin(pi/8)

ans =

    0.3827

>> log(10)

ans =

    2.3026

>> log10(2)

ans =

    0.3010
```

The trigonometric functions all take radian arguments; and `pi` is a built-in constant. The functions `log` and `log10` are the natural logarithm and logarithms to base 10.

A.3   Variables and the Workspace

When using any sort of computer system, we need to store things with appropriate names. In the context of MATLAB, we use *variables* to store values. Here are some examples:

```
>> a = 5^(7/2)

a =

   279.5085

>> b = sin(pi/9)-cos(pi/9)

b =

   -0.5977
```
MATLAB/Octave

Note that although **a** and **b** are displayed using the **short** format, MATLAB in fact stores their full values. We can see this with:

```
>> format long; a, format

a =

     2.795084971874737e+02
```
MATLAB/Octave

We can now use these new variables in further calculations:

```
>> log(a^2)/log(5)

ans =

     7

> atan(1/b)

ans =

   -1.0321
```
MATLAB/Octave

The Workspace

If you are using a windowed version of MATLAB, you may find a **Workspace** item in the **View** menu. This lists all your currently defined variables, their numeric data types, and their sizes in bytes. To open the workspace window, use the **View** menu, and choose **Workspace**. The same information can be obtained using the **whos** function:

```
>> whos
  Name        Size           Bytes  Class

   a          1x1               8    double array
   ans        1x1               8    double array
   b          1x1               8    double array

Grand total is 3 elements using 24 bytes
```

MATLAB/Octave

Note also that ans is variable: it is automatically created by MATLAB to store the result of the last calculation. A listing of the variable names only is obtained using who:

```
>> who

Your variables are:

a    ans  b
```

MATLAB/Octave

The numeric date type **double** is MATLAB's standard for numbers; such numbers are stored as double-precision 8-byte values.

Other data types will be discussed next.

A.4 Dealing with Matrices

MATLAB and Octave have an enormous number of commands for generating and manipulating matrices. Since a grayscale image is a matrix, we can use some of these commands to investigate aspects of the image.

We can enter a small matrix by listing its elements row by row, using spaces or commas as delimiters for the elements in each row, and using semicolons to separate the rows. Thus, the matrix

$$\begin{bmatrix} 4 & -2 & -4 & 7 \\ 1 & 5 & -3 & 2 \\ 6 & -8 & -5 & -6 \\ -7 & 3 & 0 & 1 \end{bmatrix}$$

can be entered as

```
>> a = [4 -2 -4 7;1 5 -3 2;6 -8 -5 -6;-7 3 0 1]
```

MATLAB/Octave

Matrix Elements

Matrix elements can be obtained by using the standard row, column indexing scheme. So for our image matrix a above, the command

```
>> a(2,3)

ans =

    -3
```

returns the element of the matrix in row 2 and column 3.

MATLAB also allows matrix elements to be obtained using a single number; this number being the position when the matrix is written out as a single column. Thus, in a 4×4 matrix as above, the order of elements is

$$\begin{bmatrix} 1 & 5 & 9 & 13 \\ 2 & 6 & 10 & 14 \\ 3 & 7 & 11 & 15 \\ 4 & 8 & 12 & 16 \end{bmatrix}.$$

So the element `a(2,3)` can also be obtained as `a(10)`:

```
>> a(10)

ans =

    -3
```

In general, for a matrix M with r rows and c columns, element $m(i, j)$ corresponds to $m(i + r(j - 1))$. Using the single indexing allows us to extract multiple values from a matrix:

```
>> a([1 6 11 16])

ans =

    4    5    -5    1
```

To obtain a row of values, or a block of values, we use the *colon* operator (`:`). This generates a vector of values; the command

```
>> a:b
```

where `a` and `b` are integers, lists all integers from `a` to `b`. The more general version of this command:

```
a:i:b
```

lists all values from `a` by increment `i` up to `b`. So, for example:

```
>> 2:3:16
```

generates the list

```
ans =

     2 5 8 11 14
```
MATLAB/Octave

Applied to our matrix **a**, for example:

```
>> a(2,1:3)
```

```
ans =

     6    -8    -5
```
MATLAB/Octave

lists all values in row 2 that are between columns 1 and 3 inclusive.

Similarly

```
>> a(2:4,3)
```

```
ans =

    -3
    -5
     0
```
MATLAB/Octave

lists all the values in column 3 that are between rows 2 to 4 inclusive. And we can choose a block of values such as

```
>> a(2:3,3:4)
```

```
ans =

    -3    2
    -5   -6
```
MATLAB/Octave

which lists the 2 by 2 block of values that lie between rows 2 to 3 and columns 3 to 4.

The colon operator by itself lists *all* the elements along that particular row or column. So, for example, all of row 3 can be obtained with:

```
>> a(3,:)
```

```
ans =

     6    -8    -5    -6
```
MATLAB/Octave

and all of column 2 with

```
>> a(:,2)

ans =

    -2
     5
    -8
     3
```
MATLAB/Octave

Finally, the colon on its own lists all the matrix elements as a single column:

```
>> a(:)
```

shows all 16 elements of a.

Matrix Operations

All the standard operations are supported. We can add, subtract, multiply and invert matrices, and take matrix powers. For example, with the matrix a from above, and a new matrix b defined by

```
>> b=[2 4 -7 -4;5 6 3 -2;1 -8 -5 -3;0 -6 7 -1]
```
MATLAB/Octave

we can have, for example:

```
>> 2*a-3*b

ans =

     2    -16     13     26
   -13     -8    -15     10
     9      8      5     -3
   -14     24    -21      1
```
MATLAB/Octave

As an example of matrix powers:

```
>> a^3*b^4

ans =

      103788     2039686     1466688      618345
      964142     2619886     2780222      345543
    -2058056    -2327582      721254     1444095
     1561358     3909734    -3643012    -1482253
```
MATLAB/Octave

Inversion is performed using the inv function:

```
>> inv(a)

ans =

    -0.0125     0.0552    -0.0231    -0.1619
    -0.0651     0.1456    -0.0352    -0.0466
    -0.0406    -0.1060    -0.1039    -0.1274
     0.1082    -0.0505    -0.0562     0.0064
```

MATLAB/Octave

A transpose can be obtained by using the apostrophe:

```
>> a'

ans =

     4     1     6    -7
    -2     5    -8     3
    -4    -3    -5     0
     7     2    -6     1
```

MATLAB/Octave

As well as these standard arithmetic operations, MATLAB supports some geometric operations on matrices; `flipud` and `fliplr` flip a matrix up/down and left/right, respectively, and `rot90` rotates a matrix by 90 degrees:

```
>> flipud(a)

ans =

    -7     3     0     1
     6    -8    -5    -6
     1     5    -3     2
     4    -2    -4     7

>> fliplr(a)

ans =

     7    -4    -2     4
     2    -3     5     1
    -6    -5    -8     6
     1     0     3    -7
>> rot90(a)

ans =

     7     2    -6     1
    -4    -3    -5     0
    -2     5    -8     3
     4     1     6    -7
```

MATLAB/Octave

The `reshape` function produces a matrix with elements taken column by column from the given matrix. Thus:

```
>> c = [1 2 3 4 5;6 7 8 9 10;11 12 13 14 15;16 17 18 19 20]

c =

     1     2     3     4     5
     6     7     8     9    10
    11    12    13    14    15
    16    17    18    19    20

>> reshape(c,2,10)

ans =

     1    11     2    12     3    13     4    14     5    15
     6    16     7    17     8    18     9    19    10    20

>> reshape(c,5,4)

ans =

     1     7    13    19
     6    12    18     5
    11    17     4    10
    16     3     9    15
     2     8    14    20
```

MATLAB/Octave

Reshape produces an error if the product of the two values is not equal to the number of elements of the matrix. Note that we could have produced the original matrix above with

```
>> c = reshape([1:20],5,4)'
```

MATLAB/Octave

All these commands work equally well on vectors. In fact, MATLAB makes no distinction between matrices and vectors; a vector merely being a matrix with the number of rows or columns equal to 1.

The Dot Operators

A very distinctive class of operators in MATLAB are those that use dots; these operate in an element-wise fashion. For example, the command

```
a*b
```

performs the usual matrix multiplication of **a** and **b**. But the corresponding dot operator:

```
a.*b
```

produces the matrix whose elements are the products of the corresponding elements of **a** and **b**. That is, if

```
c=a.*b
```

then $c(i,j) = a(i,j) \times b(i,j)$:

```
>> a.*b

ans =

     8    -8    28   -28
     5    30    -9    -4
     6    64    25    18
     0   -18     0    -1
```

MATLAB/Octave

We have dot division and dot powers. The command `a.^2` produces a matrix each element of which is a square of the corresponding elements of `a`:

```
>> a.^2

ans =

    16     4    16    49
     1    25     9     4
    36    64    25    36
    49     9     0     1
```

MATLAB/Octave

Similarly we can produce a matrix of reciprocals by writing `1./a`:

```
>> 1./a

ans =

    0.2500   -0.5000   -0.2500    0.1429
    1.0000    0.2000   -0.3333    0.5000
    0.1667   -0.1250   -0.2000   -0.1667
   -0.1429    0.3333       Inf    1.0000
```

MATLAB/Octave

The value `Inf` is MATLAB's version of infinity; it is returned for the operation $1/0$.

Operators on Matrices

Many functions in MATLAB, when applied to a matrix, work by applying the function to each element in turn. Such functions are the trigonometric and exponential functions, and logarithms. The use of functions in this way means that in MATLAB many iterations and repetitions can be done with *vectorization* rather than by using loops. We will explore this next.

Constructing Matrices

We have seen that we can construct matrices by listing all their elements. However, this can be tedious if the matrix is large, or if it can be generated by a function of its indices.

Two special matrices are the matrix consisting of all zeros, and the matrix consisting of all ones. These are generated by the **zero** and **ones** functions, respectively. Each function can be used in several different ways:

`zeros(n)`	if n is a number, will produce a zeros matrix of size $n \times n$
`zeros(m,n)`	if m and n are numbers, will produce a zeros matrix of size $m \times n$
`zeros(m,n,p,...)`	where m, n, p and so on are numbers, will produce an $m \times n \times p \times \cdots$ multidimensional array of zeros
`zeros(a)`	where a is a matrix, will produce a matrix of zeros of the same size as a.

Matrices of random numbers can be produced using the **rand** and **randn** functions. They differ in that the numbers produced by **rand** are taken from a uniform distribution on the interval $[0, 1]$, and those produced by **randn** are taken from a normal distribution with mean zero and standard deviation one. For creating matrices, the syntax of each is the same as the first three options of **zeros** above. The **rand** and **randn** functions on their own produce single numbers taken from the appropriate distribution.

We can construct random integer matrices by multiplying the results of **rand** or **randn** by an integer and then using the **floor** function to take the integer part of the result:

```
>> floor(10*rand(3))

ans =

     8     4     6
     8     8     8
     5     8     6

>> floor(100*randn(3,5))

ans =

  -134   -70  -160   -40    71
    71    85  -145    68   129
   162   125    57    81    66
```
MATLAB/Octave

The **floor** function will be automatically applied to every element in the matrix.

Suppose we wish to create a matrix every element of which is a function of one of its indices. For example, the 10×10 matrix A for which $A_{ij} = i + j - 1$. In most programming languages, such a task would be performed using nested loops. We can use nested loops in MATLAB, but it is easier here to use dot operators. We can first construct two matrices: one containing all the row indices, and one containing all the column indices:

```
>> rows=(1:10)'*ones(1,10)

rows =

     1     1     1     1     1     1     1     1     1     1
     2     2     2     2     2     2     2     2     2     2
     3     3     3     3     3     3     3     3     3     3
     4     4     4     4     4     4     4     4     4     4
     5     5     5     5     5     5     5     5     5     5
     6     6     6     6     6     6     6     6     6     6
     7     7     7     7     7     7     7     7     7     7
     8     8     8     8     8     8     8     8     8     8
     9     9     9     9     9     9     9     9     9     9
    10    10    10    10    10    10    10    10    10    10

>> cols=ones(10,1)*(1:10)

cols =

     1     2     3     4     5     6     7     8     9    10
     1     2     3     4     5     6     7     8     9    10
     1     2     3     4     5     6     7     8     9    10
     1     2     3     4     5     6     7     8     9    10
     1     2     3     4     5     6     7     8     9    10
     1     2     3     4     5     6     7     8     9    10
     1     2     3     4     5     6     7     8     9    10
     1     2     3     4     5     6     7     8     9    10
     1     2     3     4     5     6     7     8     9    10
     1     2     3     4     5     6     7     8     9    10
```

MATLAB/Octave

Now we can construct our matrix using `rows` and `cols`:

```
>> A=rows+cols-1

A =

     1     2     3     4     5     6     7     8     9    10
     2     3     4     5     6     7     8     9    10    11
     3     4     5     6     7     8     9    10    11    12
     4     5     6     7     8     9    10    11    12    13
     5     6     7     8     9    10    11    12    13    14
     6     7     8     9    10    11    12    13    14    15
     7     8     9    10    11    12    13    14    15    16
     8     9    10    11    12    13    14    15    16    17
     9    10    11    12    13    14    15    16    17    18
    10    11    12    13    14    15    16    17    18    19
```

MATLAB/Octave

The construction of `rows` and `cols` can be done automatically with the `meshgrid` function:

```
[cols,rows]=meshgrid(1:10,1:10)
```

will produce the two index matrices above.

The size of our matrix `a` can be obtained by using the `size` function:

```
>> size(a)

ans =

    4    4
```

which returns the number of rows and columns of `a`.

Vectorization

Vectorization refers to an operation carried out over an entire matrix or vector. We have seen examples of this already, in our construction of the 10×10 matrix A above, and in our use of the dot operators. In most programming languages, applying an operation to elements of a list or array will require the use of a loop, or a sequence of nested loops. Vectorization in MATLAB allows us to dispense with loops in almost all instances, and is a very efficient replacement for them.

For example, suppose we wish to calculate the sine values of all the integer radians one to one million. We can do this with a `for` loop:

```
>> for i = 1:10^6, sin(i); end
```

and we can measure the time of the operation with MATLAB's `tic`, `toc` timer: `tic` starts a stopwatch timer, and `toc` stops it and prints out the elapsed time in seconds. Thus, on my computer:

```
>> tic, for i = 1:10^6, sin(i); end, toc

elapsed_time =

    27.4969
```

We can perform the same calculation with:

```
>> i = 1:10^6; sin(i);
```

and print out the elapsed time with:

```
>> tic, i=1:10^6; sin(i); toc

elapsed_time =

    1.3522
```

Note that the second command applies the sine function to *all* the elements of the vector `1:10^6`, whereas with the `for` loop, sine is only applied to each element of the loop in turn.

As another example, we can easily generate the first 10 square numnbers with:

```
>> [1:10].^2

ans =

1 4 9 16 25 36 49 64 81 100
```

MATLAB/Octave

What happens here is that `[1:10]` generates a vector consisting of the numbers 1 to 10; and the dot operator `.^2` squares each element in turn.

Vectorization can also be used with logical operators; we can obtain all positive elements of the matrix **a** above with:

```
>> a>0

ans =

     1    0    0    1
     1    1    0    1
     1    0    0    0
     0    1    0    1
```

MATLAB/Octave

The result consists of 1's only in the places where the elements are positive.

MATLAB and Octave are designed to perform vectorized commands very quickly, and whenever possible such a command should be used instead of a `for` loop.

Cell Arrays

Cell arrays are able to hold different objects: matrices of different sizes, images, or any other data. They are delineated with braces, and can be defined by listing all values:

```
>> C = {[37],ones(2),randi([10,20],3,3),magic(4)}
C =
{
  [1,1] =  37
  [1,2] =

     1   1
     1   1

  [1,3] =

     15   14   13
     19   18   12
     15   17   16

  [1,4] =

     16    2    3   13
      5   11   10    8
      9    7    6   12
      4   14   15    1

}
```
MATLAB/Octave

Elements of C can be obtained again by using braces:

```
>> C{3}
ans =

   15   14   13
   19   18   12
   15   17   16
```
MATLAB/Octave

A cell array can be extended simply by putting a value into an index; any intermediate values will be filled with empty arrays:

```
>> C{10} = "Now is the winter of our discontent";

>> C{5:10}
ans = [](0x0)
ans = [](0x0)
ans = [](0x0)
ans = [](0x0)
ans = [](0x0)
ans = Now is the winter of our discontent
```
MATLAB/Octave

For another example, suppose we create a cell array each of whose elements is a list of the binomial coefficients:

```
>> B = {[1]}
>> for i = 2:11, B{i} = [B{i-1} 0]+[0,B{i-1}]; end
>> B{7}
ans =

    1    6   15   20   15    6    1
```

MATLAB/Octave

The entire cell array can be displayed with `celldisp`.

The `cellfun` function maps a function onto each value of a cell array:

```
>> cellfun(@(x) sum(x), B)
ans =

    1     2     4     8    16    32    64   128   256   512
       1024
```

MATLAB/Octave

There are several other functions for managing cell arrays; "`lookfor cell`" will display a list which includes them.

A.5   Plots

MATLAB and Octave have outstanding graphics capabilities. But we shall just look at some simple plots. The idea is straightforward: we create two vectors x and y of the same size, then the command

```
>> plot(x,y)
```

will plot y against x. If y has been created from x by using a vectorized function $f(x)$, the plot will show the graph of $y = f(x)$. Here is a simple example:

```
>> x = [0:0.1:2*pi]is;
>> plot(x,sin(x))
```

MATLAB/Octave

and the result is shown in Figure A.3. The `plot` function can be used to produce many different plots. We can, for example, plot two functions simultaneously with different colors or plot symbols. For example:

```
>> plot(x,sin(x),'.',x,cos(x),'--r','linewidth',2)
```

MATLAB/Octave

produces the graph shown in Figure A.4.

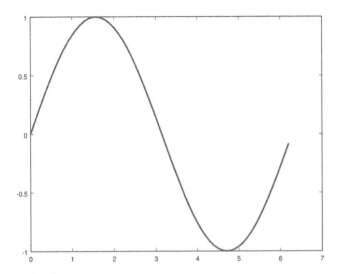

FIGURE A.3: A simple plot in MATLAB/Octave

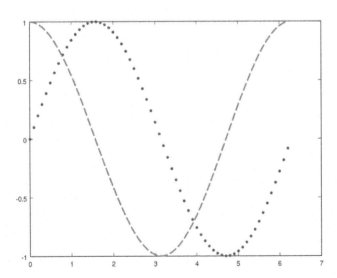

FIGURE A.4: A different plot in MATLAB/Octave

A.6   Online Help

A vast amount of online help and information is available. So much, in fact, that it is quite easy to use MATLAB or Octave without a manual. To obtain information on a particular command, you can use `help`. For example:

```
>> help for

 FOR    Repeat statements a specific number of times.
    The general form of a FOR statement is:

       FOR variable = expr, statement, ..., statement END

    The columns of the expression are stored one at a time in
    the variable and then the following statements, up to the
    END, are executed. The expression is often of the form X:Y,
    in which case its columns are simply scalars. Some examples
    (assume N has already been assigned a value).

          FOR I = 1:N,
              FOR J = 1:N,
                  A(I,J) = 1/(I+J-1);
              END
          END

    FOR S = 1.0: -0.1: 0.0, END steps S with increments of -0.1
    FOR E = EYE(N), ... END  sets E to the unit N-vectors.

    Long loops are more memory efficient when the colon expression appears
    in the FOR statement since the index vector is never created.

    The BREAK statement can be used to terminate the loop prematurely.

    See also IF, WHILE, SWITCH, BREAK, END.
```

MATLAB/Octave

If there is too much information, it may scroll past you too fast to see. In such cases, you can turn on the MATLAB pager with the command

```
>> more on
```

MATLAB/Octave

For more help on `help`, try:

```
>> help help
```

MATLAB/Octave

Better formatted help can be obtained with the `doc` function, that opens up a help browser which interprets HTML-formatted help files. The result of the command

```
>> doc help
```

MATLAB/Octave

is the window shown in Figure A.5.

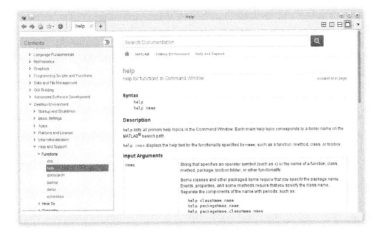

FIGURE A.5: The MATLAB help browser

You can find more about the **doc** function with any of

```
>> doc doc
>> help doc
```
MATLAB/Octave

If you want to find help on a particular topic, but do not know the function to use, the **lookfor** function is extremely helpful. The command

```
lookfor topic
```

lists all commands for which the first line of the help text contains the string **topic**. For example, suppose we want to find out if MATLAB supports the exponential function

$$e^x.$$

Here's how:

```
>> lookfor exponential
EXP     Exponential.
EXPINT Exponential integral function.
EXPM    Matrix exponential.
EXPM1   Matrix exponential via Pade approximation.
EXPM2   Matrix exponential via Taylor series.
EXPM3   Matrix exponential via eigenvalues and eigenvectors.
BLKEXP Defines a function that returns the exponential of the input.
```
MATLAB/Octave

and we now know that the function is implemented using **exp**. We could have used

```
>> lookfor exp
```
MATLAB/Octave

and this would have returned many more functions.

Note that MATLAB convention is to use uppercase names for functions in help texts, even though the function itself is called in lowercase.

A.7   Programming

MATLAB and Octave have a very rich programming language. Only a small set of functions is actually "built in" to each system, other functions are written in the system's own programming language. We consider two distinct programs: script files and functions.

Script Files

A script file is simply a list of commands to be executed. It may be that we will execute the same sequence of commands many times; in which case, it is more efficient to write a file containing those commands. If the file is called `script.m` and placed somewhere on the path, then simply entering `script` at the prompt will execute all the commands in it. Of course you can use any name you like for your script file! However, it is convention to end MATLAB files with the extension `.m`.

Functions

As we have seen, a *function* is a MATLAB command that takes an input (one or several variables) and returns one or several values. Let's look at a simple example: writing a function that returns the number of positive values of a matrix. This function will take a matrix as input and return a single number as output. We have seen that

```
>> a>0
```
MATLAB/Octave

produces a matrix with 1's in the positions of positive elements. So the sum of all elements in this new matrix is the number we require. We can obtain the sum of matrix elements using the `sum` function. If applied to a vector, `sum` produces the sum of all its elements. If applied to a matrix, however, `sum` produces a vector whose elements are the sums of the matrix columns:

```
>> sum(a)

ans =

     4    -2   -12     4

>> sum(a>0)

ans =

     3     2     0     3
```
MATLAB/Octave

So we have two options here: we can use `sum` twice:

```
>> sum(sum(a>0))

ans =

    8
```
MATLAB/Octave

or we could convert the matrix to a vector by using the colon operator before finding the positive elements:

```
>> sum(a(:)>0)

ans =

    8
```
MATLAB/Octave

Our function must have a name; let's call it `countpos`. A function file starts with the word `function`. The first line defines how the function is called. After that come some lines of help text, and finally the code. Our `countpos` function is:

```
function num = countpos(a)

% COUNTPOS finds the number of positive elements in a matrix.  The matrix
    can
% be of any data type.
%
% Usage:
%
%   n=countpos(a)

num = sum(a(:)>0);
```
MATLAB/Octave

If this file is saved as `countpos.m` somewhere on the path, we can use `countpos` exactly as we use any other MATLAB function or function:

```
>> countpos(a)

>> help countpos

>> doc countpos
```
MATLAB/Octave

and the command

```
>> lookfor count
```
MATLAB/Octave

will include a reference to `countpos`.

Finally, if we want to explore functions in more detail, we can use `type`, which lists the entire program corresponding to that function or command. So we can enter

```
>> type countpos.m
```

MATLAB/Octave

to see a listing of our new function.

Anonymous Functions

An *anonymous function* is a function that can be generated "on the go"; that is, without needing to write a complete definition in a file. Such a function can be useful as input to another function, such as `nlfilter`. For example, suppose we want to use a function that takes a square matrix as input, and produces the scalar product of the means of its rows with the means of its columns. Such a function can be written as

```
>> dotmeans = @(x) sum(mean(x).*mean(x'))
```

MATLAB/Octave

This can be applied to any square matrix:

```
>> A = randi([1,10],6,6)
A =

    8    2    6    1    9    8
    6    1    5    6    1    3
    6    7    9    2    9   10
    7    2    8    1    4    8
    7    4   10    1    7    5
    1    4   10    1    5    1

>> dotmeans(A)
ans = 167.06
```

MATLAB/Octave

Note here that `randi` is a handy function that produces a random matrix of integers with low and high values given in the first parameter, and the size of the last two values.

Exercises

1. Perform the following calculations:

$$132 + 45, \quad 235 \times 645, \quad 12.45/17.56. \quad \sin(\pi/6), \quad e^{0.5}, \quad \sqrt{2}$$

2. Now enter `format long` and repeat the above calculations.

3. Read the help file for `format`, and experiment with some of the other settings.

4. Enter the following variables: $a = 123456$, $b = 3^{1/4}$, $c = \cos(\pi/8)$. Now calculate:

$$(a + b)/c, \quad 2a - 3b, \quad c^2 - \sqrt{a - b}, \quad a/(3b + 4c), \quad \exp(a^{1/4} - b^{10})$$

5. Find the functions for the inverse trigonometric functions arcsin, arccos, and arctan. Then calculate:

$$\arcsin(0.5), \quad \arccos(\sqrt{3}/2), \quad \arctan(2)$$

Convert your answers from radians to degrees.

6. Using vectorization and the colon operator, use a single command each to generate:

   (a) The first 15 cubes
   (b) The values $\sin(n\pi/16)$ for n from 1 to 16
   (c) The values $\sqrt{n}$ for n from 10 to 20

7. Enter the following matrices:

$$A = \begin{bmatrix} 1 & 2 & 3 \\ 2 & 3 & 4 \\ 3 & 4 & 5 \end{bmatrix}, \quad B = \begin{bmatrix} -1 & 2 & -1 \\ -3 & -4 & 5 \\ 2 & 3 & -4 \end{bmatrix}, \quad C = \begin{bmatrix} 0 & -2 & 1 \\ -3 & 5 & 2 \\ 1 & 1 & -7 \end{bmatrix}$$

Now calculate:

$$2A - 3B, \quad A^T, \quad AB - BA, \quad BC^{-1}, \quad (AB)^T, \quad B^T A^T, \quad A^2 + B^3$$

8. Use the **det** function to determine the determinant of each of the matrices in the previous question. What happens if you try to find the inverse of matrix A?

9. Write a little function **issquare**, which will determine whether a given integer is a square number. Thus:

```
>> issquare(9)

ans =

    1

>> issquare(9.000)

ans =

    1

>> issquare(9.001)

ans =

    0

>> issquare([1:10])

ans =

    1   0   0   1   0   0   0   0   1   0
```

MATLAB/Octave

10. Enter the command

```
>> imshow(issquare(reshape([1:65536],256,256)))
```
MATLAB/Octave

What are you seeing here?

11. Plot the function $tan(x)$ with the following commands:

```
>> x = [0:0.1:10];
>> plot(x,tan(x))
>> figure,plot(x,tan(x)),axis([0,10,-10,10])
```
MATLAB/Octave

What does the **axis** function do? Read the help file for it. Experiment with changing the last two numbers in **axis** for the above command.

Appendix B

Introduction to Python

Python is a programming language that was initially released in 1991 by Guido Van Rossum, and which has since become deservedly popular for its elegance, ease of use, readability, portability, and extensibility. It marks code blocks using indentation, instead of parentheses (such as C), or "begin," "end" words (such as MATLAB). Since its inception Python has matured, is used worldwide for enterprise systems as much as for elementary teaching, and there are a vast number of specialized libraries—over 54,000 at the time of writing—which support particular programming tasks. Some of the most richly developed and most used libraries include:

`numpy`	"Numerical Python," routines for array manipulation and high-level functions for operating on arrays.
`scipy`	"Scientific Python," which includes modules for optimization, numerical analysis, signal and image processing, and differential equation solvers.
`matplotlib`	A plotting library, which includes an interface `pyplot` designed to emulate the plotting tools of MATLAB.

Python's *standard library* is rich and full-featured, and allows the handling of many different data types, mathematics, Internet protocols, file and data handling, system interfaces, multimedia, and graphics interface design.

Unlike MATLAB or Octave, not all of Python's functions are available to the user at the start; many functions have to be "imported", either by importing their library or by importing just the functions needed. This makes the namespace very manageable, and can allow for different functions sharing the same name.

Python has many interfaces. IDLE ("Integrated Development Environment") is bundled with the language, and may be considered a simple and uncluttered Python shell. IPython is an enhanced shell that includes syntax highlighting, automatic indentation, inline graphics, and many other enhancements. Spyder provides a graphics development environment, similar in some ways to that of MATLAB or Octave, and with integration of numpy, scipy, matplotlib, and iPython. There are numerous other interfaces, some open source, others proprietary. Unlike most programming systems, where the choice is between a graphics interface or a command line interface, the Python user is spoiled for choice.

Figure B.1 shows two standard development environments: iPython and Spyder. IPython has been configured to produce graphics "inline"; this is just one of many configuration options. Note that Spyder includes the iPython enhanced shell within it (and support for other consoles), as well as a variable and file browser, integrated editor, and interface to the Python debugger.

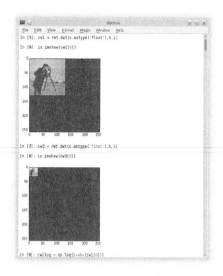
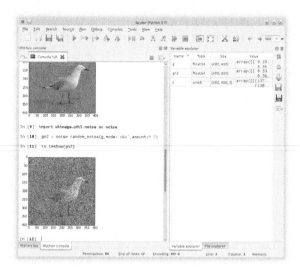

iPython   Spyder

FIGURE B.1: Python development environments

B.1   Basic Use

Suppose Python is started, and you have your prompt. Depending on the interface, it might be three "greater than" signs (IDLE):

```
>>>
```

Python

or a prompt that includes the command number (iPython):

```
In [1]:
```

Python

For ease of exposition, we shall use an unnumbered prompt

```
In :
```

Python

Python can perform standard arithmetic, with a couple of caveats: the "double division" sign: // is used for *integer division*, and exponentiation is implemented with two asterisks: **; the caret ^ is used for bitwise conjunction. Here are a few examples:

```
In : 37+42
Out: 79

In : 2**67-1
Out: 147573952589676412927L

In : 1.0/7
Out: 0.14285714285714285

In : 11/7
Out: 1

In : 11.0/7
Out: 1.5714285714285714

In : 11.0//7
Out: 1.0
```
`Python`

Python will return a value of the same type as the input: thus "**11/7**," having integer inputs, will return an integer result. Python can also perform integer arithmetic to arbitrary precision:

```
In : 37**100
Out:
66095557828843866774348296857793615320986068325257944996730965130260195627
49349063704800410525656374299407003776959988239901239717056920027944664127
58131334001L
```
`Python`

Arbitrary precision real arithmetic can be obtained through specialized libraries.

Standard mathematical functions are available in the `math` standard library, which must be loaded before its functions are available. This can be done in several ways:

```
In :  from math import *
```
`Python`

This makes all functions from the `math` library available.

```
In :  import math
```
`Python`

This imports the library; individual functions (such as `sin`) must be called as `math.sin`:

```
In : sqrt(3.0)
---------------------------------------------------------------------------
NameError                                 Traceback (most recent call last)
<ipython-input-1-d4d1770368ad> in <module>()
----> 1 sqrt(3.0)

NameError: name 'sqrt' is not defined

In : import math
```
`Python`

```
In : math.sqrt(3.0)
Out: 1.7320508075688772

In : from math import exp

In : exp(1.5)
Out: 4.4816890703380645

In : sin(pi/6)
---------------------------------------------------------------------------
NameError                                 Traceback (most recent call last)
<ipython-input-6-dbaab9b2f48b> in <module>()
----> 1 sin(pi/6)

NameError: name 'sin' is not defined

In : from math import *

In : sin(pi/6)
Out: 0.49999999999999994

In : cos(pi/12)
Out: 0.9659258262890683
```
                                                                    Python

If you want to keep the namespace functions different, you can import a library with an alias:

```
In : import math as m

In : m.tan(pi/8)
Out: 0.41421356237309503
```
                                                                    Python

To see all math functions, enter `math.` (including the period) and press the `tab` key:

```
In : math.⌐Tab⌐
math.acos        math.cos         math.factorial   math.ldexp       math.sin
math.acosh       math.cosh        math.floor       math.lgamma      math.sinh
math.asin        math.degrees     math.fmod        math.log         math.sqrt
math.asinh       math.e           math.frexp       math.log10       math.tan
math.atan        math.erf         math.fsum        math.log1p       math.tanh
math.atan2       math.erfc        math.gamma       math.modf        math.trunc
math.atanh       math.exp         math.hypot       math.pi
math.ceil        math.expm1       math.isinf       math.pow
math.copysign    math.fabs        math.isnan       math.radians
```
                                                                    Python

B.2 Arrays

Arrays are the basic data type for image processing; in Python, an array can be entered as a list of lists, using the `array` function from `numpy`

```
In : import numpy as np
In : A = np.array([[1,2,3],[4,5,6],[7,8,9]])
```
Python

Python does not return the result of a variable assignment, so we can see this new array by calling it:

```
In : A
Out:
array([[1, 2, 3],
       [4, 5, 6],
       [7, 8, 9]])

In : print A
[[1 2 3]
 [4 5 6]
 [7 8 9]]
```
Python

Python also has a matrix class, in which operations are slightly different, but there will be no need of it in this text; most image processing functions return arrays. Python arrays do not use the "dot" syntax of MATLAB for operation on every element:

```
In : A**3
Out:
array([[  1,   8,  27],
       [ 64, 125, 216],
       [343, 512, 729]])

In : A**(0.5)
Out:
array([[ 1.        ,  1.41421356,  1.73205081],
       [ 2.        ,  2.23606798,  2.44948974],
       [ 2.64575131,  2.82842712,  3.        ]])
In : np.sqrt(A)
Out:
array([[ 1.        ,  1.41421356,  1.73205081],
       [ 2.        ,  2.23606798,  2.44948974],
       [ 2.64575131,  2.82842712,  3.        ]])
```
Python

To multiply two matrices use the `dot` method:

```
In : B = A-1
In : print A.dot(B)
[[ 24  30  36]
 [ 51  66  81]
 [ 78 102 126]]
```
Python

Note that we invoked the `sqrt` function from `numpy`, which operates on arrays. The `sqrt` function from `math` is designed only to operate on single numbers:

```
In : math.sqrt(A)
----------------------------------------------------------------
TypeError                           Traceback (most recent call last)
<ipython-input-69-8214aa97e221> in <module>()
----> 1 math.sqrt(A)

TypeError: only length-1 arrays can be converted to Python scalars
```
Python

Here is an example of two functions `sqrt` with the same name, but with completely different operations. For this reason, the importing of all functions from a library:

```
In :  from math import *
```
Python

may be dangerous, as it may place the wrong version of a function into the user namespace. It is better to call functions from their libraries, as above.

Python provides an object oriented model, where an object of a particular type (such as an array) has many methods associated with it. To see them, enter `A.` (with period) followed by the Tab key:

```
In : A. Tab
A.T             A.conjugate    A.flatten      A.prod          A.sort
A.all           A.copy         A.getfield     A.ptp           A.squeeze
A.any           A.ctypes       A.imag         A.put           A.std
A.argmax        A.cumprod      A.item         A.ravel         A.strides
A.argmin        A.cumsum       A.itemset      A.real          A.sum
A.argpartition  A.data         A.itemsize     A.repeat        A.swapaxes
A.argsort       A.diagonal     A.max          A.reshape       A.take
A.astype        A.dot          A.mean         A.resize        A.tofile
A.base          A.dtype        A.min          A.round         A.tolist
A.byteswap      A.dump         A.nbytes       A.searchsorted  A.tostring
A.choose        A.dumps        A.ndim         A.setfield      A.trace
A.clip          A.fill         A.newbyteorder A.setflags      A.transpose
A.compress      A.flags        A.nonzero      A.shape         A.var
A.conj          A.flat         A.partition    A.size          A.view
```
Python

In order to see what each one does, enter it followed by a question mark:

```
In : A.max?
Type:         builtin_function_or_method
String Form:<built-in method max of numpy.ndarray object at 0x1d9c4f0>
Docstring:
a.max(axis=None, out=None)

Return the maximum along a given axis.

Refer to 'numpy.amax' for full documentation.

 See Also
 --------
numpy.amax : equivalent function
```
`Python`

Then

```
In : np.amax?
```
`Python`

will produce copious documentation about finding the maxima of arrays.

Elements can be accessed with square bracket notation, and noting the vital fact that in Python, arrays and lists are indexed **starting at zero**.

```
In : A[0,1]
Out: 2

In : A[1]
Out: array([4, 5, 6])

In : B = np.reshape(range(25),(5,5))
In : print B
[[ 0  1  2  3  4]
 [ 5  6  7  8  9]
 [10 11 12 13 14]
 [15 16 17 18 19]
 [20 21 22 23 24]]
```
`Python`

The `range` function in its simplest form `range(n)` provides a list of n integers from 0 to $n - 1$; more generally, we can have `range(<start>, <stop>, <step>)`:

```
In : range(3,10)
Out: [3, 4, 5, 6, 7, 8, 9]

In : range(2,30,3)
Out: [2, 5, 8, 11, 14, 17]
```
`Python`

Note that `range` does *not* include the final value.

Elements and rows or columns of an array can be produced by array `slicing`, in a manner similar to that of MATLAB, using a colon operator:

```
In : print B[::2]
[[ 0  1  2  3  4]
 [10 11 12 13 14]
 [20 21 22 23 24]]

In : print B[::-1]
[[20 21 22 23 24]
 [15 16 17 18 19]
 [10 11 12 13 14]
 [ 5  6  7  8  9]
 [ 0  1  2  3  4]]

In : print B[:,::2]
[[ 0  2  4]
 [ 5  7  9]
 [10 12 14]
 [15 17 19]
 [20 22 24]]

In : print B[:,::-1]
[[ 4  3  2  1  0]
 [ 9  8  7  6  5]
 [14 13 12 11 10]
 [19 18 17 16 15]
 [24 23 22 21 20]]

In : print B[[0,3]]
[[ 0  1  2  3  4]
 [15 16 17 18 19]]

In : print B[:,[3,4]]
[[ 3  4]
 [ 8  9]
 [13 14]
 [18 19]
 [23 24]]
```

Python

As you see, two colons will list from the entire array in steps of the value given. Note also that in Python elements are entered into arrays by rows, as opposed to MATLAB, where elements are entered column-wise.

B.3   Graphics and Plots

The standard library for plotting in Python is `matplotlib` in which the module `pyplot` provides plotting functionality in the style of MATLAB:

```
In : import matplotlib as mpl
In : import matplotlib.pyplot as plt
In : x = np.arange(0,2*np.pi,0.1)
In : plt.plot(x,np.sin(x),x,np.cos(x),'or')
```
Python

Depending on your setup, you may now need to enter

```
In : plt.show()
```
Python

to display the plot, which is given in Figure B.2. Note that to create the list of x values, we used the **arange** function, which returns an array instead of a list.

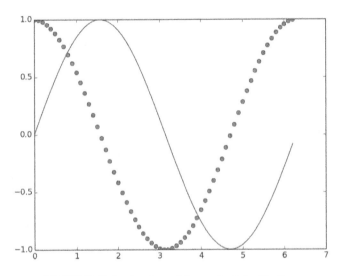

FIGURE B.2: A plot with **pyplot** in Python

B.4   Programming

Programming in Python is broadly similar to any other high level language, but with some extra flexibility and power. All the programming constructs–repetition, flow control, input and output–are very easy.

One very elegant aspect is that the looping **for** function allows you to iterate over pretty much any list; the list may contain numbers, arrays, graphics, or other lists.

Suppose we aim to write a program that will set to zero all elements in an array that are equal to the largest or second largest absolute value. For example:

```
In : A = np.random.randint(-100,100,size=(10,10))
```
Python

The `randint` function produces values between the first (low) and second (high) values, in an array of size given by the `size` parameter.

The largest absolute value can be found by

```
In : maxA = abs(A).max()

In : maxA
Out: 98
```
Python

Now we need to set to zero all values equal to ±98:

```
In : B = np.copy(A)

In : B[np.where(abs(A)==maxA)] = 0
```
Python

The `where` function acts like MATLAB's `find`: it produces lists of indices of all elements satisfying a condition.

In order to work on a new array, the `copy` function creates a new version of the original array. Without `copy`, B and A would simply be two names for the same array.

Now we simply do the same thing again, to eliminate elements with the next highest absolute value:

```
In : C = np.copy(B)

In : maxB = abs(B).max()

In : maxB
Out: 97

In : C[np.where(abs(B)==maxB)] = 0
```
Python

We can see the difference with

```
In : print abs(C-A)
[[ 0  0  0  0  0  0  0  0  0  0]
 [ 0  0  0  0  0  0  0  0  0  0]
 [ 0  0  0  0 97  0  0  0  0  0]
 [ 0  0  0  0  0 98  0  0  0  0]
 [ 0  0  0  0  0  0  0  0  0  0]
 [ 0  0  0 97  0  0  0  0  0  0]
 [ 0  0  0  0  0  0  0  0  0  0]
 [ 0  0  0  0  0  0  0  0  0  0]
 [ 0  0  0  0  0  0  0 97  0  0]
 [ 0  0  0  0  0  0  0  0  0  0]]
```
Python

Now for the program:

```
def biggish(A):
    B = np.copy(A)
    B[np.where(abs(A) == abs(A).max())] = 0
    C = np.copy(B)
    C[np.where(abs(B) == abs(B).max())] = 0
    return C
```

Python

Note that there is no need for "begin" or "end" statements, or braces, as the block of statements that comprise the program is delineated by indentation. Save this program in a file, say `myfile.py`, and it can then be read into your running Python with any of

```
In : execfile('myfile.py')

In : import myfile

In : from myfile import *
```

Python

Those three statements allow the following usages respectively:

```
In : biggish(A)

In : myfile.biggish(A)

In : biggish(A)
```

Python

It is impossible to give more than a brief "taste" of Python in a few pages, but it is an extraordinarily powerful language, with a huge user base, constantly being extended, and well worth the effort of learning it.

Appendix C

The Fast Fourier Transform

We have discussed the Fourier transform and its uses in Chapter 7. However, as we know, the Fourier transform gains much of its usefulness by the existence of a fast algorithm to compute it. We look briefly at one version of the fast Fourier transform here. More details can be found in [13] or [54].

To start, we shall look at the very simple DFT for a two-element vector, where for convenience we shall omit the scaling factor:

$$\begin{bmatrix} X_0 \\ X_1 \end{bmatrix} = \begin{bmatrix} 1 & 1 \\ 1 & -1 \end{bmatrix} \begin{bmatrix} x_0 \\ x_1 \end{bmatrix}$$

$$= \begin{bmatrix} x_0 + x_1 \\ x_0 - x_1 \end{bmatrix}.$$

We can express this combination with a "butterfly diagram" as shown in Figure C.1. We

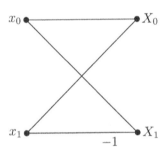

FIGURE C.1: A butterfly diagram

shall see how this butterfly diagram can be extended to vectors of greater length. In general, a butterfly diagram will consist of nodes joined as shown in Figure C.1, with scaling factors as shown in Figure C.2. In this figure, we have

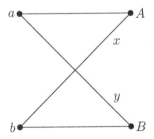

FIGURE C.2: A general butterfly

$$A = a + xb,$$
$$B = a + yb.$$

By convention, if a scaling factor is not specified, it is assumed to be 1.

We can extend this to the DFT for a four-element vector, starting with its matrix definition.

$$
\begin{bmatrix} X_0 \\ X_1 \\ X_2 \\ X_3 \end{bmatrix} =
\begin{bmatrix} 1 & 1 & 1 & 1 \\ 1 & -i & -1 & i \\ 1 & -1 & 1 & -1 \\ 1 & i & -1 & -i \end{bmatrix}
\begin{bmatrix} x_0 \\ x_1 \\ x_2 \\ x_3 \end{bmatrix}
$$

$$
= \begin{bmatrix}
x_0 + x_1 + x_2 + x_3 \\
x_0 - ix_1 - x_2 + ix_3 \\
x_0 - x_1 + x_2 - x_3 \\
x_0 + ix_1 - x_2 - ix_3
\end{bmatrix}
$$

$$
= \begin{bmatrix}
(x_0 + x_2) + (x_1 + x_3) \\
(x_0 - x_2) - i(x_1 - x_3) \\
(x_0 + x_2) - (x_1 + x_3) \\
(x_0 - x_2) + i(x_1 - x_3)
\end{bmatrix}
$$

$$
= \begin{bmatrix}
x_0' + x_1' \\
x_2' - ix_3' \\
x_0' - x_1' \\
x_2' + ix_3'
\end{bmatrix}
$$

This means we can obtain a four-element DFT by a two-stage process; first we create "intermediate" values x_i':

$$x_0' = x_0 + x_2$$
$$x_1' = x_1 + x_3$$
$$x_2' = x_0 - x_2$$
$$x_3' = x_1 - x_3$$

and use these new values to calculate the final values:

$$X_0 = x_0' + x_1'$$
$$X_1 = x_2' - ix_3'$$
$$X_2 = x_0' - x_1'$$
$$X_3 = x_2' + ix_3'$$

The butterfly diagram for this is shown in Figure C.3. Note that it consists of two simple butterfly diagrams on the left, joined together with more butterflies on the right.

Note that the order of the original elements x_i has been changed; we shall discuss this more below.

To apply the same idea to an eight-element vector, we consider the general term of the transform:

$$X_k = \omega^0 x_0 + \omega^k x_1 + \omega^{2k} x_2 + \omega^{3k} x_3 + \cdots + \omega^{7k} x_7 \qquad \text{where } \omega = e^{-2\pi i/8}$$

$$= \underbrace{\omega^0 x_0 + \omega^{2k} x_2 + \omega^{4k} x_4 + \omega^{6k} x_6}_{\text{even values}} + \underbrace{\omega^k x_1 + \omega^{3k} x_3 + \omega^{5k} x_5 + \omega^{7k} x_7}_{\text{odd values}}$$

$$= \omega^0 x_0 + \omega^{2k} x_2 + \omega^{4k} x_4 + \omega^{6k} x_6 + \omega^k (x_1 + \omega^{2k} x_3 + \omega^{4k} x_5 + \omega^{6k} x_7).$$

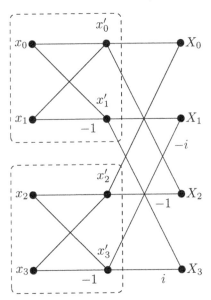

FIGURE C.3: A butterfly diagram for a four-element FFT

Now, let $z = \omega^2 = e^{-2\pi i/4}$. Then

$$X_k = (z^0 x_0 + z^k x_2 + z^{2k} x_4 + z^{3k} x_6) + \omega^k (z^0 x_1 + z^k x_3 + z^{2k} x_5 + z^{3k} x_7) \qquad \text{(C.1)}$$

If we examine at the brackets in this expression, we note that the first bracket is the kth term of the DFT of (x_0, x_2, x_4, x_6), and the second bracket is the kth term of the DFT of (x_1, x_3, x_5, x_7). So we can write:

$$X_k = \text{DFT}(x_0, x_2, x_4, x_6)_k + \omega^k \text{DFT}(x_1, x_3, x_5, x_7)_k.$$

To make the argument easier to follow, let us write

$$(Y_0, Y_1, Y_2, Y_3) = \text{DFT}(x_0, x_2, x_4, x_6)$$
$$(Y_0', Y_1', Y_2', Y_3') = \text{DFT}(x_1, x_3, x_5, x_7)$$

Thus, we have

$$X_k = Y_k + \omega^k Y_k'.$$

There is a slight problem here: the DFT of a four-element vector only has four terms, so that k here only takes values 0 to 3, and we need values 0 to 7. However, if we look back at Equation C.1, we can see that if k takes values 4 to 7, the powers of z just cycle around again, since $z^4 = 1$. Thus, for these values of k we have:

$$X_k = Y_{k-4} + \omega^k Y_{k-4}'.$$

This means that the indices for X_k can take all values between 0 and 7, but the indices of Y_k and Y_k' only take on values 0 to 3.

We can check this from first principles. Recall that in MATLAB and Octave, the `fft` function, to obtain a correct result, must be applied to a *column* vector.

Let us first create an eight-element vector, and obtain its Fourier transform:

```
>> x = 2:9

x =

    2    3    4    5    6    7    8    9

>> fx=fft(x')

fx =

   44.0000
   -4.0000 + 9.6569i
   -4.0000 + 4.0000i
   -4.0000 + 1.6569i
   -4.0000
   -4.0000 - 1.6569i
   -4.0000 - 4.0000i
   -4.0000 - 9.6569i
```

MATLAB/Octave

Now we shall split up the vector into even and odd parts, and put their Fourier transforms together using the above formula:

```
>> even = [2 4 6 8]; odd = [3 5 7 9];
>> feven = fft(even');
>> fodd = fft(odd');
>> X = zeros(8,1);
>> omega = exp(-2*pi*sqrt(-1)/8);
>> for i = 0:3,X(i+1) = feven(i+1)+omega^i*fodd(i+1);end
>> for i = 4:7,X(i+1) = feven(i-3)+omega^i*fodd(i-3);end
>> X

X =

   44.0000
   -4.0000 + 9.6569i
   -4.0000 + 4.0000i
   -4.0000 + 1.6569i
   -4.0000
   -4.0000 - 1.6569i
   -4.0000 - 4.0000i
   -4.0000 - 9.6569i
```

MATLAB/Octave

and this result agrees with the Fourier transform obtained from above. Note that in the for loops we used indices $i + 1$ and $i - 3$. This is because the theory starts indices at 0, but MATLAB starts indices at 1.

In Python it is equally straightforward:

```
In : from numpy.fft import fft
In : x = np.arange(2,10)
In : fx = fft(x)
In : print fx
[ 44.+0.j    -4.+9.65685425j  -4.+4.j    -4.+1.65685425j
  -4.+0.j    -4.-1.65685425j  -4.-4.j    -4.-9.65685425j]
```

Python

This agrees—as it should!—with the computations above. Now, for even and odd subarrays:

```
In : evenx = x[::2]
In : oddx = x[1::2]
In : feven =fft(evenx)
In : fodd = fft(oddx)
In : omega = np.exp(-2.0*np.pi*1j/8)
In : X1 = feven + omega**np.arange(4)*fodd
In : X2 = feven + omega**np.arange(4,8)*fodd

In : np.set_printoptions(precision=4)
In : print np.hstack([X1,X2])
[ 44. +0.0000e+00j  -4. +9.6569e+00j  -4. +4.0000e+00j  -4. +1.6569e+00j
  -4. -1.0658e-14j  -4. -1.6569e+00j  -4. -4.0000e+00j  -4. -9.6569e+00j]
```

Python

Note that because of Python's array indexing starting at zero, the computation of X1 and X2 is very straightforward.

As with the four-element transform, we can express the eight-element transform by using a butterfly diagram. Such a diagram, treating the four-element FFT as a "black box," is shown in Figure C.4. Note that as in Figure C.3 the order of the original elements is changed.

If we now replace the black box FFTs from Figure C.4 with the four-element butterfly diagram of Figure C.3 we obtain a "complete" butterfly diagram for eight elements, which is given in Figure C.5. To show the structure of the diagram, we have left out the multiplication factors, which are simply inherited from their respective diagrams.

We have noted that the input values in our butterfly diagrams are not given in their correct order. The ordering we require can be obtained by *binary bit reversal*. Suppose we list the input elements x_i in order but replace the indices with their binary expansions:

$$
\begin{array}{ccccccccc}
& x_0 & x_1 & x_2 & x_3 & x_4 & x_5 & x_6 & x_7 \\
= & x_{000} & x_{001} & x_{010} & x_{011} & x_{100} & x_{101} & x_{110} & x_{111}
\end{array}
$$

Now we replace each index with the reversal of its bits, and go back to decimal:

$$
\begin{array}{ccccccccc}
& x_{000} & x_{100} & x_{010} & x_{110} & x_{001} & x_{101} & x_{011} & x_{111} \\
= & x_0 & x_4 & x_2 & x_6 & x_1 & x_5 & x_3 & x_7
\end{array}
$$

and now we have the new order for input into the FFT.

The particular form of the FFT we have developed here is known as the "decimation in time, 2 radix" FFT. We can describe its general workings as follows:

1. Re-order the initial 2^n vector elements by binary bit reversal

2. Butterfly the elements two at a time, using scaling factors 1 and -1

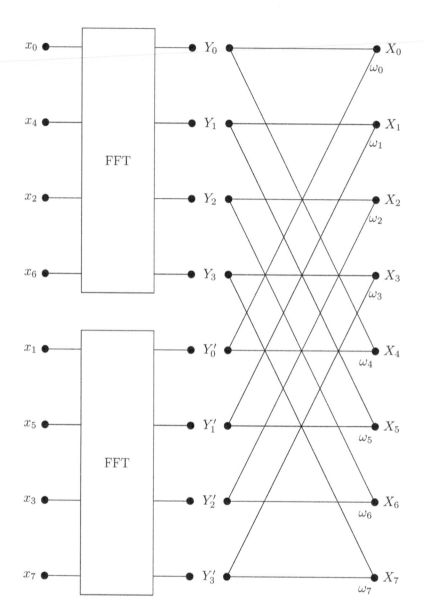

FIGURE C.4: A butterfly diagram for an eight-element FFT

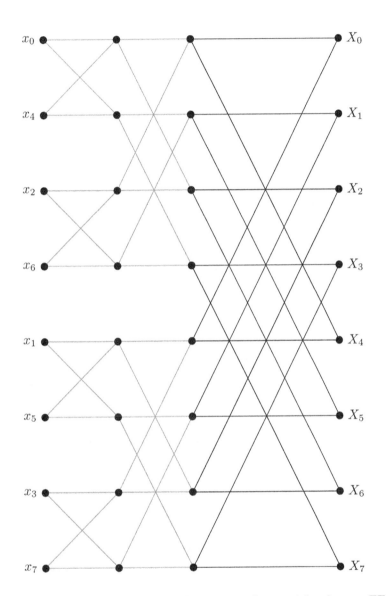

FIGURE C.5: A complete butterfly diagram for an eight-element FFT

3. Butterfly the resulting elements four at a time, using the scaling factors 1, ω, ω^2 and ω^3, with $\omega = \exp(2i\pi/4)$

4. Butterfly the resulting elements eight at a time, using the scaling factors $1, \omega, \omega^2, \ldots, \omega^7$, with $\omega = \exp(2i\pi/8)$

5. And so on, until we butterfly all elements together using scaling factors ω^k, with $\omega = \exp(2i\pi/2^n)$.

In practice, an FFT program will use a divide and conquer strategy: the initial vector is broken up into smaller vectors, the algorithm is applied recursively to these smaller vectors, and the results "butterflied" together.

There are many other forms of the FFT, but they all work by the same basic principle of dividing the vector into shorter lengths. Clearly we will obtain the greatest speed for vectors whose lengths are a power of 2 (such as in the previous examples), but we can also apply similar schemes to other lengths.

Bibliography

[1] Adobe Developers Association. *TIFF Revision 6.0.* available at `http://partners.adobe.com/public/developer/en/tiff/TIFF6.pdf`. Adobe Systems, Inc., 1992.

[2] Stormy Attaway. *Matlab: A Practical Introduction to Programming and Problem Solving.* 2nd Edition. Butterworth-Heinemann, 2013.

[3] Gregory A. Baxes. *Digital Image Processing: Principles and Applications.* John Wiley & Sons, 1994.

[4] Gary Bradski and Adrian Kaehler. *Learning OpenCV: Computer Vision with the OpenCV Library.* O'Reilly Media, Inc., 2008.

[5] Wayne C. Brown and Barry J. Shepherd. *Graphics File Formats: Reference and Guide.* Manning Publications, 1995.

[6] John Canny. "A computational approach to edge detection". In: *IEEE Transactions in Pattern Analysis and Machine Intelligence* 6 (1986), pp. 679–698.

[7] Kenneth R. Castleman. *Digital Image Processing.* Prentice Hall, 1996.

[8] Ashley R. Clark and Colin N Eberhardt. *Microscopy Techniques for Materials Science.* CRC Press, Boca Raton, Fl, 2002.

[9] Luis Pedro Coelho. "Mahotas: Open source software for scriptable computer vision". In: *arXiv preprint arXiv:1211.4907* (2012).

[10] Octave community. *GNU Octave 3.8.1.* 2014. URL: `\url{www.gnu.org/software/octave/}`.

[11] James D. Foley et al. *Introduction to Computer Graphics.* Addison-Wesley, 1994.

[12] Python Software Foundation. *Python Language Reference, v2.7.* available at `http://www.python.org`.

[13] Rafael Gonzalez and Richard E. Woods. *Digital Image Processing.* Third Edition. Addison-Wesley, 2007.

[14] Robert M. Haralick and Linda G. Shapiro. *Computer and Robot Vision.* Addison-Wesley, 1993.

[15] Stephen Hawley. "Ordered dithering". In: *Graphics Gems.* Ed. by Andrew S. Glassner. Academic Press, 1990, pp. 176–178.

[16] M. D. Heath et al. "A robust visual method for assessing the relative performance of edge-detection algorithms". In: *IEEE Transactions on Pattern Analysis and Machine Intelligence* 19.2 (1997), pp. 1338–1359.

[17] Robert V. Hogg and Allen T. Craig. *Introduction to Mathematical Statistics.* Fifth. Prentice-Hall, 1994.

[18] Gerard J. Holzmann. *Beyond Photography: the Digital Darkroom.* Prentice Hall, 1988.

[19] Joseph Howse. *OpenCV Computer Vision with Python.* Packt Publishing Ltd, 2013.

[20] William J. Palm III. *Introduction to MATLAB for Engineers.* Third edition. McGraw-Hill, 2010.

[21] Adobe Systems Incorporated. *PostScript(R) Language Reference.* Third Edition. Addison-Wesley Publishing Co., 1999.

[22] Anil K. Jain. *Fundamentals of Digital Image Processing.* Prentice Hall, 1989.

[23] Glyn James and David Burley. *Advanced Modern Engineering Mathematics.* Second Edition. Addison-Wesley, 1999.

[24] Arne Jensen and Anders la Cour-Harbo. *Ripples in Mathematics: the Discrete Wavelet Transform*. Springer Science & Business Media, 2001.

[25] Daniel J Jobson, Z-U Rahman, and Glenn A Woodell. "Properties and performance of a center/surround retinex". In: *Image Processing, IEEE Transactions on* 6.3 (1997), pp. 451–462.

[26] David C. Kay and John R. Levine. *Graphics File Formats*. Windcrest/McGraw-Hill, 1995.

[27] Edwin H. Land. "The Retinex Theory of Color Vision". In: *Scientific American* 237.6 (1977), pp. 108–128.

[28] Hans Petter Langtangen. *A Primer on Scientific Programming with Python*. Springer, 2011.

[29] Jae S. Lim. *Two-Dimensional Signal and Image Processing*. Prentice Hall, 1990.

[30] Wes McKinney. *Python for data analysis: Data wrangling with Pandas, NumPy, and IPython*. "O'Reilly Media, Inc.", 2012.

[31] Andrew Mertz and William Slough. "Graphics with TikZ". In: *The PracTEX Journal* 1 (2007).

[32] James R. Parker. *Algorithms for Image Processing and Computer Vision*. John Wiley and Sons, 1997.

[33] Jim R Parker. *Algorithms for Image Processing and Computer Vision*. John Wiley & Sons, 2010.

[34] Maria Petrou and Pangiota Bosdogianni. *Image Processing: the Fundamentals*. John Wiley and Sons, 1999.

[35] William K Pratt. *Introduction to Digital Image Processing*. CRC Press, 2013.

[36] Alfio Quarteroni, Fausto Saleri, and Paola Gervasio. *Scientific Computing with MATLAB and Octave*. Springer, 2006.

[37] Majid Rabbani and Paul W. Jones. *Digital Image Compression Techniques*. SPIE Optical Engineering Press, 1991.

[38] Greg Roelofs. *PNG: The Definitive Guide*. O'Reilly and Associates, 1999.

[39] Steven Roman. *Introduction to Coding and Information Theory*. Springer-Verlag, 1997.

[40] Azriel Rosenfeld and Avinash C. Kak. *Digital Picture Processing*. Second Edition. Academic Press, 1982.

[41] David Salomon. *A Concise Introduction to Data Compression*. Springer Science & Business Media, 2007.

[42] David Salomon and Giovanni Motta. *Handbook of Data Compression*. Springer Science & Business Media, 2010.

[43] Khalid Sayood. *Introduction to Data Compression*. Newnes, 2012.

[44] Dale A. Schumacher. "A comparison of digital halftoning techniques". In: *Graphics Gems II*. Ed. by James Arvo. Academic Press, 1991, pp. 57–71.

[45] Jean Paul Serra. *Image Analysis and Mathematical Morphology*. Academic Press, 1982.

[46] Frank Y Shih. *Image Processing and Mathematical Morphology: Fundamentals and Applications*. CRC press, 2009.

[47] Frank Y Shih. *Image Processing and Pattern Recognition: Fundamentals and Techniques*. John Wiley & Sons, 2010.

[48] Melvin P. Siedband. "Medical Imaging Systems". In: *Medical instrumentation : application and design*. Ed. by John G. Webster. John Wiley and Sons, 1998, pp. 518–576.

[49] Milan Sonka, Vaclav Hlavac, and Roger Boyle. *Image Processing, Analysis and Machine Vision*. Second Edition. PWS Publishing, 1999.

[50] Steven Tanimoto. *Structured Computer Vision: Machine Perception Through Hierarchical Computation Structures*. Elsevier, 2014.

[51] David S. Taubman and Michael W. Marcellin. *Jpeg2000: Image Compression Fundamentals, Standards, and Practice*. Kluwer Academic Publishers, 2001.

[52] Scott E. Umbaugh. *Computer Vision and Image Processing: A Practical Approach Using CVIPTools*. Prentice-Hall, 1998.

[53] Stefan Van Der Walt et al. "scikit-image: image processing in Python". In: *PeerJ* 2 (2014), e453.

[54] James S. Walker. *Fast Fourier Transforms*. Second Edition. CRC Press, 1996.

[55] Dominic Welsh. *Codes and Cryptography*. Oxford University Press, 1989.

[56] Joan L. Mitchell William B. Pennebaker. *JPEG Still Image Data Compression Standard*. Van Nostrand Reinhold, 1993.

Index